NONLINEAR CONTROL UNDER NONCONSTANT DELAYS

Advances in Design and Control

SIAM's Advances in Design and Control series consists of texts and monographs dealing with all areas of design and control and their applications. Topics of interest include shape optimization, multidisciplinary design, trajectory optimization, feedback, and optimal control. The series focuses on the mathematical and computational aspects of engineering design and control that are usable in a wide variety of scientific and engineering disciplines.

Series Volumes

Bekiaris-Liberis, Nikolaos and Krstic, Miroslav, *Nonlinear Control Under Nonconstant Delays*
Osmolovskii, Nikolai P. and Maurer, Helmut, *Applications to Regular and Bang-Bang Control: Second-Order Necessary and Sufficient Optimality Conditions in Calculus of Variations and Optimal Control*
Biegler, Lorenz T., Campbell, Stephen L., and Mehrmann, Volker, eds., *Control and Optimization with Differential-Algebraic Constraints*
Delfour, M. C. and Zolésio, J.-P., *Shapes and Geometries: Metrics, Analysis, Differential Calculus, and Optimization, Second Edition*
Hovakimyan, Naira and Cao, Chengyu, *$\mathcal{L}_1$ Adaptive Control Theory: Guaranteed Robustness with Fast Adaptation*
Speyer, Jason L. and Jacobson, David H., *Primer on Optimal Control Theory*
Betts, John T., *Practical Methods for Optimal Control and Estimation Using Nonlinear Programming, Second Edition*
Shima, Tal and Rasmussen, Steven, eds., *UAV Cooperative Decision and Control: Challenges and Practical Approaches*
Speyer, Jason L. and Chung, Walter H., *Stochastic Processes, Estimation, and Control*
Krstic, Miroslav and Smyshlyaev, Andrey, *Boundary Control of PDEs: A Course on Backstepping Designs*
Ito, Kazufumi and Kunisch, Karl, *Lagrange Multiplier Approach to Variational Problems and Applications*
Xue, Dingyü, Chen, YangQuan, and Atherton, Derek P., *Linear Feedback Control: Analysis and Design with MATLAB*
Hanson, Floyd B., *Applied Stochastic Processes and Control for Jump-Diffusions: Modeling, Analysis, and Computation*
Michiels, Wim and Niculescu, Silviu-Iulian, *Stability and Stabilization of Time-Delay Systems: An Eigenvalue-Based Approach*
Ioannou, Petros and Fidan, Barış, *Adaptive Control Tutorial*
Bhaya, Amit and Kaszkurewicz, Eugenius, *Control Perspectives on Numerical Algorithms and Matrix Problems*
Robinett III, Rush D., Wilson, David G., Eisler, G. Richard, and Hurtado, John E., *Applied Dynamic Programming for Optimization of Dynamical Systems*
Huang, J., *Nonlinear Output Regulation: Theory and Applications*
Haslinger, J. and Mäkinen, R. A. E., *Introduction to Shape Optimization: Theory, Approximation, and Computation*
Antoulas, Athanasios C., *Approximation of Large-Scale Dynamical Systems*
Gunzburger, Max D., *Perspectives in Flow Control and Optimization*
Delfour, M. C. and Zolésio, J.-P., *Shapes and Geometries: Analysis, Differential Calculus, and Optimization*
Betts, John T., *Practical Methods for Optimal Control Using Nonlinear Programming*
El Ghaoui, Laurent and Niculescu, Silviu-Iulian, eds., *Advances in Linear Matrix Inequality Methods in Control*
Helton, J. William and James, Matthew R., *Extending H^∞ Control to Nonlinear Systems: Control of Nonlinear Systems to Achieve Performance Objectives*

NONLINEAR CONTROL UNDER NONCONSTANT DELAYS

Nikolaos Bekiaris-Liberis
University of California at San Diego
La Jolla, California

Miroslav Krstic
University of California at San Diego
La Jolla, California

Society for Industrial and Applied Mathematics
Philadelphia

10 9 8 7 6 5 4 3 2 1

Library of Congress Cataloging-in-Publication Data
Bekiaris-Liberis, Nikolaos.
Nonlinear control under nonconstant delays / Nikolaos Bekiaris-Liberis, University of California at San Diego, La Jolla, California, Miroslav Krstic, University of California at San Diego, La Jolla, California.
pages cm – (Advances in design and control ; 25)
Includes bibliographical references and index.
ISBN 978-1-611973-17-4
1. Automatic control. I. Krstic, Miroslav. II. Title.
TJ213.B4175 2013
629.8'312–dc23
2013020352

Contents

Preface

Consider the time periods required

- for a driver to react in response to large disturbances in road traffic;
- for a signal sent from Houston to reach a satellite in orbit;
- for a coolant to be distributed among all air-conditioners in a residential building;
- for a job to be executed on a server;
- for a cell to reach a certain maturation level;
- for a relativistic particle to feel the electromagnetic force from another particle;
- for a cutting tool to perform two succeeding cuts.

The common attribute of all these time periods is that they do not remain constant.

Another feature in common to the dynamics of traffic flow, cooling systems, networks and queuing systems, population growth, and cutting processes is that they are all nonlinear. Although a plethora of techniques exist for the control of nonlinear systems without delays, control design for nonlinear systems in the presence of *long* delays with *large and rapid variation* in the actuation or sensing path, or delays affecting the internal states of a system, introduces significant feedback design challenges that have, heretofore, remained largely untackled.

In this book we present systematic design techniques applicable to general nonlinear systems with long, nonconstant delays. While there is a nearly inexhaustible number of combinations in which one or multiple delays (as well as discrete or distributed delays) can enter a dynamical system, we focus our attention on problems with input delays. Arguably, if the system has only a single discrete delay, the case where the delay affects the input (rather than some of the state components that appear on the system model's right-hand side) is the most challenging case for control, and in particular for stabilization. Hence, our focus on input delay problems is without much loss of generality.

In ODE systems with input delays, the overall state of the dynamical system consists of the vector state of the ODE and the functional state of the input delay. (If the delay is nonconstant, the support of the functional part of the state is nonconstant as well.) For a problem with such a (relatively) "unusual" state, the control design can be approached—in principle—in an abstract setting where the particular structure of the system and of the state is deemphasized and the design is performed in an abstract infinite-dimensional setting. However, at present, methods that fit such an abstract approach exist only when the ODE plant is linear (and when the delay is constant), but not when the plant is nonlinear.

To develop designs that are applicable to both linear and nonlinear plants, a much more structure-specific approach is needed. This approach exploits the structure of the system with input delay. In this approach the delay is "compensated for." On the surface, the approach appears very simple: the idea is to give up on controlling the current state over the immediate near term—because systems with input delays are not "small-time" controllable but only controllable over time intervals longer than the delay—and to, instead, design the feedback law to control the future state. To control the future state, the feedback law requires the value of the future state, rather than merely of the current state.

Knowledge of the future state only seems like an impossible thing to ask. It is not. The value of the state in the future can be expressed, using the system model, in terms of the current state and of the past inputs. Such a future state value is called a prediction and the formula for the state is called a "predictor." A feedback law employing the predictor—a formula for the future state—achieves successful control of the ODE system in the future, after an initial period of time equal to the input delay. Hence, in this approach one only needs to design a feedback law for the delay-free system and to construct the predictor.

The predictor construction for linear plants is straightforward (thanks to the variation of constants formula, using the current state as the initial condition). In the nonlinear case the predictor is not given explicitly, but the approach is conceptually the same, employing a predictor that depends on the current state and past inputs.

The *predictor-based* approach outlined above applies not only to systems (linear and nonlinear) with constant delays but also to systems with time-varying delays. It even applies to systems whose delays are time varying as a result of being dependent on the system state.

This book guides the reader from the basic idea of predictor feedback for linear systems with constant delays only on the input all the way through to nonlinear systems with state-dependent delays on the input as well as on system states.

What Does the Book Cover? While the most useful part of the book is the design of the feedback laws for systems with input delays and certain system structures involving state delays, the design is the easy part.

The key challenge is in the analysis of stability. While the ODE state is trivially endowed with stability-like properties after the initial time equal to the delay, the analysis of stability requires not only the quantification of the ODE state over this initial period, but also the quantification of the infinite-dimensional delay state over the entire time period, from zero to infinity.

To conduct such an analysis, we employ the recently introduced techniques based on infinite-dimensional backstepping transformations. These transformations employ linear or nonlinear Volterra operators of the delayed input state. In addition, the transformations involve the ODE state. The stability analysis of the predictor-based feedback laws employs Lyapunov functionals that incorporate the backstepping transformations, the inverses of the backstepping transformations, and the complex nonlinear relationships between the Lyapunov functionals and the norms of the overall system state (combining the vector state and the functional state).

Although the book's emphasis is on heretofore intractable problems involving nonlinear systems with time-varying and state-dependent delays, we also provide designs of predictor feedback laws for linear systems with constant distributed delays and known or unknown plant parameters, and for linear systems with simultaneous known or unknown constant delays on the input and the state.

Our results are always accompanied by a stability analysis which we perform by constructing Lyapunov–Krasovskii functionals for each particular problem. With our Lyapunov functionals we provide, explicitly, performance measures of the closed-loop system such as convergence rate and overshoot. In addition, the Lyapunov functionals that we construct allow us to quantify the robustness properties of our control laws to plant uncertainties (including delays), as well as to exogenous disturbances.

This book's most advanced results are the ones for state-dependent delays. State-dependent delays introduce a puzzling challenge. For input delays that are time varying—whether as a result of a direct dependence of the delay on time or of an indirect dependence of the delay on time thanks to the delay's dependence on the system's state—the challenge in constructing the predictor is that the time horizon over which prediction should be conducted is not in general equal to the length of the (time-varying) delay. This is illustrated in the cover art for our book. The prediction horizon depends on an inverse function of the function—which we refer to as the "delayed time"—which is given as a difference between the current time and the current delay. When the delay is state dependent, the "delayed time" function is not known a priori in the future. As a result, the "delayed time's" inverse function, which determines the prediction horizon, is not known at present time. In fact, the prediction horizon depends on the future predictor state. This gives rise to a seemingly intractable, seemingly circuitous situation, in which the predictor state is calculated over a time period that depends on the predictor state itself. We resolve this quandary and give a design formula for the predictor state even for state-dependent delays. We also provide a stability analysis, in which we overcome challenges associated with the fact that the support of the functional state (the input delay state) depends on the value of the vector state of the ODE plant.

Who Is This Book For? This book should be of interest to researchers working on control of delay systems, including engineers, mathematicians, and students. They may find many parts of it quite fascinating since it provides elegant and systematic treatments of long-standing problems that arise in many applications.

First and foremost among research communities that may benefit from the book, mathematicians working on nonlinear ordinary and functional differential equations, as well as on partial differential equations, may be stimulated by the wealth of mathematical challenges that arise in the systems considered in this book, particularly by systems with simultaneous input and state delays and systems with state-dependent delays.

All of our designs are given by explicit formulae. Therefore, the book should be of interest to any engineer who has faced delay-related challenges and is concerned with actual implementations: electrical and computer engineers who encounter varying delays imposed by communication networks; mechanical and other manufacturing engineers that are forced to operate their machinery within conservative bounds, since otherwise the uncompensated delay can lead to instability; and aerospace engineers working on combustion engines. Even civil engineers come across challenges due to the presence of long, varying delays in terms of traffic flow dynamics, or water and gas distribution dynamics. All of them may find this book useful because it provides systematic control synthesis techniques, as well as analysis tools for establishing stability and performance guarantees.

Chemical engineers and engineers working in automotive industry may significantly benefit from this book, since we devote a whole chapter to the control of gas emissions in automotive catalysts. They are going to gain insight into the mechanisms due to which the current production strategies operate effectively, despite such strategies actually being heuristic.

Graduate classes in engineering and applied mathematics could also use this as a supplemental textbook. Reading parts of the book is a viable alternative to homework exercises or finals in classes such as nonlinear systems, nonlinear control, adaptive control (Sections 2.3, 3.1.3, 3.2.1), control of distributed parameter systems, robust control

(Chapters 7, 14), linear systems and linear control (Chapters 2, 3, 6, 7), and ordinary or partial differential equations with applications.

The reader is assumed to have a basic graduate-level background on differential equations and calculus. All required notions, such as Lyapunov stability, as well as basic inequalities and lemmas, such as Young's inequality and Barbalat's lemma, used in this book are summarized in appendices for the reader's convenience.

Acknowledgments. We would like to thank Mrdjan Jankovic for his contributions in Chapters 3, 4, and 13.

We are grateful to Mrdjan Jankovic, Iasson Karafyllis, Frederic Mazenc, Michael Malisoff, and Pierdomenico Pepe for their inspiring and pioneering contributions to control of nonlinear delay systems.

We thank the Ford Motor Company and the National Science Foundation for their support of this research.

Nikolaos Bekiaris-Liberis dedicates this book to his mother Παγώνα, father Κώστα, and sister Δέσποινα (Pagona, Kosta, and Despoina). If it weren't for their continuous support and encouragement, this book would have never been written. Miroslav Krstic warmly thanks his wife, Angela, his daughter Victoria, and especially his daughter Alexandra, for providing daily inspiration for research on delay compensation.

La Jolla, California
October 2012

Nikolaos Bekiaris-Liberis
Miroslav Krstic

Chapter 1
Introduction

1.1 ▪ Delay Systems

Delays are ubiquitous in applications. A nonexhaustive list includes network controlled systems, teleoperation, milling processes, rolling mills, cooling systems, chemical processes, traffic flow, supply networks, automotive engines, 3D printing/additive manufacturing, irrigation channels, and population dynamics. The presence of delays can severely degrade the performance of such systems and in some cases even lead to instability. The resulting poor performance in the former case and the catastrophic consequences for the physical system in the latter create the need for developing control techniques that explicitly take into account the delay.

In addition to their practical importance, delay systems carry a wealth of mathematical challenges, since they are infinite-dimensional. For this reason, delay systems have drawn the attention not only of engineers but also of mathematicians.

Over the past 65 years major breakthroughs have been reported in the control of delay systems. The first systematic treatment of delays goes back to 1946 in the paper by Tsypkin [164]. A breakthrough in analysis is reported in the late 1950s/early 1960s with the seminal papers by Razumikhin in 1956 [144] and by Krasovskii in 1963 [82].

Meanwhile, Smith introduced in 1959 the celebrated Smith Predictor [155]. The next well-known landmark in the control of delay systems is the introduction of the finite spectrum assignment technique (also known as model reduction or predictor-like technique) by Manitius and Olbrot in 1979 [102] which overcomes the limitations of the Smith Predictor. However, a 1968 paper by Mayne [103] predates Manitius and Olbrot by over ten years in introducing a "predictor-based" technique that is applicable to unstable systems with input delays.

In subsequent years numerous papers appeared dealing with the control of delay systems by adopting techniques coming from the control of finite-dimensional systems. These techniques include sliding mode techniques, optimal control, H_∞ control, and LMI techniques. However, it wasn't until the 2000s that systematic nonlinear control designs based on recursive techniques, originally developed for nonlinear ODEs, such as forwarding and backstepping started to appear for nonlinear delay systems.

Some excellent books and survey articles summarizing the achievements in analysis and control of nonlinear delay systems are the books by Dugard and Verriest [37], Gu and coauthors [45], Karafyllis and Jiang [68], Kolmanovskii and Myshkis [80], Krstic [87], Michiels and Niculescu [117], Niculescu [127], and Zhong [182], and the surveys by Gu and Niculescu [46] and Richard [145].

The most challenging problems in the control of delay systems are arguably those in which the system has to be controlled through a delayed input, even when the system is linear and even when the delay is constant. Since a large portion of the available results are finite-dimensional control laws, their applicability is limited to either a narrow class of systems or to systems with short delays. Infinite-dimensional control laws are required for the stabilization of general systems with long delays.

When it comes to nonlinear systems with input delays, one is faced not only with limitations due to the applicability and complexity of the control designs, but also with limitations that are inherent to the problem. Nonlinear systems in certain classes can exhibit a finite escape time. For this reason, global stabilization of nonlinear systems, under input delays, is not possible in general, even for systems that are globally stabilizable in the absence of the delay. In this book we focus on systems whose solutions remain bounded for all times and that are globally stabilizable in the absence of the delay. For this class of systems our designs achieve global stabilization for arbitrarily long delays. Yet, for completeness we devote a chapter to systems that are only locally stabilizable in the absence of the delay, for which we achieve local stabilization for long delays.

The control challenges grow further with the simultaneous presence of input and state delays. Control of systems with simultaneous input and state delays is an underdeveloped area in the field of delay systems. In this book we present solutions to problems of this kind.

1.2 ▪ Control via Delay Compensation

Finite spectrum assignment and *reduction method* are synonyms for the control design approach that we refer to as "predictor feedback." We adopt this name to emphasize that feedback laws in this class are exactly what their name implies—feedback of the future (predicted) values of the state. Since predictor feedback laws use the future values of the state, they are able to compensate for the delay—after the control signal reaches the state of the plant, the state evolves as if there were no delay at all. There are two key challenges in developing predictor feedback laws. The first challenge is the determination of an implementable form for the future values of the state. Having determined the predictor state, the control law is then obtained by replacing the state in a nominal state feedback law (which stabilizes the delay-free plant) by the state's predictor. The second challenge is the stability analysis of the closed-loop system under predictor feedback, since the overall state of the system consists of the finite-dimensional internal states of the plant and the infinite-dimensional actuator or sensor state.

1.2.1 ▪ Overview of the Literature on Predictor Feedback

Arguably, the most mathematically and practically significant development in delay systems is the introduction of the Smith Predictor by Otto Smith in 1959 [155]. The Smith Predictor is in widespread use in industry, and perhaps second in popularity to PID controllers, especially among process control practitioners. More importantly, the Smith Predictor is the first design dealing with the *compensation* of delays. Smith's original development was applicable only to linear, stable systems with constant input delays. The open-loop stability limitation was removed by Mayne in 1968 [103], and then in 1979 by Manitius and Olbrot [102], who introduced the finite spectrum assignment technique for the compensation of input delays. In 1982 Artstein [4] systematized this methodology. Artstein used the name *reduction method* for his approach to emphasize that the stabilization problem for linear systems with input delays can be *reduced* to the stabilization problem of a delay-free (reduced) system. Later, Fiagbedzi and Pearson [41] and Watanabe [171] in 1986 and Watanabe and coauthors [172] in 1992 extended the finite spectrum assignment technique to linear systems with a special structure also having state delays.

However, the introduction of the predictor feedback design for unstable linear systems with time-varying input delay did not occur until 1992 and the paper by Nihtila [131], who explicitly constructed the predictor feedback using the variation of constants formula. During the 1990s further studies on improving the performance and robustness properties of predictor-based techniques appeared; they are summarized in the paper by Palmor [136] in 1996 and the book by Wang and coauthors [170] in 1998. In the early and mid-2000s interest in studying the numerical implementation of predictor-based control laws developed. Mondie and Michiels in 2003 [122] pointed out drawbacks related to a class of numerical implementations of predictor-based control laws. Yet, Mirkin in 2004 [119] and Zhong in 2006 [181] showed that these drawbacks can be avoided with a more suitable alternative implementation.

Despite the numerous contributions to the design and analysis of predictor-based control laws, two major research paths had remained unexplored. The first was the construction of a Lyapunov–Krasovskii functional, and the second was the design of predictor feedback laws for nonlinear systems. In 2008 Krstic and Smyshlyaev [93] introduced an infinite-dimensional backstepping transformation which enabled the construction of a Lyapunov–Krasovskii functional for linear systems with constant input delays under predictor feedback. Krstic in 2008 [86] introduced the first predictor feedback design for nonlinear systems. The book [87] by Krstic gives an introductory idea of nonlinear predictor feedback but does not deal with time-varying, state-dependent, or distributed delays, or on delays on the state.

1.2.2 ▪ Results in This Book

In this book we design predictor feedback laws for the compensation of time-varying and state-dependent delays on the input and on the state for general nonlinear systems. We also provide designs of predictor feedback laws for linear systems with constant distributed delays and known or unknown plant parameters, and for linear systems with simultaneous known or unknown constant delays on the input and the state. Moreover, we introduce infinite-dimensional backstepping transformations for each particular problem, which enables us to construct Lyapunov–Krasovskii functionals. With the available Lyapunov–Krasovskii functionals we study stability, as well as robustness, of our control laws to plant uncertainties (including delays) and exogenous disturbances.

1.3 ▪ Varying vs. Constant Delays

In this book we put emphasis on stabilization of plants of the form

$$\dot{X}(t) = f(X(t - D_1(t, X(t))), U(t - D_2(t, X(t)))),$$

where f is the vector field, X is the state, U is the input, and D_1, D_2 are arbitrary nonnegative-valued functions of time t and the state X. Depending on whether D_1 or D_2 is constant, time-varying, or state-dependent, one can formulate various combinations of problems, each one possessing different challenges. The degree of difficulty in designing controllers for the various cases arising is proportional to the complexity of the dependence of the delays on their arguments, and increases with the simultaneous presence of input and state delays.

In this book we focus on the case in which the input is delayed, i.e., the delay D_2 is not identically equal to zero, which is the most challenging case for control. In particular, stabilization of *general* nonlinear systems under *long* delays requires predictor feedback.

The challenges for predictor feedback design, as well as for stability analysis in nonlinear systems with time-varying or state-dependent delays, grow significantly as compared to the case of constant delays. In the case of time-varying delays (i.e., $D_1 = D_1(t)$ and $D_2 = D_2(t)$) the main challenge in control design is to construct the predictor state—the prediction horizon is not in general equal to the length of the delay, but it depends on an inverse function of the function which is given as a difference between the current time and the current delay. In addition, in the case of systems with delays both on the input and on the state, the prediction horizons are different for different states and depend on the inverse functions and on the function which is given as a difference between the current time and the current delay. Also, in contrast to the case of only input delays, the knowledge of a nominal delay-free design does not suffice. One has to design a feedback law directly based on the delay systems.

The predictor feedback design in the case of state-dependent delays (i.e., $D_1 = D_1(t,X(t))$ and $D_2 = D_2(t,X(t))$) inherits the challenges of the time-varying case and introduces an additional challenge. The prediction horizon depends on the future state of the system, that is, the time period over which the predictor state is calculated depends on the predictor state itself.

As compared to constant delays, nonconstant delays induce greater challenges also in the analysis. In the case of time-varying delays, the support of the functional part of the state is also time varying, and in the case of state-dependent delays depends also on the value of the vector state of the ODE plant. Because of the time or state dependency of the support, a much more delicate stability analysis, in comparison to constant delays, is required.

1.4 ▪ Future Paths

Robust Control of Systems with Input Delays

Consider a scalar linear system with a parametric uncertainty within a priori known bounds. How would one design a robust stabilizing control law in the presence of an arbitrarily long but known input delay? If one attempts to employ a high gain predictor feedback law, this approach would fail since the model of the system is not known for computing the predictor of the state. This simple example reveals the fact that the control problems that one can solve with the extant design tools for delay systems are limited as compared to the problems that have been solved for finite-dimensional systems. For attacking such problems in delay systems one should first look into extending techniques originally developed for finite-dimensional systems to infinite dimension. The continuum (infinite-dimensional) version of backstepping used in this book is an example of such an extension. In addition, since our designs are always accompanied with a Lyapunov functional, one could try to design robust control laws for delay systems using Lyapunov redesign techniques.

Nonlinear Systems with Distributed Input Delay

For systems with distributed delays the concepts of prediction and backstepping are not directly applicable, and we go beyond the classical predictor approach in this book. The reason for the inapplicability of predictors/backstepping is that for systems with distributed delays the infinite-dimensional control law cannot be based simply on the predictor of the state for a single prediction time because the same input affects the plant over an interval of times. The transformations that we introduce for linear systems with distributed delays are not of the backstepping type, i.e., they are not of a purely Volterra type, but are of a mixed Fredholm–Volterra type. In this book we transition from the backstepping transformation for linear systems with lumped delays, which is explicitly defined in terms of the state of the plant and the Volterra integral of the history of the input, to the nonlinear backstepping transformation defined in terms of the state of the plant and the Volterra integral of the history of the input through an intermediate variable, that is, through the predictor state. For achieving an analogous transition in the distributed delay case one has to appropriately define an intermediate transformation (which is the predictor state in the case of lumped delays) defined implicitly through a relation that incorporates the state of the plant and the Fredholm–Volterra integral of the history of the input.

Systems with Input-Dependent Delays

Consider a plant in which the larger the control effort that is applied, the larger the time that the control signal needs to reach the plant. This is an example of an input delay which grows with the magnitude of the input itself, i.e., an input-dependent input delay. Systems with an input-dependent delay are similar to systems with a state-dependent delay. While the control design that compensates for an input-dependent delay would employ a predictor feedback law, the procedure that one has to follow for actually computing the predictor at each time instant would be much more involved. In contrast to the case of state-dependent delays, in which the predictor state satisfies an ODE in the spatial variable (over a state-dependent domain), in the case of input-dependent delays the predictor would be given by some nonstandard DDE (delay differential equation) of neutral type.

Nonlinear Systems with PDE Actuator Dynamics

Nonlinear systems with input delays can be viewed as a particular form of PDE/nonlinear ODE cascade. It is of interest to consider problems of other PDE/nonlinear ODE cascades. We have launched an effort to tackle such problems in Bekiaris-Liberis and Krstic [10], where we consider a wave PDE/nonlinear ODE cascade. Our design methodology was based on converting the problem of compensation of the wave actuator dynamics to a stabilization problem of a 2×2 system of transport PDEs coupled with a nonlinear ODE. This transformation enabled us to use some of the techniques that we develop in this book. One could attack problems in which the PDE is of some other type, for example, of parabolic type. In our opinion there will be two possible distinct lines of attack. The first is to try to convert the problem of compensating for the, say, diffusive actuator dynamics to the stabilization problem of transport-like PDEs/nonlinear ODE cascades. The second is to try to develop novel backstepping transformations directly on the PDE/nonlinear ODE cascade, with a starting point being the backstepping design for nonlinear parabolic PDEs in Vazquez and Krstic [166, 167].

PDEs with State-Dependent Coefficients

Nonlinear systems with state-dependent delays are a special form of transport PDE/nonlinear ODE cascade in which the transport coefficient or the size of the spatial domain of the PDE depends, through a particular relation, on the state of the ODE. One could try to extend our developments to other PDE/nonlinear ODE cascades or to transport

PDE/transport PDE cascades, with state-dependent coefficients. An important problem, both practically and mathematically, in the former category are diffusive-type PDE/nonlinear ODE cascades in which the diffusion coefficient (or the spatial domain) of the PDE depends on the state of the nonlinear ODE. In the latter category are the first order hyperbolic PDE systems in which the coefficients depend on the PDE states.

1.5 ▪ Organization of the Book

The book is composed of three parts.

Part I: In the first part of the book we deal with constant delays.

In Chapter 2 we provide the introductory ideas of predictor feedback and infinite-dimensional backstepping through the study of linear systems with only input delays. We then design predictor feedback laws for linear systems with simultaneous known delays on the input and on the state. In addition, we consider systems with simultaneous but unknown delays on the input and the state, for which we design a delay-adaptive feedback law.

In Chapter 3 we consider linear systems with distributed input delays and known or unknown plant parameters. In particular, we construct a Lyapunov–Krasovskii functional for multi-input systems under predictor feedback. We also design an adaptive version of the predictor feedback to compensate for constant, matched disturbances on the input, and an infinite-dimensional observer for multi-output systems with distributed sensor delays. The availability of our Lyapunov–Krasovskii functionals allows us to design an adaptive version of the predictor feedback when the plant parameters are unknown.

Chapter 4 is devoted to control of automotive catalysts. The model of the catalyst is composed of coupled, nonlinear first order hyperbolic PDEs, which are in turn coupled with an infinite-dimensional ODE. In particular, we reduce the stability analysis of the infinite-dimensional ODE to the stability analysis of an autonomous infinite-dimensional ODE which incorporates integral terms of the ODE state on its right-hand side.

In Chapter 5 we introduce the concepts of nonlinear predictor feedback and nonlinear infinite-dimensional backstepping through the study of nonlinear systems with input delays.

Part II: In the second part of the book we consider systems with time-varying delays.

In Chapter 6 we introduce the basic idea of predictor feedback for time-varying delays by studying linear systems with time-varying input delays.

In Chapter 7 we study robustness of nominal, constant-delay predictor feedback laws to time-varying delay perturbations in linear systems with input delays.

We then proceed to Chapters 8 and 9, in which we consider nonlinear systems with time-varying input delays, and simultaneous input and state delays.

Part III: The third part of the book is devoted to systems with state-dependent delays.

In Chapter 10 we design the predictor feedback law for nonlinear systems with state-dependent input delays.

We analyze the stability properties of our design for forward complete, globally stabilizable in the absence of the delay nonlinear systems (Chapter 11), as well as for nonlinear systems that are locally stabilizable in the delay-free case (Chapter 12).

In Chapter 13 we consider nonlinear systems with state-dependent delays on the state.

In Chapter 14 we study robustness of nonlinear, constant-delay predictor feedbacks to time- and state-dependent perturbations on the delay.

Chapter 15 is devoted to the stabilization problem of nonlinear systems with delays that do not depend on the current but on past values of the state of the system.

1.6 ▪ Spaces and Solutions

- We denote by $\mathbb{R}^n$ the space of n-dimensional vectors taking real values. For a vector $X \in \mathbb{R}^n$ we denote by $|X|$ its usual Euclidean norm. For a matrix $A \in \mathbb{R}^{m\times n}$ we denote by $|A| = \sup\{|AX| : X \in \mathbb{R}^n, |X| = 1\}$ its induced norm.
- We denote by $\mathbb{R}_+$ the set of nonnegative real numbers.
- The definitions of class $\mathcal{K}$, $\mathcal{K}_\infty$, $\mathcal{KL}$, $\mathcal{KC}$, and $\mathcal{KC}_\infty$ functions are given in Appendix C.1.
- Let $[a, b]$ be an interval on the real line. We denote by $L_p[a, b]$, where $p \in [1, \infty)$, the space of measurable functions $U(\theta)$, $a \le \theta \le b$, such that $\int_a^b |U(\theta)|^p d\theta < \infty$. We denote by $L_\infty[a, b]$ the space of measurable and bounded functions $U(\theta)$, $a \le \theta \le b$.
- We denote by $H_l[a, b]$, where $l \in [1, \infty)$ is an integer, the space of functions that belong to $L_2[a, b]$ whose (weak) derivatives up to order l exist and belong to $L_2[a, b]$.
- We denote by $C^j(A;\Omega)$, where $j \ge 0$ is an integer, the class of functions taking values in Ω that have continuous derivatives of order j on A.

In the first two parts of the book we do not belabor the issue of existence and uniqueness of solutions. Our closed-loop system has a unique solution, which is continuously differentiable, provided that the initial internal states and the initial actuator state are compatible with the feedback law. Otherwise, the closed-loop system has a unique solution which continuous.

Yet, the issue of existence and uniqueness is less obvious in the third part of the book, in which we deal with state-dependent delays. In this case we discuss existence and uniqueness of solutions in more detail.

Part I

Constant Delays

Chapter 2

Linear Systems with Input and State Delays

Following the invention of the Smith Predictor in the late 1950s for dead compensation in stable linear systems, the first systematic feedback design for potentially unstable, linear plants with arbitrarily long input delays was introduced by Mayne [103] in 1968. This predictor-based approach reemerged, under the name finite spectrum assignment, in the widely known work by Manitius and Olbrot [102] in the late 1970s. Yet, the question of Lyapunov-based stability analysis of the plant under predictor feedback remained open for decades after these early contributions. In Section 2.1 we lay out some of the main ideas of predictor feedback. With the introduction of the infinite-dimensional backstepping transformation we construct a Lyapunov functional for the closed-loop system consisting of the plant and the predictor feedback law.

In our development we provide two alternative representations for the infinite- dimensional actuator state: one using standard delay notation and one using a representation with a transport PDE for the actuator state. Consequently, we provide two alternative representations for the backstepping transformation. The infinite-dimensional backstepping transformation was initially developed for the stabilization of PDEs. So we keep the PDE representation to emphasize that this technique was inspired by the boundary control of PDEs.

However, the reasons of keeping, in some of our developments, the PDE representation of the actuator state are not only historical. The PDE representation allows a linear parametrization of the delay in the case in which we are designing adaptive control laws for systems with unknown delays (as in Section 2.3). In addition, the PDE representation is used also when dealing with time-varying delays since it reveals an explicit dependence of the transport speed coefficient on both time and space. Finally, the PDE notation allows a reader who is not an expert on delay systems to digest more easily the details of our analysis.

The advantages of using the standard delay representation is not obvious until we deal with state-dependent delays in Part III of the book. This advantage has to do with the fact that, when one uses a PDE representation for the actuator state in the case of a state-dependent delay, the state-dependency is hidden in the transport coefficient, and it appears as if the delay were only time-dependent.

Although the problem of compensation of input delays for unstable linear systems was solved in the late 1960s, and further advanced in the late 1970s, systems with simultaneous input and state delay have remained a challenge. Exponential stabilization has been solved for systems that are not exponentially unstable, such as chains of delayed integrators and systems in the "feedforward" form. In Section 2.2 we consider a general system in strict-feedback form with delayed integrators, which is an example of a particularly challenging class of exponentially unstable systems with simultaneous input and state delays.

An even more challenging problem is the adaptive control of systems with simultaneous input and state delays. From a practical point of view, controllers for delay systems should be robust to parametric uncertainties, including plant parameters and delays. On the other hand, since there already exists a rich literature for the control of time delay systems, adaptive control schemes that are based on existing control techniques are of interest. Since predictor techniques are an indispensable part of the control design toolbox for plants with input and state delays of significant size (which are known to be very sensitive to delay uncertainties [145]), designing adaptive versions of these control schemes is crucial for making them usable in scenarios with uncertain delays. In Section 2.3 we develop a delay-adaptive predictor feedback design for linear feedforward systems with simultaneous, unknown state and input delays.

2.1 ▪ Predictor Feedback Design for Systems with Input Delay

We introduce the basic predictor feedback design for linear systems with long input delays in Section 2.1.1. In Section 2.1.2 we construct an infinite-dimensional backstepping transformation for the actuator state, and we provide two alternative representations for it: one based on standard delay representation of delayed states and one using a representation of the actuator state with a transport PDE. In Section 2.1.3 we construct a Lyapunov functional and use it to prove exponential stability of the closed-loop system.

2.1.1 ▪ The Main Principle of Predictor Feedback

We consider the following linear system with input delay:

$$\dot{X}(t) = AX(t) + BU(t-D), \tag{2.1}$$

where $X \in \mathbb{R}^n$, (A,B) is a controllable pair, and $D > 0$ is arbitrary long. When $D = 0$ a stabilizing control law for system (2.1) is given as

$$U(t) = KX(t), \tag{2.2}$$

where the gain K is such that it renders the matrix $A + BK$ Hurwitz. The predictor feedback design is based on the nominal delay-free design (2.2). The main idea is to replace the state X by its prediction over a D-time-unit horizon, namely, the signal

$$P(t) = X(t+D), \tag{2.3}$$

so that $U(t-D) = KX(t)$. The key challenge is how to derive an implementable form of the signal P, i.e., a form that does not incorporate the future values of the state X as they are not available for feedback. Having determined the predictor P of the state X, the control law for the system with delay is given by

$$U(t) = KP(t). \tag{2.4}$$

This control law completely compensates for the input delay since for all $t \geq D$, $U(t-D) = KP(t-D) = KX(t)$, and hence the closed-loop system behaves as if there were no delay at all (after an initial transient period of D time units). An implementable form of the predictor signal is derived as follows. Performing the change of variables $t=\theta+D$, for all $t-D\leq\theta\leq t$, in (2.1) and using the fact that $\frac{d\theta}{dt}=1$, we get

$$\frac{dX(\theta+D)}{d\theta}=AX(\theta+D)+BU(\theta)\quad\text{for all } t-D\leq\theta\leq t. \tag{2.5}$$

Defining the new signal

$$P(\theta)=X(\theta+D)\quad\text{for all } t-D\leq\theta\leq t \tag{2.6}$$

and solving the resulting ODE in θ for P, with initial condition $P(t-D)=X(t)$ we arrive at

$$P(\theta)=e^{A(\theta-t+D)}X(t)+\int_{t-D}^{\theta}e^{A(\theta-s)}BU(s)ds\quad\text{for all } t-D\leq\theta\leq t, \tag{2.7}$$

and hence

$$P(t)=e^{AD}X(t)+\int_{t-D}^{t}e^{A(t-\theta)}BU(\theta)d\theta. \tag{2.8}$$

Representation (2.8) for the predictor signal is directly implementable since it is given in terms of the measured state $X(t)$ and the history of $U(\theta)$ for all $t-D\leq\theta\leq t$. Note also that $P(\theta)$ in (2.7) should be viewed as the output of an operator, parametrized by t, acting on $U(s)$, $t-D\leq s\leq\theta$, in the same way that the solution $X(t)$ to a linear ODE (i.e., $X(t)=e^{A(t-t_0)}X(t_0)+\int_{t_0}^{t}e^{A(t-s)}BU(s)ds$) can be viewed as the output of an operator, parametrized by t_0, acting on $U(s)$, $t_0\leq s\leq t$.

The final predictor feedback law is given by combining (2.4) and (2.8) as

$$U(t)=K\left(e^{AD}X(t)+\int_{t-D}^{t}e^{A(t-\theta)}BU(\theta)d\theta\right). \tag{2.9}$$

One can view the feedback law (2.9) as implicit, since U appears both on the left and on the right. However, one should observe that the input memory $U(\theta)$, $\theta\in[t-D,t]$, is a part of the state of the overall infinite-dimensional system, so the control law is in fact given by an explicit full-state feedback formula.

2.1.2 ▪ Backstepping Transformation of the Actuator State

Backstepping Transformation in Standard Delay Notation

The main idea of backstepping is to find an invertible transformation of the original state variables, such that the system in the new variables has some desirable properties (such as stability) or it is easier to analyze than the original system. We refer from now on to this transformed system as the "target system." For proving the stability of the closed-loop system consisting of the plant (2.1) and the control law (2.9), one has to find an appropriate backstepping transformation of the state variables (X,U). It turns out that such a transformation is needed only for the infinite-dimensional actuator state (and not also for the finite-dimensional state of the plant X) $U(\theta)$ for all $t-D\leq\theta\leq t$, and it has the following form:

$$W(\theta)=U(\theta)-KP(\theta), \tag{2.10}$$

where $P(\theta)$ is given in (2.7). Using the facts that $U(t)=KP(t)$ and that $P(t-D)=X(t)$ for all $t \geq 0$, transformation (2.10) maps the closed-loop system consisting of the plant (2.1) and the control law (2.9) to the following target system:

$$\dot{X}(t)=(A+BK)X(t)+BW(t-D), \tag{2.11}$$

$$W(t)=0 \quad \text{for all } t \geq 0. \tag{2.12}$$

Since the matrix $A+BK$ is Hurwitz and the "disturbance" $W(t-D)$ vanishes in finite time, it is evident that the target system (2.11), (2.12) possesses some desirable properties in terms of stability. However, even if the system in the transformed variables (X,W) is stable, this do not directly imply that the system in the original variables (X,U) is stable. One has to find the inverse backstepping transformation of W. This transformation is

$$U(\theta)=W(\theta)+K\Pi(\theta) \quad \text{for all } t-D \leq \theta \leq t, \tag{2.13}$$

where for all $t-D \leq \theta \leq t$

$$\Pi(\theta)=e^{(A+BK)(\theta-t+D)}X(t)+\int_{t-D}^{\theta} e^{(A+BK)(\theta-s)}BW(s)ds. \tag{2.14}$$

One can see that the inverse backstepping transformation is given by (2.13), (2.14) as follows. First note from (2.10) that $U(\theta)=W(\theta)+KP(\theta)$. Second, and most importantly, observe that $\Pi(\theta)=P(\theta)$ for all $t-D \leq \theta \leq t$ (one can see this fact by showing that $\Pi(\theta)=X(\theta+D)$ using the ODE (2.11) in the same way that one can show that $P(\theta)=X(\theta+D)$ using ODE (2.1)). However, P is given as a function of X and U through relation (2.7), whereas Π is given as a function of X and W through relation (2.14).

Backstepping Transformation in PDE Representation

An alternative representation of system (2.1) can be written, by representing the actuator state $U(\theta)$, $\theta \in [t-D,t]$, with a transport PDE, as

$$\dot{X}(t)=AX(t)+Bu(0,t), \tag{2.15}$$

$$u_t(x,t)=u_x(x,t) \quad \text{for all } x \in [0,D], \tag{2.16}$$

$$u(D,t)=U(t). \tag{2.17}$$

Note that with this representation, $u(x,t)=U(t+x-D)$ for all $x \in [0,D]$. Analogously, the PDE representation of the predictor state given in (2.7) is

$$p(x,t)=e^{Ax}X(t)+\int_0^x e^{A(x-y)}Bu(y,t)dy \quad \text{for all } x \in [0,D]. \tag{2.18}$$

With this representation

$$P(t)=p(D,t). \tag{2.19}$$

The backstepping transformation (2.10) in the PDE notation is written as

$$w(x,t)=u(x,t)-Kp(x,t), \tag{2.20}$$

and together with the control law $U(t)=KP(t)=Kp(D,t)$ transforms system (2.15)–(2.17) into the target system (2.11), (2.12), which is written now as

$$\dot{X}(t)=(A+BK)X(t)+Bw(0,t), \tag{2.21}$$

$$w_t(x,t)=w_x(x,t) \quad \text{for all } x\in[0,D], \tag{2.22}$$

$$w(D,t)=0. \tag{2.23}$$

To see this, first note that one gets relation (2.21) using (2.20) and the fact that from (2.18) it follows that $p(0,t) = X(t)$. Relation (2.23) follows from (2.20) and the fact that $U(t)=KP(t)=Kp(D,t)$. For deriving relation (2.22) one has only to show that $p_t(x,t) = p_x(x,t)$ for all $x\in[0,D]$, since already $u_t(x,t) = u_x(x,t)$ for all $x\in[0,D]$. Differentiating (2.18) with respect to t and using (2.15), (2.16) we get that

$$p_t(x,t)=e^{Ax}X(t)+e^{Ax}Bu(0,t)+\int_0^x e^{A(x-y)}Bu_y(y,t)dy. \tag{2.24}$$

Using integration by parts in the integral we arrive at

$$p_t(x,t)=e^{Ax}X(t)+Bu(x,t)+A\int_0^x e^{A(x-y)}Bu(y,t)dy. \tag{2.25}$$

Differentiating (2.18) with respect to x we conclude that

$$p_t(x,t)=p_x(x,t) \quad \text{for all } x\in[0,D]. \tag{2.26}$$

The inverse backstepping transformation in PDE notation, i.e., the analogue of (2.13), (2.14), is given by

$$u(x,t)=w(x,t)+K\pi(x,t), \tag{2.27}$$

where for all $x\in[0,D]$

$$\pi(x,t)=e^{(A+BK)x}X(t)+\int_0^x e^{(A+BK)(x-y)}Bw(y,t)dy. \tag{2.28}$$

This fact can be proved analogously with the proof of the fact that transformation (2.13), (2.14) is the inverse of (2.10), (2.7).

2.1.3 ▪ Stability Analysis with the Construction of a Lyapunov Functional

Construction of a Lyapunov Functional in Standard Delay Notation

One of the benefits of the backstepping transformation is that it allows for the construction of a Lyapunov functional for the target system (2.11), (2.12), which is less complicated from the closed-loop system consisting of the plant (2.1) and the control law (2.9). The Lyapunov functional for the target system is given as

$$V(t)=X(t)^TPX(t)+b\int_{t-D}^t e^{\theta+D-t}W(\theta)^2d\theta, \tag{2.29}$$

where the matrix $P=P^T>0$ is such that

$$(A+BK)^TP+P(A+BK)=-Q \tag{2.30}$$

for some matrix $Q=Q^T>0$. We have the following result.

Theorem 2.1. *The closed-loop system consisting of the plant* (2.1) *with the controller* (2.9) *is exponentially stable at the origin in the sense that there exist positive constants λ and ρ such that*

$$|X(t)|+\sqrt{\int_{t-D}^{t} U(\theta)^2 d\theta} \leq \rho\left(|X(0)|+\sqrt{\int_{-D}^{0} U(\theta)^2 d\theta}\right) e^{-\lambda t} \tag{2.31}$$

for all $t \geq 0$.

Proof. Taking the time derivative of V along (2.11), (2.12) we get that

$$\begin{aligned} \dot{V}(t) \leq & -\lambda_{\min}(Q)|X(t)|^2+2\left|X(t)^T PBW(t-D)\right|-bW(t-D)^2 \\ & -b\int_{t-D}^{t} e^{\theta+D-t} W(\theta)^2 d\theta. \end{aligned} \tag{2.32}$$

With Young's inequality we get

$$\begin{aligned} \dot{V}(t) \leq & -\frac{1}{2}\lambda_{\min}(Q)|X(t)|^2+\left(\frac{2|PB|^2}{\lambda_{\min}(Q)}-b\right) W(t-D)^2 \\ & -b\int_{t-D}^{t} e^{\theta+D-t} W(\theta)^2 d\theta. \end{aligned} \tag{2.33}$$

Choosing

$$b=2\frac{|PB|^2}{\lambda_{\min}(Q)}, \tag{2.34}$$

we get that

$$\dot{V}(t) \leq -\min\left\{\frac{\lambda_{\min}(Q)}{2\lambda_{\max}(P)},1\right\} V(t), \tag{2.35}$$

and hence, using the comparison principle (see also Lemma B.4 from Appendix B), we obtain

$$V(t) \leq e^{-\mu t} V(0) \quad \text{for all } t \geq 0, \tag{2.36}$$

where

$$\mu=\min\left\{\frac{\lambda_{\min}(Q)}{2\lambda_{\max}(P)},1\right\}. \tag{2.37}$$

To conclude the exponential stability in the (X,W) variables, we observe from relation (2.29) that

$$\mu_1\left(|X(t)|^2+\int_{t-D}^{t} W(\theta)^2 d\theta\right) \leq V(t) \leq \mu_2\left(|X(t)|^2+\int_{t-D}^{t} W(\theta)^2 d\theta\right), \tag{2.38}$$

where

$$\mu_1 = \min\left\{\lambda_{\min}(P), \frac{2|PB|^2}{\lambda_{\min}(Q)}\right\}, \tag{2.39}$$

$$\mu_2 = \max\left\{\lambda_{\max}(P), \frac{2|PB|^2}{\lambda_{\min}(Q)} e^D\right\}. \tag{2.40}$$

Therefore,

$$|X(t)|^2 + \int_{t-D}^{t} W(\theta)^2 d\theta \leq \frac{\mu_2}{\mu_1}\left(|X(0)|^2 + \int_{-D}^{0} W(\theta)^2 d\theta\right) e^{-\mu t}. \tag{2.41}$$

For proving stability in the original variables (X,U) one has to relate the norm of the system in the transformed variables with the norm of the system in the original variables. For doing so we use the direct and inverse backstepping transformations. Using the direct backstepping transformation (2.10) and Young's inequality we get that

$$\int_{t-D}^{t} W(\theta)^2 d\theta \leq 2\left(\int_{t-D}^{t} U(\theta)^2 d\theta + |K|^2 \int_{t-D}^{t} |P(\theta)|^2 d\theta\right). \tag{2.42}$$

Using (2.7) together with the Young and Cauchy–Schwarz inequalities, we arrive at

$$\int_{t-D}^{t} |P(\theta)|^2 d\theta \leq 2De^{2|A|D}\left(|X(t)|^2 + D|B|^2 \int_{t-D}^{t} U(\theta)^2 d\theta\right), \tag{2.43}$$

and hence

$$|X(t)|^2 + \int_{t-D}^{t} W(\theta)^2 d\theta \leq \nu_1\left(|X(t)|^2 + \int_{t-D}^{t} U(\theta)^2 d\theta\right), \tag{2.44}$$

where

$$\nu_1 = 2\max\left\{1 + 2|K|^2 De^{2|A|D} D|B|^2, 2|K|^2 De^{2|A|D}\right\}. \tag{2.45}$$

Analogously, using the inverse backstepping transformation and Young's inequality, we get that

$$\int_{t-D}^{t} U(\theta)^2 d\theta \leq 2\left(\int_{t-D}^{t} W(\theta)^2 d\theta + |K|^2 \int_{t-D}^{t} |\Pi(\theta)|^2 d\theta\right). \tag{2.46}$$

With relation (2.14), and the Young and Cauchy–Schwarz inequalities, we get that

$$\int_{t-D}^{t} |\Pi(\theta)|^2 d\theta \leq 2De^{2|A+BK|D}\left(|X(t)|^2 + D|B|^2 \int_{t-D}^{t} W(\theta)^2 d\theta\right), \tag{2.47}$$

and hence

$$|X(t)|^2 + \int_{t-D}^{t} U(\theta)^2 d\theta \leq \nu_2\left(|X(t)|^2 + \int_{t-D}^{t} W(\theta)^2 d\theta\right), \tag{2.48}$$

where

$$\nu_2 = 2\max\left\{1 + 2|K|^2 De^{2|A+BK|D} D|B|^2, 2|K|^2 De^{2|A+BK|D}\right\}. \tag{2.49}$$

Combining (2.41), (2.44), (2.48), we get (2.31) with

$$\rho = \sqrt{\frac{2\mu_1\nu_1\nu_2}{\mu_1}}, \tag{2.50}$$

$$\lambda = \frac{\mu}{2}. \qquad \square \tag{2.51}$$

Construction of a Lyapunov Functional in PDE Representation

Using the backstepping transformation in PDE representation (2.20) one can, analogously with the construction of a Lyapunov functional using the delay notation, construct a Lyapunov functional using PDEs. The analogue of the Lyapunov functional given in (2.29) is

$$V(t) = X(t)^T PX(t) + b\int_0^D e^x w(x,t)^2 dx, \tag{2.52}$$

where b is defined in (2.34). Using (2.52) one can show that for the closed-loop system consisting of the plant (2.15)–(2.17) with the controller (2.9) the following holds:

$$|X(t)| + \sqrt{\int_0^D u(x,t)^2 dx} \le \rho\left(|X(0)| + \sqrt{\int_0^D u(x,0)^2 dx}\right) e^{-\lambda t} \tag{2.53}$$

for all $t \ge 0$, where ρ and λ are given in (2.50) and (2.51), respectively. To see this, start by differentiating V and using relations (2.21)–(2.23), (2.30) to get

$$\begin{aligned}\dot{V}(t) \le {} & -\lambda_{\min}(Q)|X(t)|^2 + 2\left|X(t)^T PBw(0,t)\right| \\ & + 2b\int_0^D e^x w(x,t) w_x(x,t) dx.\end{aligned} \tag{2.54}$$

Using integration by parts together with relation (2.23) we get

$$\begin{aligned}\dot{V}(t) \le {} & -\lambda_{\min}(Q)|X(t)|^2 + 2\left|X(t)^T PBw(0,t)\right| - bw(0,t)^2 \\ & - b\int_0^D e^x w(x,t)^2 dx.\end{aligned} \tag{2.55}$$

With Young's inequality and choosing b as in (2.34) we arrive at (2.36). Using the fact that for all $x \in [0,D]$, $w(x,t) = W(t+x-D)$, which follows from (2.22), one can conclude from (2.38) that

$$\mu_1\left(|X(t)|^2 + \int_0^D w(x,t)^2 dx\right) \le V(t) \le \mu_2\left(|X(t)|^2 + \int_0^D w(x,t)^2 dx\right), \tag{2.56}$$

where μ_1 and μ_2 are given in (2.39) and (2.40), respectively. Using also the fact that for all $x \in [0,D]$, $u(x,t) = U(t+x-D)$, which follows from (2.16), we get using (2.44), (2.48) that

$$|X(t)|^2 + \int_0^D w(x,t)^2 dx \leq \nu_1 \left(|X(t)|^2 + \int_0^D u(x,t)^2 dx \right), \tag{2.57}$$

$$|X(t)|^2 + \int_0^D u(x,t)^2 dx \leq \nu_2 \left(|X(t)|^2 + \int_0^D w(x,t)^2 dx \right). \tag{2.58}$$

Estimate (2.53) then follows by combining (2.36), (2.56)–(2.58).

2.2 ▪ Strict-Feedback Systems with Delayed Integrators

We develop a predictor-based controller for linear strict-feedback systems with delays in the virtual inputs in Section 2.2.1. Specifically, an infinite-dimensional backstepping procedure is developed, which together with a control law converts the system to an exponentially stable target system. Using the invertibility of the backstepping transformations, we then prove exponential stability of the closed-loop system using a suitably weighted Lyapunov–Krasovskii functional in Section 2.2.2. The effectiveness of the proposed controller is illustrated by a simulation example of a second order unstable system in Section 2.2.3.

2.2.1 ▪ Predictor Feedback Design

We consider the following n-dimensional linear system:

$$\dot{\bar{X}}_1(t) = \bar{a}_{11}\bar{X}_1(t) + b_1\bar{X}_2(t-D_1), \tag{2.59}$$

$$\dot{\bar{X}}_2(t) = \bar{a}_{21}\bar{X}_1(t) + \bar{a}_{22}\bar{X}_2(t) + b_2\bar{X}_3(t-D_2), \tag{2.60}$$

$$\vdots$$

$$\dot{\bar{X}}_n(t) = \bar{a}_{n1}\bar{X}_1(t) + \cdots + \bar{a}_{nn}\bar{X}_n(t) + b_n\bar{U}(t-D_n), \tag{2.61}$$

where $\bar{X}_i(t), \bar{a}_{ij}, \bar{U}(t) \in \mathbb{R}$, $b_i \neq 0$, and $D_i \in \mathbb{R}_+$. We start by redefining the states of system (2.59)–(2.61) such that the coefficients in front of the delayed terms are unity. That is, we define

$$X_1(t) = \bar{X}_1(t), \tag{2.62}$$

$$X_2(t) = b_1\bar{X}_2(t), \tag{2.63}$$

$$X_3(t) = b_1 b_2 \bar{X}_3(t), \tag{2.64}$$

$$\vdots$$

$$X_n(t) = b_1 b_2 \dots b_{n-1}\bar{X}_n(t). \tag{2.65}$$

Moreover, for notational consistency we define

$$U(t) = b_1 b_2 \dots b_n \bar{U}(t), \tag{2.66}$$

$$a_{ij} = \left\{ \begin{array}{cc} \bar{a}_{ij} & \text{if} \quad i=j \\ b_j \dots b_{i-1}\bar{a}_{ij} & \text{if} \quad i>j \end{array} \right\}. \tag{2.67}$$

In the new variables, system (2.59)–(2.61) is transformed to

$$\dot{X}_1(t)=a_{11}X_1(t)+X_2(t-D_1), \tag{2.68}$$
$$\dot{X}_2(t)=a_{21}X_1(t)+a_{22}X_2(t)+X_3(t-D_2), \tag{2.69}$$
$$\vdots$$
$$\dot{X}_n(t)=a_{n1}X_1(t)+\ldots+a_{nn}X_n(t)+U(t-D_n). \tag{2.70}$$

We state here our controller, and in Section 3 we analyze the stability properties of the closed-loop system. The controller for the system (2.68)–(2.70) is given by

$$\begin{aligned} U(t) &= u(D_n,t) \\ &= \alpha_n(D_n,t) \\ &= -a_{n1}P_1\left(t-\sum_{k=1}^{n-1}D_k\right)-\cdots-a_{nn}P_n(t)-c_n\left(P_n(t)-\alpha_{n-1}\left(D_{n-1}+D_n,t\right)\right) \\ &\quad +\frac{\partial\alpha_{n-1}\left(D_{n-1}+D_n,t\right)}{\partial x}, \end{aligned} \tag{2.71}$$

where

$$\begin{aligned} \alpha_i(x,t) &= -a_{i1}P_1\left(t-\sum_{k=1}^{n}D_k+x\right)-\cdots-a_{ii}P_i\left(t-\sum_{k=i}^{n}D_k+x\right) \\ &\quad -c_i\left(P_i\left(t-\sum_{k=i}^{n}D_k+x\right)-\alpha_{i-1}\left(D_{i-1}+x,t\right)\right) \\ &\quad +\frac{\partial\alpha_{i-1}\left(D_{i-1}+x,t\right)}{\partial x}, \quad x\in\left[0,\sum_{k=i}^{n}D_k\right], \end{aligned} \tag{2.72}$$

$$\alpha_1(x,t)=-(a_{11}+c_1)P_1\left(t+x-\sum_{k=1}^{n}D_k\right), \quad x\in\left[0,\sum_{k=1}^{n}D_k\right], \tag{2.73}$$

and the $c_i, i=1,2,\ldots,n$, are arbitrary positive constants. In the above control scheme we use the $P_i(t)$ signals, the $\sum_{k=i}^{n}D_k$ seconds ahead predictors of the $X_i(t)$ state (this fact becomes clear later on). That is, it holds that $P_i(t)=X_i(t+\sum_{k=i}^{n}D_k)$. These signals are given by

$$P_1(t)=X_1(t)+\int_{t-\sum_{k=1}^{n}D_k}^{t}(a_{11}P_1(\theta)+P_2(\theta))d\theta, \tag{2.74}$$
$$P_2(t)=X_2(t)+\int_{t-\sum_{k=2}^{n}D_k}^{t}(a_{21}P_1(\theta-D_1)+a_{22}P_2(\theta)+P_3(\theta))d\theta, \tag{2.75}$$
$$\vdots$$
$$\begin{aligned} P_n(t) &= X_n(t)+\int_{t-D_n}^{t}\left(a_{n1}P_1\left(\theta-\sum_{k=1}^{n-1}D_k\right)+a_{n2}P_2\left(\theta-\sum_{k=2}^{n-1}D_k\right)\ldots\right. \\ &\quad \left.+a_{nn}P_n(\theta)+U(\theta)\right)d\theta, \end{aligned} \tag{2.76}$$

with initial conditions

$$P_1(\theta)=X_1(0)+\int_{-\sum_{k=1}^{n}D_k}^{\theta}(a_{11}P_1(\sigma)+P_2(\sigma))\,d\sigma, \tag{2.77}$$

$$P_2(\theta)=X_2(0)+\int_{-\sum_{k=2}^{n}D_k}^{\theta}(a_{21}P_1(\sigma-D_1)+a_{22}P_2(\sigma)+P_3(\sigma))\,d\sigma, \tag{2.78}$$

$$\vdots$$

$$P_n(\theta)=X_n(0)+\int_{-D_n}^{\theta}\left(a_{n1}P_1\left(\sigma-\sum_{k=1}^{n-1}D_k\right)+a_{n2}P_2\left(\sigma-\sum_{k=2}^{n-1}D_k\right)\dots\right.$$
$$\left.+a_{nn}P_n(\sigma)+U(\sigma)\right)d\sigma, \tag{2.79}$$

where θ is defined in each $P_i(\theta)$ as $\theta\in\left[-\sum_{k=i}^{n}D_k,0\right]$. Note here that the notation $\frac{\partial\alpha_{i-1}(D_{i-1}+x,t)}{\partial x}$ corresponds to $\frac{\partial\alpha_{i-1}(x',t)}{\partial x'}\big|_{x'=x+D_{i-1}}$, which includes the time derivatives of the signals $P_1(t),\dots,P_{i-1}(t)$. These derivatives are obtained from (2.68)–(2.70) and (2.74)–(2.76).

2.2.2 ▪ Stability analysis

We first state a theorem describing our main stability result and then prove it using a series of technical lemmas.

Theorem 2.2. *System* (2.68)–(2.70) *with the controller* (2.71) *is exponentially stable in the sense that there exist constants $\varkappa$ and λ such that*

$$\Omega(t)\le\varkappa\Omega(0)e^{-\lambda t}, \tag{2.80}$$

where

$$\Omega(t)=\frac{1}{2}\sum_{i=1}^{n}X_i^2(t)+\frac{1}{2}\sum_{i=2}^{n}\int_{t-D_{i-1}}^{t}X_i^2(\theta)d\theta+\frac{1}{2}\int_{t-D_n}^{t}U^2(\theta)d\theta \tag{2.81}$$

and

$$\int_{t-D_{i-1}}^{t}X_i^2(\theta)d\theta=\int_0^{D_{i-1}}\xi_i^2(x,t)dx=\|\xi_i(t)\|^2, \tag{2.82}$$

$$\int_{t-D_n}^{t}U^2(\theta)d\theta=\int_0^{D_n}u^2(x,t)dx=\|u(t)\|^2, \tag{2.83}$$

$$\xi_i(x,t)=X_i(t+x-D_{i-1}),\quad x\in[0,D_{i-1}], \tag{2.84}$$

$$u(x,t)=U(t+x-D_n),\quad x\in[0,D_n]. \tag{2.85}$$

We first give and prove the following lemmas.

Lemma 2.3. *The signals $P_i(t)$ defined in* (2.74)–(2.76) *are, respectively, the $\sum_{k=i}^{n}D_k$ seconds ahead predictors of the $X_i(t)$ states. Moreover an equivalent representation for* (2.74)–(2.76) *is given by*

$$p_1\left(\sum_{k=1}^{n} D_k, t\right) = X_1(t) + \int_0^{\sum_{k=1}^{n} D_k} (a_{11}p_1(y,t) + p_2(y-D_1,t))\,dy, \tag{2.86}$$

$$p_2\left(\sum_{k=2}^{n} D_k, t\right) = X_2(t) + \int_0^{\sum_{k=2}^{n} D_k} (a_{21}p_1(y,t) + a_{22}p_2(y,t) + p_3(y-D_2,t))\,dy, \tag{2.87}$$

$$\vdots$$

$$p_n(D_n,t) = X_n(t) + \int_0^{D_n} (a_{n1}p_1(y,t) + \cdots + a_{nn}p_n(y,t) + u(y,t))\,dy, \tag{2.88}$$

where

$$p_i(x,t) = P_i\left(t + x - \sum_{k=i}^{n} D_k\right), \quad x \in \left[0, \sum_{k=i}^{n} D_k\right]. \tag{2.89}$$

Proof. Consider the equivalent representation of system (2.68)–(2.70) using transport PDEs for the delayed states and control:

$$\dot{X}_1(t) = a_{11}X_1(t) + \xi_2(0,t), \tag{2.90}$$
$$\xi_{2_t}(x,t) = \xi_{2_x}(x,t), \tag{2.91}$$
$$\xi_2(D_1,t) = X_2(t), \tag{2.92}$$
$$\dot{X}_2(t) = a_{21}X_1(t) + a_{22}X_2(t) + \xi_3(0,t), \tag{2.93}$$
$$\xi_{3_t}(x,t) = \xi_{3_x}(x,t), \tag{2.94}$$
$$\xi_3(D_2,t) = X_3(t), \tag{2.95}$$
$$\vdots$$
$$\dot{X}_n(t) = a_{n1}X_1(t) + \cdots + a_{nn}X_n(t) + u(0,t), \tag{2.96}$$
$$u_t(x,t) = u_x(x,t), \tag{2.97}$$
$$u(D_n,t) = U(t). \tag{2.98}$$

Consider the following ODEs in x (to make clear that these are ODEs in x, consider the time t acting as a parameter rather than as a running variable):

$$p_{1_x}(x,t) = a_{11}p_1(x,t) + p_2(x-D_1,t), \tag{2.99}$$
$$p_{2_x}(x,t) = a_{21}p_1(x,t) + a_{22}p_2(x,t) + p_3(x-D_2,t), \tag{2.100}$$
$$\vdots$$
$$p_{n_x}(x,t) = a_{n1}p_1(x,t) + \cdots + a_{nn}p_n(x,t) + u(x,t), \tag{2.101}$$

where, for each $p_i(x,t)$, x varies in $\left[0, \sum_{k=i}^{n} D_k\right]$. The initial conditions for the above system of ODEs are given by

$$p_i(0,t) = X_i(t) \quad \text{for all } i \tag{2.102}$$

and

$$p_i(\theta_i,t)=X_i(t+\theta_i),\quad \theta_i\in[-D_{i-1},0],\quad i=2,\dots,n. \tag{2.103}$$

Since system (2.99)–(2.101) is driven by the input $u(x,t)$, which satisfies a transport PDE, the same holds for all the $p_i(x,t)$ (see, for example, [86], [89]). Thus,

$$\frac{\partial p_i(x,t)}{\partial t}=\frac{\partial p_i(x,t)}{\partial x},\quad x\in\left[0,\sum_{k=i}^{n}D_k\right],\quad \text{for all } i. \tag{2.104}$$

By taking into account (2.102) we have that

$$p_i(x,t)=X_i(t+x),\quad x\in\left[0,\sum_{k=i}^{n}D_k\right],\quad \text{for all } i. \tag{2.105}$$

To see that (2.104) holds it is sufficient to prove that (2.105) is the unique solution of the ODEs in x given by (2.99)–(2.101) with the initial conditions (2.102)–(2.103). Thus, the $p_i(x,t)$ are functions of only one variable, namely, $x+t$, and consequently (2.104) holds. Therefore, it remains to prove that (2.105) is the unique solution of the initial value problem (2.99)–(2.103). Toward this end, by taking into account (2.85) we point out that (2.105) satisfies the initial value problem (2.99)–(2.103). Then, assuming that $X_i(t+\theta_i), i=2,\dots,n$, are continuous for all $\theta_i\in[-D_{i-1},0]$, using Theorem 2.1 from [49] we can conclude that (2.105) is the unique solution of the ODEs in x given by (2.99)–(2.101) with the initial conditions (2.102)–(2.103). Thus, (2.104) holds.

From relation (2.105) it becomes clear that the $p_i(x,t)$ are the x seconds ahead predictors of the states. By defining

$$p_i\left(\sum_{k=i}^{n}D_k,t\right)=P_i(t)\quad \text{for all } i, \tag{2.106}$$

we get (2.89). By integrating from 0 to x (2.99)–(2.101) we get

$$p_1(x,t)=X_1(t)+\int_0^x(a_{11}p_1(y,t)+p_2(y-D_1,t))\,dy, \tag{2.107}$$

$$p_2(x,t)=X_2(t)+\int_0^x(a_{21}p_1(y,t)+a_{22}p_2(y,t)+p_3(y-D_2,t))\,dy, \tag{2.108}$$

$$\vdots$$

$$p_n(x,t)=X_n(t)+\int_0^x(a_{n1}p_1(y,t)+\cdots+a_{nn}p_n(y,t)+u(y,t))\,dy. \tag{2.109}$$

By setting in each $p_i(x,t)$, $x=\sum_{k=i}^{n}D_k$, and using (2.89) we get (2.86)–(2.88). □

It is important here to observe that the total delay from the input to each state $X_i(t)$ is $\sum_{k=i}^{n}D_k$. This explains the fact that our predictor intervals are different for each state and specifically must be $\sum_{k=i}^{n}D_k$ seconds for each state $X_i(t)$. Our controller design is based on a recursive procedure that transforms system (2.68)–(2.70) into a target system which is exponentially stable with the controller (2.71). Then, using the invertibility of this transformation, we prove exponential stability of the original system. We now state this transformation, along with its inverse.

Lemma 2.4. *The state transformation defined by*

$$Z_1(t) = X_1(t), \tag{2.110}$$

$$Z_{i+1}(t) = X_{i+1}(t) - \alpha_i(D_i, t), \quad i = 1, 2, \ldots, n-1, \tag{2.111}$$

along with the transformation of the actuator state

$$w(x,t) = u(x,t) - \alpha_n(x,t), \quad x \in [0, D_n], \tag{2.112}$$

where the $\alpha_i(x,t)$ are defined as in (2.72)–(2.73), transforms the system (2.68)–(2.70) into the target system with the control law given by (2.71). The target system is given by

$$\dot{Z}_1(t) = -c_1 Z_1(t) + Z_2(t - D_1), \tag{2.113}$$

$$\dot{Z}_2(t) = -c_2 Z_2(t) + Z_3(t - D_2), \tag{2.114}$$

$$\vdots$$

$$\dot{Z}_n(t) = -c_n Z_n(t) + W(t - D_n), \tag{2.115}$$

where

$$W(\theta) = 0, \quad \theta \geq 0. \tag{2.116}$$

Proof. Before we start our recursive procedure, we rewrite the target system using transport PDEs as

$$\dot{Z}_1(t) = -c_1 Z_1(t) + \zeta_2(0,t), \tag{2.117}$$

$$\zeta_{2_t}(x,t) = \zeta_{2_x}(x,t), \tag{2.118}$$

$$\zeta_2(D_1,t) = Z_2(t), \tag{2.119}$$

$$\dot{Z}_2(t) = -c_2 Z_2(t) + \zeta_3(0,t), \tag{2.120}$$

$$\zeta_{3_t}(x,t) = \zeta_{3_x}(x,t), \tag{2.121}$$

$$\zeta_3(D_2,t) = Z_3(t), \tag{2.122}$$

$$\vdots$$

$$\dot{Z}_n(t) = -c_n Z_n(t) + w(0,t), \tag{2.123}$$

$$w_t(x,t) = w_x(x,t), \tag{2.124}$$

$$w(D_n,t) = 0. \tag{2.125}$$

Note that

$$\zeta_i(x,t) = Z_i(t + x - D_{i-1}), \quad x \in [0, D_{i-1}]. \tag{2.126}$$

Step 1. Following the backstepping procedure we first stabilize $X_1(t)$ with the virtual input $\alpha_1(D_1,t)$. We define

$$\zeta_2(x,t) = \xi_2(x,t) - \alpha_1(x,t); \tag{2.127}$$

then using (2.90) we get

$$\dot{X}_1(t) = a_{11} X_1(t) + \zeta_2(0,t) + \alpha_1(0,t). \tag{2.128}$$

By choosing $\alpha_1(x,t) = -(a_{11}+c_1)p_1(x,t)$ (note the equivalent representation of $\alpha_1(x,t)$ using (2.89)) and by using (2.102), we get

$$\dot{Z}_1(t) = -c_1 Z_1(t) + \zeta_2(0,t). \tag{2.129}$$

From (2.127) with $x = D_1$ and (2.119) it follows that

$$Z_2(t) = X_2(t) - \alpha_1(D_1,t). \tag{2.130}$$

By setting now

$$\zeta_3(x,t) = \xi_3(x,t) - \alpha_2(x,t) \tag{2.131}$$

and using (2.127), (2.91), and (2.93), we have

$$\begin{aligned}\dot{Z}_2(t) &= \zeta_{2_x}(x,t)|_{x=D_1}\\ &= a_{21}X_1(t) + a_{22}X_2(t) + \zeta_3(0,t) + \alpha_2(0,t) - \frac{\partial \alpha_1(D_1,t)}{\partial x},\end{aligned} \tag{2.132}$$

where $\frac{\partial \alpha_1(D_1,t)}{\partial x}$ corresponds to $\frac{\partial \alpha_1(x,t)}{\partial x}|_{x=D_1}$ and we use the fact that $\alpha_{1_t}(x,t) = \alpha_{1_x}(x,t)$ (which is a consequence of relation (2.73)).

Step 2. By choosing

$$\begin{aligned}\alpha_2(x,t) &= -a_{21}p_1(x,t) - a_{22}p_2(x,t) - c_2(p_2(x,t) - \alpha_1(D_1+x,t))\\ &\quad + \frac{\partial \alpha_1(x+D_1,t)}{\partial x}\\ &= -a_{21}p_1(x,t) - a_{22}p_2(x,t) - c_2(p_2(x,t) + (a_{11}+c_1)\\ &\quad \times p_1(D_1+x,t)) - (a_{11}+c_1)\\ &\quad \times (a_{11}p_1(x+D_1,t) + p_2(x,t)),\end{aligned} \tag{2.133}$$

we get from (2.132) (with the help of (2.102)) that

$$\dot{Z}_2(t) = -c_2 Z_2(t) + \zeta_3(0,t). \tag{2.134}$$

By setting now $x = D_2$ in (2.131) and using (2.122), we get

$$Z_3(t) = X_3(t) - \alpha_2(D_2,t). \tag{2.135}$$

If we now define

$$\zeta_4(x,t) = \xi_4(x,t) - \alpha_3(x,t), \tag{2.136}$$

then with the help of (2.94) we get

$$\begin{aligned}\dot{Z}_3(t) &= a_{31}X_1(t) + a_{32}X_2(t) + a_{33}X_3(t) + \zeta_4(0,t)\\ &\quad + \alpha_3(0,t) - \frac{\partial \alpha_2(D_2,t)}{\partial x}.\end{aligned} \tag{2.137}$$

Step i. Assume now that

$$\dot{Z}_{i-1}(t) = -c_{i-1}Z_{i-1}(t) + \zeta_i(0,t), \tag{2.138}$$

and define $\zeta_{i+1}(x,t)$ as

$$\zeta_{i+1}(x,t) = \xi_{i+1}(x,t) - \alpha_i(x,t). \tag{2.139}$$

Then from (2.111) with $x = D_{i-1}$ we have that

$$\begin{aligned}\dot{Z}_i(t) = {}& a_{i1}X_1(t) + \cdots + a_{ii}X_i(t) + \zeta_{i+1}(0,t)\\ &+ \alpha_i(0,t) - \frac{\partial \alpha_{i-1}(D_{i-1},t)}{\partial x}.\end{aligned} \tag{2.140}$$

Hence, with

$$\begin{aligned}\alpha_i(x,t) = {}& -a_{i1}p_1(x,t) - \cdots - a_{ii}p_i(x,t) - c_i\left(p_i(x,t) - \alpha_{i-1}(D_{i-1}+x,t)\right)\\ &+ \frac{\partial \alpha_{i-1}(D_{i-1}+x,t)}{\partial x},\end{aligned} \tag{2.141}$$

we get

$$\dot{Z}_i(t) = -c_iZ_i(t) + \zeta_{i+1}(0,t). \tag{2.142}$$

Step n. In the last step we choose the controller $U(t)$. Since

$$\dot{Z}_n(t) = a_{n1}X_1(t) + \cdots + a_{nn}X_n(t) + u(0,t) - \frac{\partial \alpha_{n-1}(D_{n-1},t)}{\partial x}, \tag{2.143}$$

then using (2.72) for $i = n$ we have that

$$\dot{Z}_n(t) = -c_nZ_n(t) + w(0,t), \tag{2.144}$$

and using (2.71),

$$w_t(x,t) = w_x(x,t), \tag{2.145}$$

$$w(D_n,t) = 0. \tag{2.146}$$

Assuming an initial condition for (2.145) as

$$w(x,0) = w_0(x), \tag{2.147}$$

and by defining a new variable $W(\cdot)$ as

$$w_0(x) = W(x-D), \quad x \in [0,D_n], \tag{2.148}$$

we get that

$$w(x,t) = \left\{\begin{array}{cc} W(t+x-D), & -D \le t+x-D \le 0\\ 0, & t+x-D \ge 0\end{array}\right\}. \tag{2.149}$$

Defining $\theta = t+x-D$ one gets (2.116). Note here that based on (2.112), $w_0(x)$ is given by

$$w_0(x) = u(x,0) - \alpha_n(x,0), \quad x \in [0,D_n]. \qquad \square \tag{2.150}$$

We now define the inverse transformation of (2.110)–(2.112).

Lemma 2.5. *The inverse transformation of* (2.110)–(2.112) *is defined as*

$$X_1(t) = Z_1(t), \tag{2.151}$$

$$X_{i+1}(t) = Z_{i+1}(t) + \beta_i(D_i, t), \quad i = 1, 2, \ldots, n-1, \tag{2.152}$$

$$u(x,t) = w(x,t) + \beta_n(x,t), \quad x \in [0, D_n], \tag{2.153}$$

where the $\beta_i(x,t)$ *are now given by*

$$\beta_1(x,t) = -(a_{11} + c_1)\epsilon_1(x,t), \quad x \in \left[0, \sum_{k=1}^{n} D_k\right], \tag{2.154}$$

$$\begin{aligned} \beta_i(x,t) = &-a_{i1}\epsilon_1(x,t) - a_{i2}(\epsilon_2(x,t) + \beta_1(D_1 + x, t)) - \cdots \\ &- a_{ii}\left(\epsilon_i(x,t) + \beta_{i-1}(D_{i-1} + x, t)\right) - c_i\epsilon_i(x,t) \\ &+ \frac{\partial \beta_{i-1}(D_{i-1} + x, t)}{\partial x}, \quad x \in \left[0, \sum_{k=i}^{n} D_k\right], \quad \textit{for all } i = 2, \ldots, n, \end{aligned} \tag{2.155}$$

and the $\epsilon_i(x,t)$ *(the predictors of the transformed states) are given by the relations*

$$\epsilon_1(x,t) = Z_1(t) + \int_0^x (-c_1\epsilon_1(y,t) + \epsilon_2(y - D_1, t))dy, \tag{2.156}$$

$$\epsilon_2(x,t) = Z_2(t) + \int_0^x (-c_2\epsilon_2(y,t) + \epsilon_3(y - D_2, t))dy, \tag{2.157}$$

$$\vdots$$

$$\epsilon_n(x,t) = Z_n(t) + \int_0^x (-c_n\epsilon_n(y,t) + w(y,t))dy, \tag{2.158}$$

where, in each $\epsilon_i(x,t)$*,* x *varies in* $\left[0, \sum_{k=i}^{n} D_k\right]$.

Proof. Applying arguments similar to those in Lemma 2, we prove that the inverse transformation of (2.110)–(2.112) and (2.72)–(2.73) is given by (2.151)–(2.155). □

We now prove stability of the transformed system.

Lemma 2.6. *The target system is exponentially stable in the sense that there exist constants* M_1*,* m_1*, and* m_2 *such that*

$$\Xi(t) \le \frac{M_1(1 + D_{\max})}{m_2}\Xi(0)e^{-\frac{m_1}{M_1}t}, \tag{2.159}$$

where

$$\Xi(t) = \frac{1}{2}\sum_{i=1}^{n} Z_i^2(t) + \frac{1}{2}\sum_{i=2}^{n}\int_{t-D_{i-1}}^{t} Z_i^2(\theta)d\theta + \frac{1}{2}\int_{t-D_n}^{t} W^2(\theta)d\theta, \tag{2.160}$$

$$D_{\max} = \max\{D_i\} \quad \textit{for all } i \tag{2.161}$$

and

$$\int_{t-D_{i-1}}^{t} Z_i^2(\theta)d\theta = \int_0^{D_{i-1}} \zeta_i^2(x,t)dx = \|\zeta_i(t)\|^2, \tag{2.162}$$

$$\int_{t-D_n}^{t} W^2(\theta)d\theta = \int_0^{D_n} w^2(x,t)dx = \|w(t)\|^2. \tag{2.163}$$

Proof. We consider the following Lyapunov-like function:

$$V(t)=\frac{1}{2}\sum_{i=1}^{n}k_i Z_i^2(t)+\frac{1}{2}\sum_{i=2}^{n}\lambda_i\int_0^{D_{i-1}}(1+x)\zeta_i^2(x,t)dx$$
$$+\frac{\lambda_{n+1}}{2}\int_0^{D_n}(1+x)w^2(x,t)dx. \tag{2.164}$$

Note that the above functional can be considered as a control Lyapunov functional in the sense of [67]. This fact reinforces the strength of the present result: a control Lyapunov functional is actually constructed. By taking the time derivative of the above function along the solutions of the $Z(t)$ system and by exploiting the fact that $\zeta_i(x,t)$ and $w(x,t)$ satisfy transport PDEs (based on (2.118), (2.121), and (2.124)), it follows that

$$\dot V(t)=-\sum_{i=1}^{n}c_i k_i Z_i^2(t)+\sum_{i=1}^{n-1}k_i Z_i(t)\zeta_{i+1}(0,t)+k_n Z_n(t)w(0,t)$$
$$+\frac{1}{2}\sum_{i=2}^{n}\lambda_i\left(1+D_{i-1}\right)Z_i^2(t)-\frac{1}{2}\sum_{i=2}^{n}\lambda_i\zeta_i^2(0,t)$$
$$-\frac{1}{2}\sum_{i=2}^{n}\lambda_i\int_0^{D_{i-1}}\zeta_i^2(x,t)dx-\frac{\lambda_{n+1}}{2}w^2(0,t)$$
$$-\frac{\lambda_{n+1}}{2}\int_0^{D_n}w^2(x,t)dx, \tag{2.165}$$

where we used integration by parts in the above integrals. By choosing the weights as

$$k_i=2\frac{\lambda_i}{c_i}\left(1+D_{i-1}\right),\quad i=2,\dots,n,$$
$$k_1=2, \tag{2.166}$$
$$\lambda_i=4\frac{\lambda_{i-1}\left(1+D_{i-2}\right)}{c_{i-1}^2},\quad i=3,\dots,n+1,$$
$$\lambda_2=\frac{1}{2c_1}, \tag{2.167}$$

and after some manipulations that incorporate completion of squares, we get

$$\dot V(t)\le-\frac{1}{2}\sum_{i=1}^{n}c_i k_i Z_i^2(t)-\frac{1}{2}\sum_{i=2}^{n}\lambda_i\int_0^{D_{i-1}}\zeta_i^2(x,t)dx$$
$$-\frac{\lambda_{n+1}}{2}\int_0^{D_n}w^2(x,t)dx. \tag{2.168}$$

Defining

$$M_1=\max\left\{k_i,\lambda_{i+1}\right\},\quad i=1,2,\dots,n, \tag{2.169}$$
$$m_1=\min\left\{\frac{c_i k_i}{2},\frac{\lambda_{i+1}}{2(1+D_i)}\right\},\quad i=1,2,\dots,n, \tag{2.170}$$

it follows that

$$\dot V(t)\le-\frac{m_1}{M_1}V(t). \tag{2.171}$$

If we now define

$$m_2 = \min\left\{\frac{k_i}{2}, \frac{\lambda_{i+1}}{2}\right\}, \quad i = 1,2,\dots,n, \tag{2.172}$$

then

$$\Xi(t) \leq \frac{V(0)}{m_2} e^{-\frac{m_1}{M_1}t} \leq \frac{M_1(1+D_{\max})}{m_2}\Xi(0)e^{-\frac{m_1}{M_1}t}. \qquad \square \tag{2.173}$$

We give now the following lemma.

Lemma 2.7. *There exist constants G_i such that*

$$p_i(x,t)^2 \leq G_i\left(|X(t)|^2 + \sum_{i=2}^{n}\int_0^{D_{i-1}} \xi_i^2(y,t)dy + \int_0^{D_n} u^2(y,t)dy\right)$$
$$\text{for all } x \in \left[0, \sum_{k=i}^{n} D_k\right], \tag{2.174}$$

where

$$|X(t)|^2 = \sum_{i=1}^{n} X_i^2(t), \tag{2.175}$$

and the bound (2.174) *is independent of x.*

Proof. By solving (2.99)–(2.101), and by taking into account that this ODE in the x system is in strict-feedback form, we get

$$p_1(x,t) = \int_{-D_1}^{x-D_1} v_{11}(x-y-D_1)p_2(y,t)dy + v_{11}(x)X_1(t),$$
$$x \in \left[0, \sum_{k=1}^{n} D_k\right], \tag{2.176}$$
$$p_2(x,t) = \sum_{i=1}^{2}\int_{-D_i}^{x-D_i} v_{2i}(x-y-D_i)p_{i+1}(y,t)dy + \sum_{i=1}^{2} v_{2i}(x)X_i(t),$$
$$x \in \left[0, \sum_{k=2}^{n} D_k\right], \tag{2.177}$$
$$\vdots$$
$$p_n(x,t) = \sum_{i=1}^{n-1}\int_{-D_i}^{x-D_i} v_{ni}(x-y-D_i)p_{i+1}(y,t)dy + \sum_{i=1}^{n} v_{ni}(x)X_i(t)$$
$$+ \int_0^x v_{nn}(x-y)u(y,t)dy, \quad x \in [0, D_n], \tag{2.178}$$

where

$$e^{A_0 x} = \begin{bmatrix} v_{11}(x) & 0 & 0 & \dots & 0 \\ v_{21}(x) & v_{22}(x) & 0 & \dots & 0 \\ \vdots & \vdots & \vdots & \vdots & \vdots \\ v_{n1}(x) & v_{n2}(x) & \dots & \dots & v_{nn}(x) \end{bmatrix}, \tag{2.179}$$

$$A_0 = \begin{bmatrix} a_{11} & 0 & 0 & \dots & 0 \\ a_{21} & a_{22} & 0 & \dots & 0 \\ \vdots & \vdots & \vdots & \vdots & \vdots \\ a_{n1} & a_{n2} & \dots & \dots & a_{nn} \end{bmatrix}. \tag{2.180}$$

By applying the Young and Cauchy-Schwarz inequalities to equations (2.176)–(2.178), we get

$$p_1^2(x,t) \le A_1 \left(X_1^2(t) + \int_{-D_1}^{x-D_1} p_2^2(y,t)dy \right), \tag{2.181}$$

$$p_2^2(x,t) \le A_2 \left(X_1^2(t) + X_2^2(t) + \sum_{i=1}^{2} \int_{-D_i}^{x-D_i} p_{i+1}^2(y,t)dy \right), \tag{2.182}$$

$$\vdots$$

$$p_n^2(x,t) \le A_n \left(\sum_{i=1}^{n} X_i^2(t) + \sum_{i=1}^{n-1} \int_{-D_i}^{x-D_i} p_{i+1}^2(y,t)dy + \int_0^x u^2(y,t)dy \right), \tag{2.183}$$

where, in each of the above bounds, $x \in \left[0, \sum_{k=i}^{n} D_k\right]$, respectively. Also

$$\begin{aligned} A_i = 2i \max \Bigg[& \sup_{x\in\left[0,\sum_{k=i}^{n} D_k\right]} v_{i1}^2(x), \dots, \sup_{x\in\left[0,\sum_{k=i}^{n} D_k\right]} v_{ii}^2(x), \\ & \sup_{x\in\left[0,\sum_{k=i}^{n} D_k\right]} \int_{-D_1}^{x-D_1} v_{i1}(x-y-D_1)^2 dy, \dots, \\ & \sup_{x\in\left[0,\sum_{k=i}^{n} D_k\right]} \int_{-D_i}^{x-D_i} v_{ii}(x-y-D_i)^2 dy \Bigg]. \end{aligned} \tag{2.184}$$

If we now take into account that $p_i(x-D_{i-1},t) = X_i(t+x-D_{i-1}) = \xi_i(x,t)$, we can rewrite (2.181)–(2.183) as

$$p_1^2(x,t) \le A_1 \left(X_1^2(t) + ||\xi_2(t)||^2 + \int_0^{x-D_1} p_2^2(y,t)dy \right), \tag{2.185}$$

$$p_2^2(x,t) \le A_2 \left(\sum_{k=1}^{2} X_i^2(t) + \sum_{i=1}^{2} ||\xi_{i+1}(t)||^2 + \sum_{i=1}^{2} \int_0^{x-D_i} p_{i+1}^2(y,t)dy \right), \tag{2.186}$$

$$\vdots$$

$$p_n^2(x,t) \le A_n\left(\sum_{k=1}^{n} X_i^2(t) + \sum_{i=1}^{n-1} ||\xi_{i+1}(t)||^2 + \sum_{i=1}^{n-1}\int_0^{x-D_i} p_{i+1}^2(y,t)dy\right.$$
$$\left. + \int_0^x u^2(y,t)dy\right), \tag{2.187}$$

where, in each of the above relations, $x \in \left[0, \sum_{k=i}^{n} D_k\right]$, respectively. From the above equations, recursively, we can take the upper bound of the lemma. To see this, we start from relation (2.185) and observe that the boundedness of $p_1^2(x,t)$ depends only on the boundedness of $X_1(t)$ and $\xi_2(x,t)$ (that is, $p_1^2(x,t)$ remains bounded for all $x \in \left[0, \sum_{k=1}^{n} D_k\right]$) if, for all $x \in \left[0, \sum_{k=2}^{n} D_k\right]$, $p_2^2(x,t)$ is upper bounded. We proceed now by proving that the boundedness of $p_2^2(x,t)$ depends only on the boundedness of $X_1(t)$, $X_2(t)$, $\xi_2(x,t)$, and $\xi_3(x,t)$ (that is, $p_2^2(x,t)$ remains bounded for all $x \in \left[0, \sum_{k=2}^{n} D_k\right]$) if, for all $x \in \left[0, \sum_{k=3}^{n} D_k\right]$, $p_3^2(x,t)$ is upper bounded. From relation (2.186) (and by noting that $\int_0^{x-D_1} p_2^2(y,t)dy \le \int_0^x p_2^2(y,t)dy$ for all x for which this equation holds, i.e., for all $x \in \left[0, \sum_{k=2}^{n} D_k\right]$) by using the comparison principle (Lemma B.7 in Appendix B) and by exploiting the fact that $e^{A_2 x} \le e^{|A_2|\sum_{k=2}^{n} D_k}$ for all $x \in \left[0, \sum_{k=2}^{n} D_k\right]$, we get that

$$\int_0^x p_2^2(y,t)dy \le A_2 e^{|A_2|\sum_{k=2}^{n} D_k}\left(\sum_{k-2}^{n} D_k\left(\sum_{k=1}^{2} X_i^2(t) + \sum_{i=1}^{2} ||\xi_{i+1}(t)||^2\right)\right.$$
$$\left.\int_0^x\int_0^{y-D_2} p_3^2(r,t)drdy\right). \tag{2.188}$$

Plugging the above bound into relation (2.186) we get a bound of $p_2^2(x,t)$ that depends on $p_3^2(x,t)$. Moreover, using the relation

$$p_3^2(x,t) \le A_3\left(\sum_{k=1}^{3} X_i^2(t) + \sum_{i=1}^{3} ||\xi_{i+1}(t)||^2 + \sum_{i=1}^{3}\int_0^{x-D_i} p_{i+1}^2(y,t)dy\right) \tag{2.189}$$

and the previous bound, we get

$$p_3^2(x,t) \le A_3\left(\sum_{k=1}^{3} X_i^2(t) + \sum_{i=1}^{2} ||\xi_{i+1}(t)||^2\right) + A_3 A_2 e^{|A_2|\sum_{k=2}^{n} D_k}$$
$$\times \sum_{k=2}^{n} D_k\left(\sum_{k=1}^{2} X_i^2(t) + \sum_{i=1}^{2} ||\xi_{i+1}(t)||^2\right) + A_3 A_2 e^{|A_2|\sum_{k=2}^{n} D_k}$$
$$\times \int_0^x\int_0^y p_3^2(r,t)drdy + A_3\int_0^x p_3^2(x,t)dy + A_3\int_0^{x-D_3} p_4^2(x,t),$$
$$x \in \left[0, \sum_{k=3}^{n} D_k\right]. \tag{2.190}$$

Note that the delayed terms in the integral for $p_3^2(x,t)$ can be removed since now $x \in \left[0, \sum_{k=3}^{n} D_k\right]$, which is the domain of definition for $p_3(x,t)$ (and of course this integral is larger than the delayed one). By changing the order of integration in the double integral of the previous relation, we can rewrite

$$\int_0^x \int_0^y p_3^2(r,t)dr\,dy = \int_0^x (x-y)p_3^2(y,t)dy. \tag{2.191}$$

By observing that $\int_0^x (x-y)p_3^2(y,t)dy \le \sum_{k=3}^n D_k \int_0^x p_3^2(y,t)dy$ for all $x \in \sum_{k=3}^n D_k$, and applying again the comparison principle (Lemma B.7 in Appendix B) for $\int_0^x p_3^2(y,t)dy$, we can bound $p_3^2(x,t)$ from $p_4^2(x,t)$ and consequently also $p_2^2(x,t)$. Repeating this process until $p_n^2(x,t)$ (the boundedness of which depends only on the boundedness of $||u(x,t)||^2$), we derive the bound of the lemma. □

Lemma 2.8. *There exists a constant $\overline{M}$ such that*

$$\Xi(t) \le \overline{M}\Omega(t). \tag{2.192}$$

Proof. From (2.110)–(2.112) it follows that

$$Z_i^2(t) \le 2\left(X_i^2(t) + \alpha_{i-1}^2(D_i,t)\right), \quad i=2,\ldots,n, \tag{2.193}$$

$$\zeta_i^2(x,t) \le 2\left(\xi_i^2(x,t) + \alpha_{i-1}^2(x,t)\right), \quad x \in [0,D_{i-1}], \quad i=2,\ldots,n, \tag{2.194}$$

$$w^2(x,t) \le 2\left(u^2(x,t) + \alpha_n^2(x,t)\right), \quad x \in [0,D_n]. \tag{2.195}$$

Moreover, from relations (2.72)–(2.73) one can see that the $\alpha_i(x,t)$ are linear functions of the predictors $p_1(x,t),\ldots,p_i(x,t)$, and hence it holds that

$$\alpha_i^2(x,t) \le b_i \sum_{k=1}^{i} p_k^2(x,t), \quad x \in \left[0, \sum_{k=i}^{n} D_k\right], \tag{2.196}$$

for some constants b_i. By employing the bound of Lemma 5, the lemma is proven. □

Lemma 2.9. *There exist constants F_i such that*

$$\epsilon_i^2(x,t) \le F_i\left(|Z(t)|^2 + \sum_{i=2}^{n} \int_0^{D_{i-1}} \zeta_i^2(y,t)dy + \int_0^{D_n} w^2(y,t)dy\right),$$
$$x \in \left[0, \sum_{k=i}^{n} D_k\right]. \tag{2.197}$$

Proof. The proof is immediate by noting that the relation for the $\epsilon_i(x,t)$ is similar to the relation for $p_i(x,t)$. Note here that in this case the derivation of the explicit bound is easier due to the special form of the $\epsilon_i(x,t)$ in (2.156)–(2.158). □

Lemma 2.10. *There exists a constant $\underline{M}$ such that*

$$\underline{M}\Omega(t) \le \Xi(t). \tag{2.198}$$

Proof. Using relations (2.151)–(2.153) we get

$$X_i^2(t) \le 2\left(Z_i^2(t) + \beta_{i-1}^2(D_i, t)\right), \quad i = 2, \ldots, n, \tag{2.199}$$

$$\xi_i^2(x,t) \le 2\left(\zeta_i^2(x,t) + \beta_{i-1}^2(x,t)\right), \quad x \in [0, D_{i-1}], \quad i = 2, \ldots, n, \tag{2.200}$$

$$u^2(x,t) \le 2\left(w^2(x,t) + \beta_n^2(x,t)\right), \quad x \in [0, D_n]. \tag{2.201}$$

By observing that $\beta_i(x,t)$ are linearly dependent on $\epsilon_1(x,t), \ldots, \epsilon_i(x,t)$, we conclude that there exist constants d_i such that

$$\beta_i^2(x,t) \le d_i \sum_{k=1}^{i} \epsilon_k^2(x,t), \quad x \in \left[0, \sum_{k=i}^{n} D_k\right]. \tag{2.202}$$

Using Lemma 7 the lemma is proven. □

Proof of Theorem 2.2. Combining Lemmas 2.8 and 2.10 we have that

$$\underline{M}\Omega(t) \le \Xi(t) \le \overline{M}\Omega(t). \tag{2.203}$$

Hence,

$$\Omega(t) \le \frac{\Xi(t)}{\underline{M}}, \tag{2.204}$$

and by Lemma 2.6 we get

$$\Omega(t) \le \frac{\overline{M} M_1 (1 + D_{\max})}{\underline{M} m_2} \Omega(0) e^{-\frac{m_1}{M_1} t}. \tag{2.205}$$

Thus Theorem 2.2 is proven with

$$\varkappa = \frac{\overline{M} M_1 (1 + D_{\max})}{\underline{M} m_2}, \tag{2.206}$$

$$\lambda = \frac{m_1}{M_1}. \quad \square \tag{2.207}$$

2.2.3 ▪ Simulations

We illustrate here our control design with a second order example given by

$$\dot{X}_1(t) = a_{11} X_1(t) + X_2(t - D_1), \tag{2.208}$$

$$\dot{X}_2(t) = a_{21} X_1(t) + a_{22} X_2(t) + U(t - D_2), \tag{2.209}$$

with parameters $a_{11} = a_{21} = a_{22} = 0.2$, $D_1 = 0.4$, and $D_2 = 0.8$. For these values for the parameters of the plant, the system is unstable (to see this one can use [133]). The control law is

$$\begin{aligned} U(t) &= u(D_2, t) \\ &= \alpha_2(D_2, t) \\ &= -a_{21} p_1(D_2, t) - a_{22} p_2(D_2, t) - c_2\left(p_2(D_2, t) + (a_{11} + c_1) p_1(D_1 + D_2, t)\right) \\ &\quad - (a_{11} + c_1)\left(a_{11} p_1(D_2 + D_1, t) + p_2(D_2, t)\right) \\ &= -a_{21} P_1(t - D_1) - a_{22} P_2(t) - c_2\left(P_2(t) + (a_{11} + c_1) P_1(t)\right) \\ &\quad - (a_{11} + c_1)\left(a_{11} P_1(t) + P_2(t)\right), \end{aligned} \tag{2.210}$$

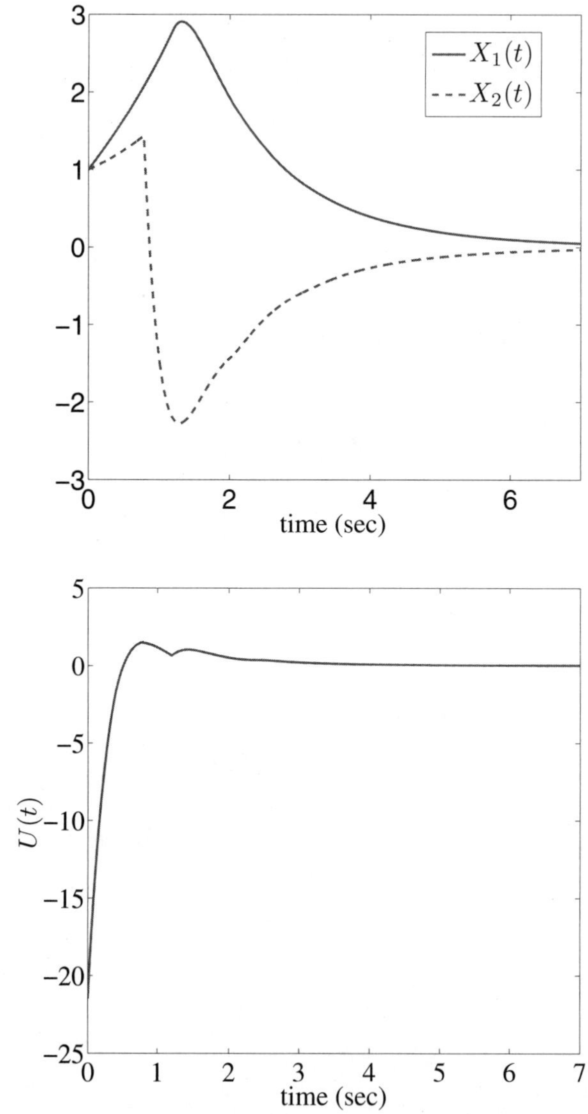

Figure 2.1. *System's response for the simulation example in Section* 2.2.3.

where $c_1 = c_2 = 2$ and

$$P_1(t) = X_1(t) + \int_{t-D_1-D_2}^{t} (a_{11}P_1(\theta) + P_2(\theta))\,d\theta, \tag{2.211}$$

$$P_2(t) = X_2(t) + \int_{t-D_2}^{t} (a_{21}P_1(\theta - D_1) + a_{22}P_2(\theta) + U(\theta))\,d\theta, \tag{2.212}$$

with initial conditions for $t = 0$. The initial conditions for the system are chosen as $X_1(0) = X_2(0) = 1$, $X_2(\theta) = 1, \theta \in [-D_1, 0]$, and $U(\theta) = 0, \theta \in [-D_2, 0]$. Note also that these integrals are computed using the trapezoidal rule.

Figure 2.1 shows that the predictor controller exponentially stabilizes the system. The control signal first reaches $X_2(t)$, since the delay from the input to $X_2(t)$ is 0.4, which is smaller than the total delay from the input to $X_1(t)$. After 1.2 seconds, which is the total delay from the input to $X_1(t)$, the controller starts stabilizing $X_1(t)$. Then both $X_1(t)$ and $X_2(t)$ converge exponentially to zero.

2.3 ▪ Adaptive Control of Feedforward Systems with Unknown Delays

In this section we develop a delay-adaptive version of the design introduced by Jankovic in [60] for linear feedforward systems with simultaneous state and input delays. In [60], for the system

$$\dot{X}_1(t) = F_1 X_1(t) + H_1 X_2(t - D_1) + B_1 U(t), \tag{2.213}$$

$$\dot{X}_2(t) = F_2 X_2(t) + B_2 U(t), \tag{2.214}$$

a predictor-based controller is designed as

$$U(t) = K_1 D_1 \int_0^1 e^{-F_1 D_1 \theta} H_1 X_2(t + D_1(\theta - 1)) d\theta + K_1 X_1(t) + K_2 X_2(t). \tag{2.215}$$

The above controller is based on a transformation that reduces the system to an equivalent system without state delay. This transformation is

$$Z_1(t) = X_1(t) + D_1 \int_0^1 e^{-F_1 D_1 \theta} H_1 X_2(t + D_1(\theta - 1)) d\theta, \tag{2.216}$$

$$Z_2(t) = X_2(t), \tag{2.217}$$

and it transforms the system (2.213)–(2.214) into

$$\dot{Z}_1(t) = F_1 Z_1(t) + e^{-F_1 D_1} H_1 Z_2(t) + B_1 U(t), \tag{2.218}$$

$$\dot{Z}_2(t) = F_2 Z_2(t) + B_2 U(t). \tag{2.219}$$

The importance of the previous transformation, besides transforming the original system to an equivalent one without state delay, is that the system can be linearly parametrized in the state delay, which is the key for designing an adaptive control law.

Then, assuming that the pair $\left[\begin{pmatrix} F_1 & e^{-D_1 F_1} H_1 \\ 0 & F_2 \end{pmatrix}, \begin{pmatrix} B_1 \\ B_2 \end{pmatrix}\right]$ is completely controllable, a state feedback controller $U(t) = K_1 Z_1(t) + K_2 Z_2(t)$ is designed such that the transformed system is asymptotically stable. It can be shown that controllability of the original system is equivalent to controllability of the transformed system [20]. This design can be also applied in the case where there is a delay in the input, say D_2. In this case, after employing the state transformation, a predictor feedback is needed for the transformed system. In this case the controller that compensates for D_2 is given by

$$U(t) = K e^{A D_2} Z(t) + K D_2 \int_0^1 e^{A D_2 (1-\theta)} B U(t - D_2(\theta - 1)) d\theta, \tag{2.220}$$

where

$$A = \begin{bmatrix} F_1 & e^{-D_1 F_1} H_1 \\ 0 & F_2 \end{bmatrix}, \tag{2.221}$$

$$B = \begin{bmatrix} B_1 \\ B_2 \end{bmatrix}, \tag{2.222}$$

$$K = \begin{bmatrix} K_1 & K_2 \end{bmatrix}. \tag{2.223}$$

The controller (2.220) is the basis of our adaptive design.

For the case where the delays are of unknown length, using the certainty equivalence principle, we design a Lyapunov-based adaptive controller in Section 2.3.1. In Section 2.3.2 we prove the global stability and regulation of the closed-loop system for arbitrary initial estimates for the delays. We present a numerical example in Section 2.3.3.

2.3.1 ▪ Delay-Adaptive Predictor Feedback Design

We consider here the case in which both the state and input delays are unknown; that is, we consider the system

$$\dot{X}_1(t) = F_1 X_1(t) + H_1 X_2(t - D_1) + B_1 U(t - D_2), \tag{2.224}$$

$$\dot{X}_2(t) = F_2 X_2(t) + B_2 U(t - D_2), \tag{2.225}$$

with D_1 and D_2 unknown. Since D_1 and D_2 are unknown, in addition to the predictor-based controller, we must design two estimators, one for each of the delays. We employ projector operators and assume a bound on the length of the delays to be known.

Assumption 2.11. *There exist known constants $\underline{D}_1$, $\overline{D}_1$, and $\overline{D}_2$ such that $D_1 \in [\underline{D}_1, \overline{D}_1]$ and $D_2 \in [0, \overline{D}_2]$.*

Our controller is based on the transformed system (i.e., on the system without state delay). As indicated in Section 2.3, the pair $\left[\begin{pmatrix} F_1 & e^{-D_1F_1}H_1 \\ 0 & F_2 \end{pmatrix}, \begin{pmatrix} B_1 \\ B_2 \end{pmatrix}\right]$ must be completely controllable. Under this assumption we can find a stabilizing state feedback. In the case of unknown state delay D_1, we must assume that there exists a stabilizing state feedback for all values of the state delay in a given interval. We thus make the following assumption.

Assumption 2.12. *The pair $\left[\begin{pmatrix} F_1 & e^{-D_1F_1}H_1 \\ 0 & F_2 \end{pmatrix}, \begin{pmatrix} B_1 \\ B_2 \end{pmatrix}\right]$ is completely controllable for all $D_1 \in [\underline{D}_1, \overline{D}_1]$. Furthermore, we assume that there exists a triple of vector-/matrix-valued functions $(K(D_1), P(D_1), Q(D_1))$ such that $K(D_1) \in C^1\left([\underline{D}_1, \overline{D}_1]\right)$, $P(D_1) \in C^1\left([\underline{D}_1, \overline{D}_1]\right)$, $Q(D_1) \in C^0\left([\underline{D}_1, \overline{D}_1]\right)$, the matrices $P(D_1)$ and $Q(D_1)$ are positive definite and symmetric, and the following Lyapunov equation is satisfied for all $D_1 \in [\underline{D}_1, \overline{D}_1]$:*

$$\begin{aligned}&(A(D_1) + BK(D_1))^T P(D_1) + P(D_1)(A(D_1) + BK(D_1)) \\ &= -Q(D_1) \quad \textit{for all } D_1 \in [\underline{D}_1, \overline{D}_1]\end{aligned} \tag{2.226}$$

Our final assumption is needed in the choice of the normalizations coefficients in the adaptation laws for the delay estimates.

Assumption 2.13. *The quantities $\underline{\lambda} = \inf_{D_1 \in [\underline{D}_1, \overline{D}_1]} \min\{\lambda_{\min}(Q(D_1)), \lambda_{\min}(P(D_1))\}$ and $\overline{\lambda} = \sup_{D_1 \in [\underline{D}_1, \overline{D}_1]} \lambda_{\max}(P(D_1))$ exist and are known.*

We first rewrite (2.224)–(2.225) using PDE representation of the delayed states and control as

$$\dot{X}_1(t) = F_1 X_1(t) + H_1 \xi(0,t) + B_1 u(0,t), \tag{2.227}$$
$$D_1 \xi_t(x,t) = \xi_x(x,t), \tag{2.228}$$
$$\xi(1,t) = X_2(t), \tag{2.229}$$
$$\dot{X}_2(t) = F_2 X_2(t) + B_2 u(0,t), \tag{2.230}$$
$$D_2 u_t(x,t) = u_x(x,t), \tag{2.231}$$
$$u(1,t) = U(t), \tag{2.232}$$

where $x \in [0,1]$. We assume that the infinite-dimensional states $\xi(x,t), u(x,t)$, $x \in [0,1]$, are available for measurement. This assumption is not in contradiction with the assumption that the convection speeds $1/D_1$ and $1/D_2$ are unknown. As restrictive as the requirement for measurement of $\xi(x,t), u(x,t), x \in [0,1]$, may appear, we do not believe that the delay-adaptive problem without such measurements is solvable globally because it cannot be formulated as linearly parametrized in the unknown delays D_1 and D_2.

The transport PDE states can be expressed in terms of the past values of X_2 and U as

$$\xi(x,t) = X_2(t + D_1(x-1)), \tag{2.233}$$
$$u(x,t) = U(t + D_2(x-1)). \tag{2.234}$$

Using the certainty equivalence principle, the controller (2.220) is taken as

$$U(t) = K(\hat{D}_1) e^{A(\hat{D}_1)\hat{D}_2(t)} \begin{bmatrix} X_1(t) + \hat{D}_1(t) \int_0^1 e^{-F_1 \hat{D}_1(t) y} H_1 \xi(y,t) dy \\ X_2(t) \end{bmatrix}$$
$$+ K(\hat{D}_1)\hat{D}_2(t) \int_0^1 e^{A(\hat{D}_1)\hat{D}_2(t)(1-y)} B u(y,t) dy. \tag{2.235}$$

The update laws for the estimations of the unknown delays D_1 and D_2 are given by

$$\dot{\hat{D}}_1(t) = \gamma_1 \mathrm{Proj}_{[\underline{D}_1, \overline{D}_1]} \{\tau_{D_1}\}, \tag{2.236}$$
$$\dot{\hat{D}}_2(t) = \gamma_2 \mathrm{Proj}_{[0, \overline{D}_2]} \{\tau_{D_2}\}, \tag{2.237}$$

where the projector operators are defined as

$$\mathrm{Proj}_{[\underline{D}_i, \overline{D}_i]} \{\tau_{D_i}\} = \begin{cases} 0 & \text{if } \hat{D}_i = \underline{D}_i \text{ and } \tau_{D_i} < 0, \\ 0 & \text{if } \hat{D}_i = \overline{D}_i \text{ and } \tau_{D_i} > 0, \\ \tau_{D_i} & \text{else} \end{cases} \tag{2.238}$$

and where

$$\tau_{D_1} = \frac{\int_0^1 (1+x)w(x,t)K(\hat{D}_1)e^{A(\hat{D}_1)\hat{D}_2(t)x}dx - \frac{2}{a_2}Z^T(t)P(\hat{D}_1)}{\Gamma(t)} R_2(t), \tag{2.239}$$

$$\tau_{D_2} = -\frac{\int_0^1 (1+x)w(x,t)K(\hat{D}_1)e^{A(\hat{D}_1)\hat{D}_2(t)x}dx}{\Gamma(t)} \times \left(Bu(0,t)+A(\hat{D}_1)Z(t)\right), \tag{2.240}$$

$$\Gamma(t) = 1 + Z^T(t)P(\hat{D}_1)Z(t) + a_2\int_0^1 (1+x)w^2(x,t)dx + k\int_0^1 (1+x)\xi^T(x,t)\xi(x,t)dx, \tag{2.241}$$

with

$$k \leq \frac{\lambda \underline{D}_1}{8}, \tag{2.242}$$

$$a_2 \geq \frac{\overline{D}_2 \sup_{\hat{D}_1\in[\underline{D}_1,\overline{D}_1]} |P(\hat{D}_1)B|^2}{\underline{\lambda}}. \tag{2.243}$$

In the above relation we use the following signals, which are derived in the stability analysis of the closed-loop system:

$$Z_1(t) = X_1(t) + \hat{D}_1(t)\int_0^1 e^{-\hat{D}_1(t)F_1 y}H_1\xi(y,t)dy, \tag{2.244}$$

$$Z_2(t) = X_2(t), \tag{2.245}$$

$$R_2(t) = \begin{bmatrix} R_2^*(t) \\ 0 \end{bmatrix}, \tag{2.246}$$

$$R_2^*(t) = e^{-\hat{D}_1(t)F_1}H_1 Z_2(t) - H_1\xi(0,t) + \hat{D}_1(t)F_1\int_0^1 e^{-\hat{D}_1(t)F_1 y}H_1\xi(y,t)dy; \tag{2.247}$$

the transformed infinite-dimensional state of the actuator is

$$w(x,t) = u(x,t) - K(\hat{D}_1)e^{A(\hat{D}_1)\hat{D}_2(t)x}Z(t) - K(\hat{D}_1)\hat{D}_2(t)\int_0^x e^{A(\hat{D}_1)\hat{D}_2(t)(x-y)}Bu(y,t)dy. \tag{2.248}$$

2.3.2 ▪ Stability Analysis

This section is devoted to the proof of the main result. We start by giving the main theorem, and in the rest of the section we prove it using a series of technical lemmas.

Theorem 2.14. *Let Assumptions* 2.11–2.13 *hold. Then system* (2.224)–(2.225) *with the controller* (2.235) *and the update laws* (2.236)–(2.237) *is stable in the sense that there exist constants R and ρ such that*

$$\Omega(t) \leq R\left(e^{\rho\Omega(0)} - 1\right), \tag{2.249}$$

where

$$\Omega(t)=|X(t)|^2+\|\xi(t)\|^2+\|u(t)\|^2+\tilde{D}_1^2(t)+\tilde{D}_2^2(t) \tag{2.250}$$

and

$$\|\xi(t)\|^2=\int_0^1 \xi(y,t)^T\xi(y,t)dy=\frac{1}{D_1}\int_{t-D_1}^t X_2(\theta)^TX_2(\theta)d\theta, \tag{2.251}$$

$$\|u(t)\|^2=\int_0^1 u^2(y,t)dy=\frac{1}{D_2}\int_{t-D_2}^t U^2(\theta)d\theta. \tag{2.252}$$

Furthermore

$$\lim_{t\to\infty} X(t)=0, \tag{2.253}$$

$$\lim_{t\to\infty} U(t)=0. \tag{2.254}$$

We start proving the above theorem by first transforming system (2.227)–(2.232) using the transformations (2.244)–(2.245) and (2.248). By differentiating with respect to time (2.244) and (2.245) and by using (2.227) and (2.230), we get

$$\begin{aligned}\dot{Z}_1(t)=&F_1X_1(t)+H_1\xi(0,t)+B_1u(0,t)+\dot{\hat{D}}_1(t)\int_0^1 e^{-\hat{D}_1(t)F_1y}H_1\xi(y,t)dy\\&-\hat{D}_1(t)\dot{\hat{D}}_1(t)\int_0^1 F_1ye^{-\hat{D}_1(t)F_1y}H_1\xi(y,t)dy\\&+\hat{D}_1(t)\int_0^1 e^{-\hat{D}_1(t)F_1y}H_1\xi_t(y,t)dy,\end{aligned} \tag{2.255}$$

$$\dot{Z}_2(t)=F_2X_2(t)+B_2u(0,t). \tag{2.256}$$

Using relations (2.228) and (2.229) and the fact that $\frac{\hat{D}_1}{D_1}=1-\frac{\tilde{D}_1}{D_1}$, and integrating by parts the last integral in (2.255), we obtain

$$\begin{aligned}\dot{Z}_1(t)=&F_1X_1(t)+H_1\xi(0,t)+B_1u(0,t)+\dot{\hat{D}}_1(t)\int_0^1 e^{-\hat{D}_1(t)F_1y}H_1\xi(y,t)dy\\&-\hat{D}_1(t)\dot{\hat{D}}_1(t)\int_0^1 F_1ye^{-\hat{D}_1(t)F_1y}H_1\xi(y,t)dy+\left(1-\frac{\tilde{D}_1}{D_1}\right)\Bigg(e^{-\hat{D}_1(t)F_1}\\&\times H_1X_2(t)-H_1\xi(0,t)+\int_0^1 \hat{D}_1(t)F_1e^{-\hat{D}_1(t)F_1y}H_1\xi(y,t)dy\Bigg),\end{aligned} \tag{2.257}$$

$$\dot{Z}_2(t)=F_2X_2(t)+B_2u(0,t). \tag{2.258}$$

Using (2.244)–(2.245), after some algebra we arrive at

$$\begin{aligned}\dot{Z}(t)=&\begin{bmatrix}\dot{Z}_1(t)\\ \dot{Z}_2(t)\end{bmatrix}\\=&A(\hat{D}_1)Z(t)+Bu(0,t)+\dot{\hat{D}}_1(t)R_1(t)-\frac{\tilde{D}_1}{D_1}R_2(t),\end{aligned} \tag{2.259}$$

where $R_2(t)$ is defined in (2.246) and

$$A(\hat{D}_1) = \begin{bmatrix} F_1 & e^{-\hat{D}_1(t)F_1}H_1 \\ 0 & F_2 \end{bmatrix}, \tag{2.260}$$

$$B = \begin{bmatrix} B_1 \\ B_2 \end{bmatrix}, \tag{2.261}$$

$$R_1(t) = \begin{bmatrix} \int_0^1 \left(I - \hat{D}_1(t)F_1 y\right) e^{-\hat{D}_1(t)F_1 y} H_1 \xi(y,t)dy \\ 0 \end{bmatrix}. \tag{2.262}$$

Using relation (2.248) for $x = 0$ we get that

$$\dot{Z}(t) = \left(A(\hat{D}_1) + BK(\hat{D}_1)\right)Z(t) + Bw(0,t) + \dot{\hat{D}}_1(t)R_1(t) - \frac{\tilde{D}_1}{D_1}R_2(t). \tag{2.263}$$

Moreover the transformation of the actuator state w satisfies

$$\begin{aligned} D_2 w_t(x,t) = {}& w_x(x,t) + \frac{\tilde{D}_1}{D_1} D_2 p_1(x,t) - \tilde{D}_2 p_2(x,t) - D_2 \dot{\hat{D}}_1(t) q_1(x,t) \\ & - D_2 \dot{\hat{D}}_2(t) q_2(x,t), \end{aligned} \tag{2.264}$$

where

$$p_1(x,t) = K(\hat{D}_1)e^{A(\hat{D}_1)\hat{D}_2(t)x}R_2(t), \tag{2.265}$$

$$\begin{aligned} q_1(x,t) = {}& \int_0^x \left(\frac{\partial K(\hat{D}_1)}{\partial \hat{D}_1} + K(\hat{D}_1)\frac{\partial A(\hat{D}_1)}{\partial \hat{D}_1}(x-y)\right)\hat{D}_2(t)e^{\hat{D}_2(t)A(\hat{D}_1)(x-y)}Bu(y,t)dy \\ & + \left(\frac{\partial K(\hat{D}_1)}{\partial \hat{D}_1} + K(\hat{D}_1)\frac{\partial A(\hat{D}_1)}{\partial \hat{D}_1}\hat{D}_2(t)x\right)e^{\hat{D}_2(t)A(\hat{D}_1)x}Z(t) \\ & + K(\hat{D}_1)e^{\hat{D}_2(t)A(\hat{D}_1)x}R_1(t), \end{aligned} \tag{2.266}$$

$$p_2(x,t) = K(\hat{D}_1)e^{\hat{D}_2(t)A(\hat{D}_1)x}Bu(0,t) + K(\hat{D}_1)A(\hat{D}_1)e^{\hat{D}_2(t)A(\hat{D}_1)x}Z(t), \tag{2.267}$$

$$q_2(x,t) = K(\hat{D}_1)\int_0^x \left(I + \hat{D}_2(t)(x-y)A(\hat{D}_1)\right)e^{A(\hat{D}_1)\hat{D}_2(t)(x-y)}Bu(y,t)dy. \tag{2.268}$$

Thus now system (2.227)–(2.232) is mapped to the target system that comprised (2.263) and (2.264). Moreover, the inverse transformation of the state $X(t)$ is easily obtained from equations (2.244)–(2.245) and the inverse transformation of (2.248) is given by

$$\begin{aligned} u(x,t) = w(x,t) + K(\hat{D}_1)\Bigg(& e^{\left(A(\hat{D}_1)+BK(\hat{D}_1)\right)\hat{D}_2(t)x}Z(t) \\ & + \hat{D}_2(t)\int_0^x e^{\left(A(\hat{D}_1)+BK(\hat{D}_1)\right)\hat{D}_2(t)(x-y)}Bw(y,t)dy\Bigg). \end{aligned} \tag{2.269}$$

We first prove that the signals in (2.263) and (2.264) that multiply the "disturbances" $\dot{\hat{D}}_i$ and $\tilde{D}_i$, $i = 1,2$, are bounded with respect to the system's transformed states $Z(t)$, $\xi(x,t)$, and $w(x,t)$. Before doing that, we point out that boundedness of the transformed states is equivalent to boundedness of the original states.

Lemma 2.15. *There exist constants M_u, M_w, M_X, and M_Z such that*

$$\|u(t)\|^2 \leq M_u\left(\|w(t)\|^2 + |Z(t)|^2\right), \tag{2.270}$$
$$|X(t)|^2 \leq M_X\left(|Z(t)|^2 + \|\xi(t)\|^2\right), \tag{2.271}$$
$$\|w(t)\|^2 \leq M_w\left(\|u(t)\|^2 + |Z(t)|^2\right), \tag{2.272}$$
$$|Z(t)|^2 \leq M_Z\left(|X(t)|^2 + \|\xi(t)\|^2\right). \tag{2.273}$$

Proof. First observe that the signals $K(\hat{D}_1)$, $P(\hat{D}_1)$, and $A(\hat{D}_1)$ are continuously differentiable with respect to $\hat{D}_1$. Moreover, since $\hat{D}_1$ and $\hat{D}_2$ are uniformly bounded, the signals $K(\hat{D}_1)$, $P(\hat{D}_1)$, and $A(\hat{D}_1)$ and their derivatives are also uniformly bounded. Denote by M_K, M_P, and M_A the bounds of $K(\hat{D}_1)$, $P(\hat{D}_1)$, and $A(\hat{D}_1)$, respectively, and with M_K', M_P', and M_A' the bounds of their derivatives. From relations (2.244)–(2.245), (2.248), (2.269) and using the Young and Cauchy–Schwarz inequalities it is easy to show that the above bounds hold with

$$M_u = 3\max\left\{1 + M_K^2\overline{D}_2^2 e^{2\overline{D}_2(M_A+|B|M_K)}|B|^2, M_K^2 e^{2\overline{D}_2(M_A+|B|M_K)}\right\}, \tag{2.274}$$
$$M_Z = 2\max\left\{1, \overline{D}_1^2 e^{2|F_1|\overline{D}_1}|H_1|^2\right\}, \tag{2.275}$$
$$M_X = 2\max\left\{1, \overline{D}_1^2 e^{2|F_1|\overline{D}_1}|H_1|^2\right\}, \tag{2.276}$$
$$M_w = 3\max\left\{1 + M_K^2\overline{D}_2^2 e^{2M_A\overline{D}_2}|B|^2, M_K^2 e^{2M_A\overline{D}_2}\right\}. \qquad \square \tag{2.277}$$

We are now ready to state the following lemma.

Lemma 2.16. *There exist constants M_{R_1}, M_{R_2}, M_{p_1}, M_{p_2}, M_{q_1}, and M_{q_2} such that the following bounds hold:*

$$|R_1(t)|^2 \leq M_{R_1}\|\xi(t)\|^2, \tag{2.278}$$
$$|R_2(t)|^2 \leq M_{R_2}\left(|Z(t)|^2 + |\xi(0,t)|^2 + \|\xi(t)\|^2\right), \tag{2.279}$$
$$p_1^2(x,t) \leq M_{p_1}\left(|Z(t)|^2 + |\xi(0,t)|^2 + \|\xi(t)\|^2\right), \tag{2.280}$$
$$p_2^2(x,t) \leq M_{p_2}\left(|Z(t)|^2 + u^2(0,t)\right), \tag{2.281}$$
$$q_1^2(x,t) \leq M_{q_1}\left(|Z(t)|^2 + \|u(t)\|^2 + \|\xi(t)\|^2\right), \tag{2.282}$$
$$q_2^2(x,t) \leq M_{q_2}\|u(t)\|^2 \tag{2.283}$$

for all $x \in [0,1]$.

Proof. From relations (2.262) and (2.246) and using the Young and Cauchy–Schwarz inequalities we get the bounds for $R_1(t)$ and $R_2(t)$ with

$$M_{R_1} = \left(1 + \overline{D}_1|F_1|\right)^2 e^{2\overline{D}_2|F_1|}|H_1|^2, \tag{2.284}$$
$$M_{R_2} = 3|H_1|^2\max\left\{e^{2\overline{D}_2|F_1|}|F_1|^2, 1, \overline{D}_1^2|F_1|^2 e^{2\overline{D}_2|F_1|}\right\}. \tag{2.285}$$

Using relations (2.265)–(2.268) together with the Young and Cauchy–Schwarz inequalities, relations (2.278)–(2.279), and (2.270) we get the bounds of the lemma with

$$M_{p_1} = M_K^2 e^{2M_A\overline{D}_2} M_{R_2}, \tag{2.286}$$

$$M_{p_2} = 3\max\left\{M_K^2 e^{2M_A\overline{D}_2}|B|^2, M_K^2 M_A^2 e^{2M_A\overline{D}_2}\right\}, \tag{2.287}$$

$$\begin{aligned} M_{q_1} = 3\max\Big\{&\left(M_K'\overline{D}_2 + \overline{D}_2 M_K M_A'\right)^2 |B|^2 e^{2M_A\overline{D}_2}, M_K^2 e^{2M_A\overline{D}_2} M_{R_1}, \\ &\left(M_K' + M_K M_A'\overline{D}_2\right)^2 e^{2M_A\overline{D}_2}\Big\}, \end{aligned} \tag{2.288}$$

$$M_{q_2} = M_K^2\left(1+\overline{D}_2 M_A\right)^2 e^{2M_A\overline{D}_2}. \qquad \square \tag{2.289}$$

Lemma 2.17. *There exist constants k, a_2, γ_1, and γ_2 such that for the Lyapunov function*

$$V(t) = D_2\log(1+\Xi(t)) + a_2 D_2\frac{\tilde{D}_1^2(t)}{D_1\gamma_1} + a_2\frac{\tilde{D}_2^2(t)}{\gamma_2}, \tag{2.290}$$

where

$$\begin{aligned} \Xi(t) = {} & Z(t)^T P\left(\hat{D}_1\right) Z(t) + k\int_0^1 (1+x)\xi^T(x,t)\xi(x,t)dx \\ & + a_2\int_0^1 (1+x)w^2(x,t)dx, \end{aligned} \tag{2.291}$$

the following holds:

$$V(t) \leq V(0). \tag{2.292}$$

Proof. Taking the time derivative of the above function we get

$$\begin{aligned} \dot{V}(t) = {} & -2a_2 D_2\frac{\tilde{D}_1}{D_1\gamma_1}\left(\dot{\hat{D}}_1(t) - \gamma_1\tau_{D_1}\right) - 2a_2\frac{\tilde{D}_2}{\gamma_2}\left(\dot{\hat{D}}_2(t) - \gamma_2\tau_{D_2}\right) + \frac{D_2}{1+\Xi(t)} \\ & \times\Bigg(-Z^T(t)Q(\hat{D}_1)Z(t) + 2Z^T(t)P(\hat{D}_1)Bw(0,t) + \frac{2k}{D_1}Z_2^T(t)Z_2(t) \\ & - \frac{k}{D_1}|\xi(0,t)|^2 - \frac{a_2}{D_2}w^2(0,t) - \frac{k}{D_1}\int_0^1 \xi^T(x,t)\xi(x,t)dx - \frac{a_2}{D_2}\int_0^1 w^2(x,t)dx \\ & + \dot{\hat{D}}_1(t)\Bigg(Z^T(t)\frac{\partial P(\hat{D}_1)}{\partial\hat{D}_1}Z(t) + 2Z^T(t)P(\hat{D}_1)R_1(t) \\ & - 2a_2\int_0^1 (1+x)w(x,t)q_1(x,t)dx\Bigg) \\ & - 2a_2\dot{\hat{D}}_2(t)\int_0^1 (1+x)w(x,t)q_2(x,t)dx\Bigg). \end{aligned} \tag{2.293}$$

Using the properties of the projector operators and relations (2.236)–(2.237) we get

$$\begin{aligned}\dot{V}(t) \leq \frac{D_2}{1+\Xi(t)}\Bigg(&-Z^T(t)Q(\hat{D}_1)Z(t)+\frac{2k}{D_1}Z_2^T(t)Z_2(t)-\frac{k}{D_1}|\xi(0,t)|^2-\frac{a_2}{D_2}w^2(0,t)\\&+Z^T(t)P(\hat{D}_1)Bw(0,t)-\frac{k}{D_1}\int_0^1\xi^T(x,t)\xi(x,t)dx-\frac{a_2}{D_2}\int_0^1 w^2(x,t)dx\\&+\dot{\hat{D}}_1(t)\Bigg(Z^T(t)\frac{\partial P(\hat{D}_1)}{\partial \hat{D}_1}Z(t)+2Z^T(t)P(\hat{D}_1)R_1(t)\\&-2a_2\int_0^1(1+x)w(x,t)q_1(x,t)dx\Bigg)\\&-2a_2\dot{\hat{D}}_2(t)\int_0^1(1+x)w(x,t)q_2(x,t)dx\Bigg).\end{aligned}\tag{2.294}$$

Then using Young's inequality and (2.242)–(2.243) we get

$$\begin{aligned}\dot{V}(t) \leq \frac{D_2}{1+\Xi(t)}\Bigg(&-\frac{\lambda}{2}|Z(t)|^2-\frac{k}{D_1}|\xi(0,t)|^2-\frac{k}{D_1}||\xi(t)||^2-\frac{a_2}{2D_2}w^2(0,t)\\&-\frac{a_2}{D_2}||w(t)||^2+\dot{\hat{D}}_1(t)\Bigg(Z^T(t)\frac{\partial P(\hat{D}_1)}{\partial \hat{D}_1}Z(t)+2Z^T(t)P(\hat{D}_1)R_1(t)\\&-2a_2\int_0^1(1+x)w(x,t)q_1(x,t)dx\Bigg)-2a_2\dot{\hat{D}}_2(t)\\&\times\int_0^1(1+x)w(x,t)q_2(x,t)dx\Bigg).\end{aligned}\tag{2.295}$$

Using bounds (2.278)–(2.283) together with relations (2.238) and (2.236)–(2.237), and employing one more time Young's inequality, we get

$$|\dot{\hat{D}}_1(t)| \leq \gamma_1|\tau_{D_1}| \leq \gamma_1 M_1\frac{(|Z(t)|^2+|\xi(0,t)|^2+||\xi(t)||^2+||w(t)||^2)}{1+\Xi(t)},\tag{2.296}$$

$$|\dot{\hat{D}}_2(t)| \leq \gamma_2|\tau_{D_2}| \leq \gamma_2 M_2\frac{(|Z(t)|^2+w^2(0,t)+||w(t)||^2)}{1+\Xi(t)},\tag{2.297}$$

where

$$M_1=\max\left\{M_{p_1}+\frac{1}{a_2}M_P+\frac{1}{a_2}M_PM_{R_2},1\right\},\tag{2.298}$$

$$M_2=\max\left\{1,2M_{p_2},2M_{p_2}M_K^2+M_{p_2}\right\}.\tag{2.299}$$

Plugging the above bound into (2.295) (and applying once more the Young and Cauchy–Schwarz inequalities) we get

$$\begin{aligned}\dot{V}(t) \leq & \frac{D_2}{1+\Xi(t)}\Bigg(-\frac{\lambda}{2}|Z(t)|^2 - \frac{k}{D_1}|\xi(0,t)|^2 - \frac{k}{D_1}\|\xi(t)\|^2 - \frac{a_2}{2D_2}w^2(0,t) \\ & -\frac{a_2}{D_2}\|w(t)\|^2 + B_1\gamma_1 \frac{\left(|Z(t)|^2 + |\xi(0,t)|^2 + \|\xi(t)\|^2 + \|w(t)\|^2\right)}{1+\Xi(t)} \\ & \times\left(|Z(t)|^2 + \|\xi(t)\|^2 + \|w(t)\|^2\right) + B_2\gamma_2 \frac{\left(|Z(t)|^2 + w^2(0,t) + \|w(t)\|^2\right)}{1+\Xi(t)} \\ & \times\left(|Z(t)|^2 + \|w(t)\|^2\right)\Bigg), \end{aligned} \tag{2.300}$$

where

$$\begin{aligned} B_1 =& M_1 \max\left\{M_P' + M_P + a_2 M_{q_1} + 2a_2 M_{q_1} M_u + 2a_2 M_{q_1},\right. \\ & \left. M_P M_{R_1} + 2a_2 M_{q_1}, 2a_2 + 2a_2 M_{q_1} M_u\right\}, \end{aligned} \tag{2.301}$$

$$B_2 = 2M_2 a_2\left(1 + M_{q_2} M_u\right). \tag{2.302}$$

Now by defining the constants

$$m_1 = \min\left\{\frac{\lambda}{2}, \frac{k}{\overline{D}_1}, \frac{a_2}{2\overline{D}_2}\right\}, \tag{2.303}$$

$$m_2 = \frac{\max\{B_1, B_2\}}{\min\{\underline{\lambda}, k, a_2\}}, \tag{2.304}$$

we get

$$\begin{aligned}\dot{V}(t) \leq & -\frac{D_2}{1+\Xi(t)}(m_1 - m_2(\gamma_1+\gamma_2)) \\ & \times\left(|Z(t)|^2 + w^2(0,t) + \|w(t)\|^2 + |\xi(0,t)|^2 + \|\xi(t)\|^2\right). \end{aligned} \tag{2.305}$$

Thus when $\gamma_1 + \gamma_2 \leq \frac{m_1}{m_2}$, $\dot{V}(t)$ is negative definite and thus

$$V(t) \leq V(0). \qquad \square \tag{2.306}$$

To prove the stability bound of Theorem 2.14 we use the following lemma.

Lemma 2.18. *There exist constants $\underline{M}$ and $\overline{M}$ such that*

$$\underline{M}\Xi(t) \leq \Pi(t) \leq \overline{M}\Xi(t), \tag{2.307}$$

where

$$\Pi(t) = |X(t)|^2 + \|\xi(t)\|^2 + \|u(t)\|^2. \tag{2.308}$$

Proof. The proof is immediate, using (2.270)–(2.273) with

$$\overline{M} = \max\{M_u + M_X, M_X + 1\}, \tag{2.309}$$

$$\underline{M} = \max\{M_w + M_Z, M_Z + 1\}. \qquad \square \tag{2.310}$$

We are now ready to derive the stability estimate of Theorem 2.14. Using (2.290) it follows that

$$\Xi(t) \leq \left(e^{\frac{V(t)}{D_2}} - 1\right), \tag{2.311}$$

$$\tilde{D}_1^2 + \tilde{D}_2^2 \leq C_2 \frac{V(t)}{D_2} \leq C_2\left(e^{\frac{V(t)}{D_2}} - 1\right), \tag{2.312}$$

where

$$C_2 = \frac{\left(\gamma_2 \overline{D}_2 + \gamma_1 \overline{D}_1\right)}{a_2}. \tag{2.313}$$

Consequently

$$\Omega(t) \leq \left(\overline{M} + C_2\right)\left(e^{\frac{V(t)}{D_2}} - 1\right). \tag{2.314}$$

Moreover, from (2.290) we take

$$\begin{aligned} V(0) &\leq \max\left\{\overline{\lambda}, k, a_2\right\}\left(|Z(0)|^2 + \|\xi(0)\|^2 + \|w(0)\|^2\right) \\ &\quad + \max\left\{\frac{a_2}{\gamma_2}, \frac{a_2 \overline{D}_2}{\gamma_1 \underline{D}_1}\right\}\left(\tilde{D}_1^2(0) + \tilde{D}_2^2(0)\right), \end{aligned} \tag{2.315}$$

and using Lemma 2.18 we have

$$V(0) \leq C_3 \Omega(0), \tag{2.316}$$

where

$$C_3 = \max\left\{\max\left\{\frac{1}{2\gamma_2}, \frac{\overline{D}_2}{2\gamma_1 \underline{D}_1}\right\}, \frac{\max\left\{\overline{\lambda}, k, a_2\right\}}{\underline{M}}\right\}. \tag{2.317}$$

Thus, by setting

$$R = \overline{M} + C_2, \tag{2.318}$$

$$\rho = C_3, \tag{2.319}$$

we get the stability result in Theorem 2.14.

We now turn our attention to proving the convergence of $X(t)$ and $U(t)$ to zero. We first point out that from (2.306) it follows that $|Z(t)|$, $\|w(t)\|$, $\|\xi(t)\|$, $\hat{D}_1$, and $\hat{D}_2$ are uniformly bounded. Moreover, using (2.270) and (2.271) we get the uniform boundedness of $|X(t)|$ and $\|u(t)\|$. Using relation (2.235) it follows that $U(t)$ is uniformly bounded. From (2.236)–(2.237), (2.296)–(2.297), and (2.263) we conclude that $\frac{dX^2(t)}{dt}$ is uniformly bounded. Finally, since from (2.305) it turns out that $|Z(t)|$ and $\|\xi(t)\|$ are square integrable, using (2.271) and Barbalat's lemma (Lemma C.9), we conclude that $\lim_{t\to\infty} X(t) = 0$. We now turn our attention to proving convergence of $U(t)$. Using (2.305) we also get that $\|w(t)\|$ is square integrable in time. Thus, with the help of (2.270) and by the square integrability of $|Z(t)|$, we conclude using (2.235) that $U(t)$ is square integrable. It only remains to show that $\frac{dU^2(t)}{dt}$ is uniformly bounded, since then one can conclude that $\lim_{t\to\infty} U(t) = 0$ using Barbalat's lemma (Lemma C.9). Hence, it is

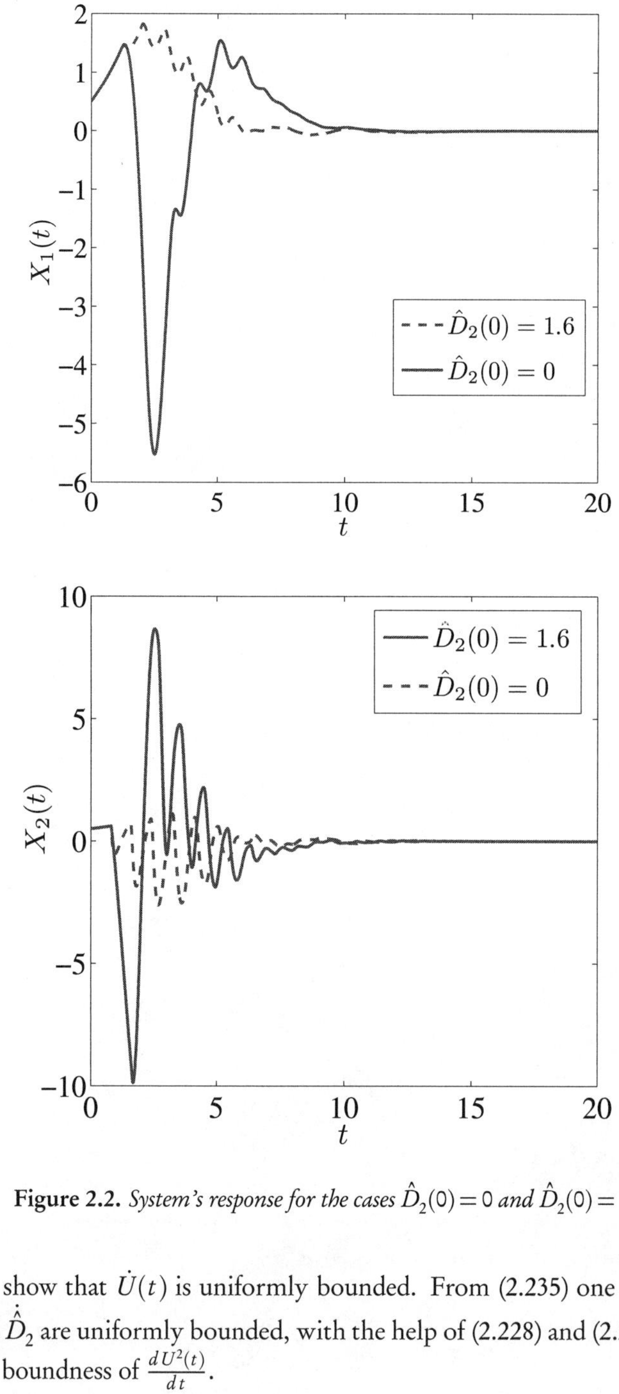

Figure 2.2. *System's response for the cases* $\hat{D}_2(0) = 0$ *and* $\hat{D}_2(0) = 1.6$.

sufficient to show that $\dot{U}(t)$ is uniformly bounded. From (2.235) one can observe that since $\dot{\hat{D}}_1$ and $\dot{\hat{D}}_2$ are uniformly bounded, with the help of (2.228) and (2.231) we conclude the uniform boundness of $\frac{dU^2(t)}{dt}$.

2.3.3 ▪ Simulations

We give here a simulation example to illustrate the effectiveness of the proposed adaptive scheme. We choose a second order feedforward system with parameters $F_1 = F_2 = 0.25$, $H_1 = 1$, $D_1 = 0.4$, and $D_2 = 0.8$. This is an unstable system with two poles at 0.25. The

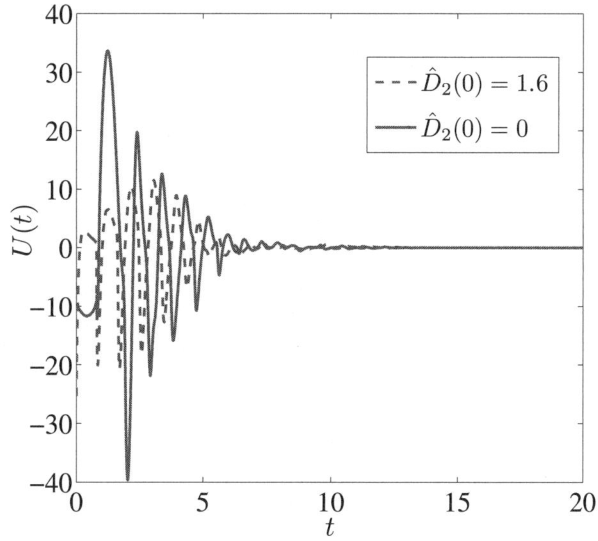

Figure 2.3. *The control effort for the cases $\hat{D}_2(0) = 0$ and $\hat{D}_2(0) = 1.6$.*

lower bounds for the unknown delays are $\underline{D}_1 = 0.1$ for D_1 and 0 for D_2. Analogously the upper bounds are chosen as twice the real values of the delays, i.e., $\overline{D}_1 = 0.8$ and $\overline{D}_2 = 1.6$. The initial conditions are chosen as $X_1(0) = 0.5$, $X_2(0) = 0.5$, and $X_2(\theta) = 0.5$ for all $\theta \in [-D_1, 0]$, and finally $\hat{D}_1(0) = \underline{D}_1 = 0.1$. The controller parameters are chosen as $a_2 = 200$, $k = 0.005$, $\gamma_1 = 25$, $\gamma_2 = 25$, and $K(\hat{D}_1) = [\ -10.0625e^{0.25\hat{D}_1} \quad -8.5\ ]$.

Figures 2.2–2.4 show two distinct simulations, starting from two extreme initial values for the input delay estimate, one at zero, and the other at twice the true delay value. In Figures 2.2 and 2.3 we observe that, as Theorem 2.14 predicts, convergence to zero is achieved for the states and the input, despite starting with initial estimate for the input delay at the two extreme values and for the state delay at the lower bound. In Figure 2.4 one can see the evolution of the estimates for the two distinct simulation cases. The estimates for the two delays, after a transient response, converge to stabilizing for the system values.

2.4 ▪ Notes and References

The predictor feedback design for linear systems with input delay presented in Section 2.1 has been studied in numerous works (potentially under the name "finite spectrum assignment" or "reduction method") [4], [41], [94], [132], [145]. However, a stability proof of the closed-loop system under predictor feedback using a Lyapunov functional has not been available until recently [93]. An alternative Lyapunov functional was provided recently in [114]. Some of the immediate benefits of having available a Lyapunov functional are summarized in [85]. Implementation issues of the predictor feedback design are discussed in [119], [122], [181], [182].

Systems with simultaneous input and state delays are considered in [41], [59], [60], [63], [98], [102], [172], [185]. In a recent work by Kharitonov [76] the predictor feedback law is constructed for linear systems, not necessarily in the strict-feedback form, with

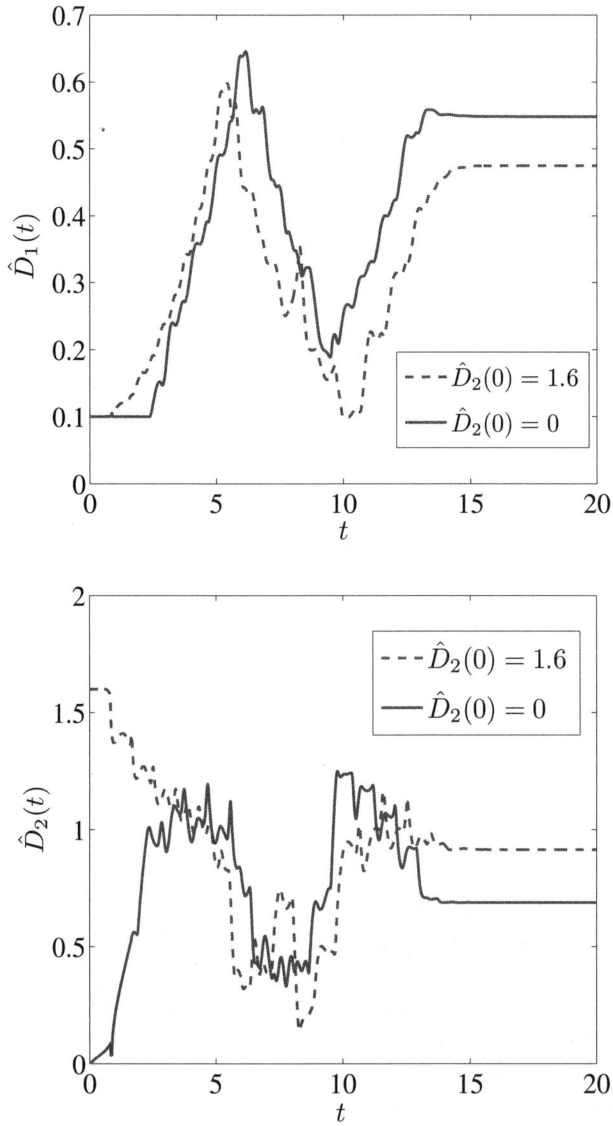

Figure 2.4. *The estimations of the unknown delays for the cases* $\hat{D}_2(0) = 0$ *and* $\hat{D}_2(0) = 1.6$.

both input and state delays. The difference with our design in Section 2.2 is that only the input delay is compensated for rather than both the input and state delays, as in our developments (which is possible due to the special structure of the systems that we considered). A backstepping-like design for linear systems with only state delay is the one considered in [59]. The major difference with the design in [59] and the one considered in Section 2.2.1 is that in [59] delays are not allowed in the virtual inputs (which is the difficult case considered here). The present procedure can be modified to incorporate state delays that are in positions other than the virtual inputs. In the case of a system with only input delay (irrespective of the form of the system, i.e., either if the system is a chain of integrators with input delay, e.g., [110], or a system in feedforward form, e.g., [60], etc.), the resulting control law is the predictor-based/finite spectrum assignment controller from [4], [41], [93], [102], with the gain K being designed using the classical backstepping procedure for linear systems from [92]. In the case of a system with simultaneous input and state delays, a backstepping-like design comparable to the one considered here is the one in [60] for the special case of a chain of delayed integrators and input delay. In this special case the resulting control law from the present work turns out to be the same as the one in [60].

The importance of designing robust controllers when the delays are the unknown parameters was highlighted in the control problems considered in [36] and [90]. Adaptive control schemes can be found in [40], [120], [128], [135], [179], [180]. Yet, the adaptive control problem when the delays are the unknown parameters had not been solved until recently with the works in [25], [26] (see also [24] for a more recent result). In [25] the problem of designing an adaptive control scheme for a linear system with unknown input delay is solved, and in [26] the result is extended to also incorporate unknown plant parameters. The aforementioned designs are based on predictor feedback together with tools that come from the adaptive control of parabolic PDEs [156].

Chapter 3

Linear Systems with Distributed Delays

As it was illustrated in Chapter 2 for single-input systems with a lumped input delay, the introduction of the infinite-dimensional backstepping transformation of the actuator state allows one to construct a Lyapunov functional for the plant under predictor feedback. Yet the backstepping methodology is applicable neither in the case of single-input systems with distributed input delay nor in the case of multi-input systems with different delays in each individual input channel. This is because the system that is composed of the finite-dimensional state X and the infinite-dimensional actuator states $U(s)$, $s \in [t-D,t]$, is not in strict-feedback form. In Section 3.1 we study multi-input multi-output linear systems, with distributed input or sensor delays that are different in each individual input or output channel. With the introduction of backstepping-forwarding transformations, we construct Lyapunov functionals which we use to prove closed-loop stability in the case of controller design, or convergence of the estimation error to zero in the case of observer design. We also design a control law for rejecting a matched, constant disturbance in the input by appropriately incorporating into the backstepping-forwarding transformations the estimation of the unknown disturbance in order to account for its effect.

In Section 3.2 we generalize the backstepping-forwarding transformations to the case where the parameters of the plant are unknown. One of the main challenges of this generalization is that one has to deal, in the case of the B matrix, with a vector of unknown functions, rather than just with a vector of unknown parameters. We resolve this challenge by constructing a Lyapunov functional with normalization and by employing an update law using projection on a projection set which can be either spherical or an infinite-dimensional hyper-rectangle. In addition, the gain kernels of these transformations are time varying, since they incorporate the estimations of the unknown parameters, and hence various technical difficulties arise when one proves that these kernels are bounded (which we need in our Lyapunov analysis).

3.1 ▪ Systems with Known Plant Parameters

We consider the system

$$\dot{X}(t) = AX(t) + \int_0^D B(\sigma)U(t-\sigma)d\sigma, \tag{3.1}$$

where $X \in \mathbb{R}^n$, $U \in \mathbb{R}^m$, and $D > 0$ is arbitrarily long. For this system and under the controllability condition of the pair

$$\left(A, \int_0^D e^{-A\sigma} B(\sigma) d\sigma\right), \tag{3.2}$$

the following controller was developed in [4] (see also [102]) and achieves asymptotic stability for any $D > 0$:

$$U(t) = Ke^{AD} X(t) + Ke^{AD} \int_{t-D}^{t} \int_{t-s}^{D} e^{A(t-s-\sigma)} B(\sigma) d\sigma \, U(s) ds, \tag{3.3}$$

where the control gain K may be designed by an LQR/Riccati approach, pole placement, or some other method that makes $A + \int_0^D e^{-A\sigma} B(\sigma) d\sigma K$ Hurwitz. The above approach works in the case of multi-input systems with different delays in each individual input channel. To see this, one can consider, for example, the case where $B(\sigma) = [B_{\delta}(\sigma) \quad B_2(\sigma)]$, with

$$B_{\delta}(\sigma) = \left\{ \begin{array}{ll} B_1(\sigma) & \text{if } \sigma \leq D_1 \\ 0 & \text{if } \sigma > D_1 \end{array} \right\}. \tag{3.4}$$

In this case system (3.1) becomes

$$\dot{X}(t) = AX(t) + \int_0^{D_1} B_1(\sigma) U_1(t-\sigma) d\sigma + \int_0^{D_2} B_2(\sigma) U_2(t-\sigma) d\sigma, \tag{3.5}$$

where $D_2 = D > D_1$. However, a Lyapunov functional for the closed-loop systems (3.1) and (3.3) is not available. The benefits of constructing a Lyapunov functional include the establishment of exponential stability as a time domain property (rather than the mere determination of the spectrum), the enablement of a study of robustness to parametric uncertainties including plant parameters and input delays, the enablement of a study of inverse optimality, and so on.

In Chapter 2 we constructed a Lyapunov functional for the plant

$$\dot{X}(t) = AX(t) + BU(t-D) \tag{3.6}$$

under the predictor feedback law

$$U(t) = Ke^{AD} X(t) + K \int_{t-D}^{t} e^{A(t-\theta)} BU(\theta) d\theta \tag{3.7}$$

using an infinite-dimensional backstepping transformation of the actuator state. However, the backstepping method is applicable neither in the case of single-input systems with distributed input delay nor in the case of multi-input systems with different delays in each individual input channel. This is because the system that is composed of the finite-dimensional state $X(t)$ and the infinite-dimensional actuator states $U(t+x-D)$, $x \in [0,D]$, is not in strict-feedback form. For this to become more clear, consider system (3.6) written using the PDE representation of the actuator state as

$$\dot{X}(t) = AX(t) + Bu(0,t), \tag{3.8}$$

$$u_t(x,t) = u_x(x,t) \quad \text{for all } x \in [0,D], \tag{3.9}$$

$$u(D,t) = U(t). \tag{3.10}$$

For system (3.8)–(3.10) the interconnection of the delay PDE with the ODE is only at one single point, namely, the output of the delay line at $x=0$. Consequently, the overall system is in strict-feedback form (the input is at the end of the integrator chain if one considers a discretization of the delay PDE). On the other hand, in (3.5) it is clear that $u(x,t)$ affects the $X(t)$ block through all of the spatial domain $x \in [0,D]$, and hence the overall system is not in strict-feedback form. For this reason, a spatially casual backstepping transformation of the infinite-dimensional actuator state which employs a Volterra integral, as the one in Chapter 2, does not apply. In the case of distributed delay the system is in the interlaced feedforward-feedback form. Hence, we use a continuum extension of a combined backstepping-forwarding transformation.

We start in Section 3.1.1 with an introduction of the predictor feedback under multiple distributed input delays and present a transformation of the finite-dimensional state $X(t)$. Then we establish exponential stability of the closed-loop system using our novel infinite-dimensional transformations of the actuator states. In Section 3.1.2 we present an example that is worked out in detail to demonstrate the construction of the Lyapunov functional. In Section 3.1.3 we design a control law that stabilizes the closed-loop system and achieves compensation of a matched, constant disturbance in the input of the plant. Finally, in Section 3.1.4 we develop a dual of the predictor-based controller and design an infinite-dimensional observer which compensates for the sensor delays.

3.1.1 ▪ Stability Analysis of Predictor Feedback for Distributed Delays

We consider the system

$$\dot{X}(t)=AX(t)+\int_0^{D_1} B_1(\sigma)U_1(t-\sigma)d\sigma+\int_0^{D_2} B_2(\sigma)U_2(t-\sigma)d\sigma, \tag{3.11}$$

where $X \in \mathbb{R}^n$, $U_1,U_2 \in \mathbb{R}$, and $D_1,D_2>0$. For notational simplicity we consider a two-input case. The same analysis can be carried out for an arbitrary number of inputs with different delays in each individual input channel. For this system, controller (3.3) achieves asymptotic stability for any $D_1,D_2>0$ under the controllability condition of the pair $\left(A,\begin{bmatrix}B_{D_1} & B_{D_2}\end{bmatrix}\right)$, where

$$B_{D_i}=\int_0^{D_i} e^{-A\sigma}B_i(\sigma)d\sigma, \quad i=1,2. \tag{3.12}$$

The predictor feedback (3.3) can be written as

$$\begin{aligned} U(t)&=\begin{bmatrix} U_1(t) \\ U_2(t) \end{bmatrix} \\ &=\begin{bmatrix} K_1e^{AD_1} \\ K_2e^{AD_2} \end{bmatrix} Z(t), \end{aligned} \tag{3.13}$$

$$Z(t)=X(t)+\sum_{i=1}^{2}\int_0^{D_i}\int_{D_i-y}^{D_i} e^{A(D_i-y-\sigma)}B_i(\sigma)d\sigma\, U_i(t+y-D_i)dy. \tag{3.14}$$

Defining

$$u_1(x,t)=U_1(t+x-D_1) \quad \text{for all } x\in[0,D_1], \tag{3.15}$$

$$u_2(z,t)=U_2(t+z-D_2) \quad \text{for all } z\in[0,D_2], \tag{3.16}$$

we obtain the main result of this section.

Theorem 3.1. *Consider the closed-loop systems consisting of the plant* (3.11) *and the controller* (3.13)*. Let the pair* $(A,[B_{D_1}\ \ B_{D_2}])$ *be completely controllable and choose* K_1 *and* K_2 *such that* $A+B_{D_1}K_1e^{AD_1}+B_{D_2}K_2e^{AD_2}$ *is Hurwitz. Then for any initial conditions* $u_i(\cdot,0)\in L^2(0,D_i)$, $i=1,2$, *the closed-loop system has a unique solution* $(X(t),u_1(\cdot,t),u_2(\cdot,t))\in C([0,\infty),\mathbb{R}^n\times L^2(0,D_1)\times L^2(0,D_2))$ *which is exponentially stable in the sense that there exist positive constants* μ *and* ρ, *such that*

$$\Omega(t)\leq \mu\Omega(0)e^{-\rho t}, \tag{3.17}$$

$$\Omega(t)=|X(t)|^2+\int_0^{D_1}U_1^2(t-\theta)d\theta+\int_0^{D_2}U_2^2(t-\theta)d\theta. \tag{3.18}$$

Moreover, if the initial conditions $u_i(\cdot,0)$, $i=1,2$, *are compatible with the control law* (3.13) *and belong to* $H^1(0,D_i)$, $i=1,2$, *then* $(X(t),u_1(\cdot,t),u_2(\cdot,t))\in C^1([0,\infty),\mathbb{R}^n\times H^1(0,D_1)\times H^1(0,D_2))$ *is the classical solution of the closed-loop system.*

Proof. We first rewrite the plant (3.11) as

$$\begin{aligned}\dot{X}(t)=AX(t)+&\int_0^{D_1}B_1(D_1-y)u_1(y,t)dy\\&+\int_0^{D_2}B_2(D_2-y)u_2(y,t)dy, &(3.19)\\ \partial_t u_1(x,t)=&\ \partial_x u_1(x,t), &(3.20)\\ u_1(D_1,t)=&\ U_1(t), &(3.21)\\ \partial_t u_2(z,t)=&\ \partial_z u_2(z,t), &(3.22)\\ u_2(D_2,t)=&\ U_2(t), &(3.23)\end{aligned}$$

where $x\in[0,D_1]$ and $z\in[0,D_2]$. Note that u_1 and u_2 satisfy (3.15), (3.16), respectively. Note also that the transformation (3.14) can be written as

$$Z(t)=X(t)+\sum_{i=1}^{2}\int_0^{D_i}\int_{D_i-y}^{D_i}e^{A(D_i-y-\sigma)}B_i(\sigma)d\sigma\, u_i(y,t)dy. \tag{3.24}$$

Consider the transformations of the infinite-dimensional actuator states $u_1(x,t)$ and $u_2(z,t)$:

$$\begin{aligned}w_1(x,t)=u_1(x,t)-\gamma_1(x)X(t)-\int_0^x p_1(x,y)u_1(y,t)dy+\int_x^{D_1}k_1(x,y)u_1(y,t)dy\\-\int_0^{D_2}q_1(x,y)u_2(y,t)dy, \qquad (3.25)\\ w_2(z,t)=u_2(z,t)-\gamma_2(z)X(t)-\int_0^z p_2(z,y)u_2(y,t)dy+\int_z^{D_2}k_2(z,y)u_2(y,t)dy\\-\int_0^{D_1}q_2(z,y)u_1(y,t)dy, \qquad (3.26)\end{aligned}$$

where the kernels $\gamma_i(\cdot)$, $p_i(\cdot)$, $k_i(\cdot)$, and $q_i(\cdot)$ for $i=1,2$ are to be specified. Using transformations (3.24)–(3.26) and relations (3.13)–(3.14) we transform the plant (3.19)–(3.23) into the "target system"

$$\dot{Z}(t)=\left(A+B_{D_1}K_1e^{AD_1}+B_{D_2}K_2e^{AD_2}\right)Z(t), \tag{3.27}$$
$$\partial_t w_1(x,t)=\partial_x w_1(x,t)-q_1(x,D_2)K_2e^{AD_2}Z(t), \tag{3.28}$$
$$w_1(D_1,t)=0, \tag{3.29}$$
$$\partial_t w_2(z,t)=\partial_z w_2(z,t)-q_2(z,D_1)K_1e^{AD_1}Z(t), \tag{3.30}$$
$$w_2(D_2,t)=0. \tag{3.31}$$

We first prove (3.27). Differentiating with respect to time (3.24) and using (3.19)–(3.20), (3.22) we obtain

$$\dot{Z}(t)=AX(t)+\sum_{i=1}^{2}\left(\int_0^{D_i} B_i(D_i-y)u_i(y,t)dy\right.$$
$$\left.+\int_0^{D_i}\int_{D_i-y}^{D_i} e^{A(D_i-y-\sigma)}B_i(\sigma)d\sigma\,\partial_y u_i(y,t)dy\right). \tag{3.32}$$

Using integration by parts together with (3.13), (3.21), (3.23) and Leibniz's differentiation rule, we obtain (3.27). We next prove (3.28)–(3.29) (the proof of (3.30)–(3.31) follows exactly the same pattern). To obtain (3.28) we differentiate (3.25) with respect to t and x. Using relations (3.20), (3.22) and integration by parts, after subtracting the resulting expressions for the time and spatial derivatives of (3.25) in order to get (3.28), we obtain a system of ODEs and PDEs which is well posed and can be solved explicitly to give

$$\gamma_1(x)=K_1e^{Ax}, \tag{3.33}$$
$$p_1(x,y)=\int_{D_1-y}^{D_1} K_1e^{A(D_1+x-y-\sigma)}B_1(\sigma)d\sigma, \tag{3.34}$$
$$k_1(x,y)=\int_0^{D_1-y} K_1e^{A(D_1+x-y-\sigma)}B_1(\sigma)d\sigma, \tag{3.35}$$
$$q_1(x,y)=\int_{D_2-y}^{D_2} K_1e^{A(D_2+x-y-\sigma)}B_2(\sigma)d\sigma. \tag{3.36}$$

Using (3.13) we get (3.28)–(3.29). Similarly, one obtains the kernels of (3.26) which can be derived from (3.33)–(3.36) by changing the index 1 to 2 and the spatial variable x to z. The next step is deriving the inverse transformations of (3.25)–(3.26). We postulate the inverse transformation of (3.25) in the form

$$u_1(x,t)=w_1(x,t)+\delta_1(x)Z(t)-\int_x^{D_1} m_1(x,y)w_1(y,t)dy. \tag{3.37}$$

Taking the time and the spatial derivatives of the above transformation, using integrations by parts together with (3.27)–(3.29) and the fact that $q_1(x,D_2)=K_1e^{Ax}B_{D_2}$, after subtracting the resulting expressions for the time and spatial derivatives, we conclude that relations

(3.20)–(3.21) hold (by taking into account also (3.13)) if

$$\delta_1(x) = K_1 e^{AD_1} e^{A_{cl}(x-D_1)} + \int_x^{D_1} \bigg(K_1 e^{Ay} B_{D_2} K_2 e^{AD_2} e^{A_{cl}(x-y)} - \int_y^{D_1} g_1(y-r) K_1 e^{Ar} B_{D_2} K_2 e^{AD_2} dr\, e^{A_{cl}(x-y)} \bigg) dy, \tag{3.38}$$

$$A_{cl} = A + B_{D_1} K_1 e^{AD_1} + B_{D_2} K_2 e^{AD_2}, \tag{3.39}$$

$$m_1(x,y) = g_1(x-y). \tag{3.40}$$

To establish that (3.37) is indeed the inverse transformation of (3.25), one has to uniquely determine $g_1(x-y)$. We find next an explicit expression for $g_1(x-y)$. To do so, we substitute $Z(t)$ and $w_1(x,t)$ from (3.24) and (3.25), respectively, into (3.37). Matching the terms for $X(t)$, $u_1(y,t)$, and $u_2(y,t)$ and performing some algebraic manipulations, we conclude that $g_1(x-y)$ must satisfy the following for all $x \in [0, D_1]$:

$$0 = -K_1 e^{Ax} + \delta_1(x) + \int_x^{D_1} g_1(x-y) K_1 e^{Ay} dy, \tag{3.41}$$

$$\begin{aligned} 0 = {} & K_1 e^{Ax} \int_x^{D_1} \int_{D_1-y}^{D_1} e^{A(D_1-y-\sigma)} B_1(\sigma) d\sigma\, u_1(y,t) dy \\ & + K_1 e^{Ax} \int_x^{D_1} \int_0^{D_1-y} e^{A(D_1-y-\sigma)} B_1(\sigma) d\sigma\, u_1(y,t) dy - \int_x^{D_1} g_1(x-y) u_1(y,t) dy \\ & + \int_0^{D_1} \left(-K_1 e^{Ax} + \delta_1(x) + \int_x^{D_1} g_1(x-r) K_1 e^{Ar} dr \right) \\ & \times \int_{D_1-y}^{D_1} e^{A(D_1-y-\sigma)} B_1(\sigma) d\sigma\, u_1(y,t) dy \\ & - \int_x^{D_1} g_1(x-y) \int_y^{D_1} K_1 e^{Ay} e^{A(D_1-r)} B_{D_1} u_1(r,t) dr\, dy. \end{aligned} \tag{3.42}$$

We can now substitute $\delta_1(x)$ from (3.38) into (3.41)–(3.42) and then find $g_1(x-y)$ that satisfies conditions (3.41)–(3.42). Instead, assuming for the moment that $g_1(x-y)$ and $\delta_1(x)$ satisfy (3.41), we remain with the following condition for all $x, y \in [0, D_1]$:

$$0 = K_1 e^{A(x-y)} B_{1e} - g_1(x-y) - \int_x^y g_1(x-r) K_1 e^{A(r-y)} dr\, B_{1e}, \tag{3.43}$$

where $B_{1e} = e^{AD_1} B_{D_1}$. Assuming a form for $g_1(x-y)$ as $g_1(x-y) = K_1 R_1(x-y) B_{1e}$, we conclude that (3.43) is satisfied if $R_1(0) = I$ together with $R_1'(w) = R_1(w)(A + B_{1e} K_1)$. Therefore

$$g_1(x) = K_1 e^{(A+B_{1e}K_1)x} B_{1e}. \tag{3.44}$$

Using (3.44) we can simplify $\delta_1(x)$. After some algebra we rewrite $\delta_1(x)$ as

$$\delta_1(x) = K_1 e^{(A+B_{1e}K_1)(x-D_1)} e^{AD_1}. \tag{3.45}$$

Finally, after some algebra one can verify that indeed relations (3.44)–(3.45) satisfy condition (3.41). Similarly, the inverse transformation of (3.26) is given by

$$u_2(z,t) = w_2(z,t) + \delta_2(z) Z(t) - \int_z^{D_2} m_2(z,y) w_2(y,t) dy, \tag{3.46}$$

where

$$m_2(z,y) = g_2(z-y) \tag{3.47}$$

$$= K_2 e^{(A+B_{2e}K_2)(z-y)} B_{2e}, \tag{3.48}$$

$$\delta_2(z) = K_2 e^{(A+B_{2e}K_2)(z-D_2)} e^{AD_2}, \tag{3.49}$$

and $B_{2e} = e^{AD_2} B_{D_2}$. Using (3.24) together with (3.37) and (3.46) we get $X(t)$ as a function of $w_1(y,t)$ and $w_2(y,t)$. We have

$$\begin{aligned} X(t) = & \left(I - \sum_{i=1}^{2} \int_0^{D_i} \int_{D_i - y}^{D_i} e^{A(D_i - y - \sigma)} B_i(\sigma) d\sigma \delta_i(y) dy \right) Z(t) \\ & - \sum_{i=1}^{2} \int_0^{D_i} \int_{D_i - y}^{D_i} e^{A(D_i - y - \sigma)} B_i(\sigma) d\sigma \\ & \times \left(w_i(y,t) - \int_y^{D_i} m_i(y,r) w_i(r,t) dr \right) dy. \end{aligned} \tag{3.50}$$

Consider now the Lyapunov functional for the target system (3.27)–(3.31):

$$\begin{aligned} V(t) = & Z(t)^T P Z(t) + \alpha \int_0^{D_1} (1+x) w_1^2(x,t) dx \\ & + \beta \int_0^{D_2} (1+z) w_2^2(z,t) dz, \end{aligned} \tag{3.51}$$

where the positive parameters α, β are to be chosen later and $P = P^T > 0$ is the solution to the Lyapunov equation

$$A_{cl}^T P + P A_{cl} = -Q \tag{3.52}$$

for some $Q = Q^T > 0$, where A_{cl} is defined in (3.39). For the time derivative of $V(t)$ along (3.27)–(3.31) we obtain

$$\begin{aligned} \dot{V}(t) \le & -\lambda_{\min}(Q) |Z(t)|^2 - \alpha w_1^2(0,t) - \alpha \int_0^{D_1} w_1^2(x,t) dx - \beta w_2^2(0,t) \\ & - 2\alpha \int_0^{D_1} w_1(x,t)(1+x) K_1 e^{Ax} B_{D_2} K_2 e^{AD_2} \sqrt{\alpha} dx Z(t) - \beta \int_0^{D_2} w_2^2(z,t) dz \\ & - 2 \int_0^{D_2} \beta w_2(z,t)(1+z) K_2 e^{Az} B_{D_1} K_1 e^{AD_1} dz Z(t), \end{aligned} \tag{3.53}$$

where $\lambda_{\min}(Q)$ is the smallest eigenvalue of Q in (3.52). Applying Young's inequality we get

$$\begin{aligned} \dot{V}(t) \le & -(\lambda_{\min}(Q) - \alpha\gamma - \beta\delta) |Z(t)|^2 - \frac{\alpha}{2} \int_0^{D_1} w_1^2(x,t) dx \\ & - \frac{\beta}{2} \int_0^{D_2} w_2^2(z,t) dz, \end{aligned} \tag{3.54}$$

where

$$\gamma = 2D_1 \left|(1+D_1)K_1 e^{AD_1} B_{D_2} K_2 e^{AD_2}\right|^2, \tag{3.55}$$

$$\delta = 2D_2 \left|(1+D_2)K_2 e^{AD_2} B_{D_1} K_1 e^{AD_1}\right|^2. \tag{3.56}$$

Hence by choosing $\alpha = \frac{\lambda_{\min}(Q)}{4\gamma}$ and $\beta = \frac{\lambda_{\min}(Q)}{4\delta}$, where $\lambda_{\max}(P)$ is the largest eigenvalue of P in (3.52), we get $\dot{V}(t) \le -\rho V(t)$, where

$$\rho = \min\left\{\frac{\lambda_{\min}(Q)}{2\lambda_{\max}(P)}, \frac{\alpha}{2(1+D_1)}, \frac{\beta}{2(1+D_2)}\right\}. \tag{3.57}$$

Using the comparison principle (see also Lemma B.4 from Appendix B) we get

$$V(t) \le V(0)e^{-\rho t}. \tag{3.58}$$

To prove stability in the original variables $X(t)$ and $u_1(x,t)$, $u_2(z,t)$ it is sufficient to show that

$$\underline{M}\Omega(t) \le V(t) \le \overline{M}\Omega(t) \tag{3.59}$$

for some positive $\overline{M}$ and $\underline{M}$, since then $\Omega(t) \le \frac{\overline{M}}{\underline{M}}\Omega(0)e^{-\rho t}$, and Theorem 3.1 is proved with $\mu = \frac{\overline{M}}{\underline{M}}$. Using (3.24) and applying the Young and Cauchy–Schwarz inequalities we get

$$|Z(t)|^2 \le m\left(|X(t)|^2 + \int_0^{D_1} u_1(x,t)^2 dx + \int_0^{D_2} u_2(z,t)^2 dz\right), \tag{3.60}$$

with $m = 3\max\{1, \max_{i=1,2}\{D_i \sup_{y\in[0,D_i]}(\int_{D_i-y}^{D_i} e^{A(D_i-y-\sigma)}B_i(\sigma)d\sigma)^2\}\}$. Using (3.25)–(3.26) together with the Young and Cauchy–Schwarz inequalities we obtain

$$\int_0^{D_i} w_i(y,t)^2 dy \le m_i\left(|X(t)|^2 + \int_0^{D_1} u_1(x,t)^2 dx + \int_0^{D_2} u_2(z,t)^2 dz\right), \tag{3.61}$$

where $i = 1,2$ and

$$\begin{aligned} m_i = 4D_i \max\Bigg\{ & 1, \sup_{y\in[0,D_i]} \gamma_i(y)^2, D_i \sup_{0\le y\le x\le D_i} p_i(x,y)^2 + D_i \sup_{0\le x\le y\le D_i} k_i(x,y)^2, \\ & D_j \sup_{x\in[0,D_i], y\in[0,D_j]} q_i(x,y)^2 \Bigg\}, \quad i,j = 1,2, \quad j \ne i. \end{aligned} \tag{3.62}$$

With $\overline{M} = \max\{\lambda_{\max}(P), \alpha(1+D_1), \beta(1+D_2)\}\max\{m, m_1, m_2\}$ the upper bound in (3.59) is proved. Similarly using (3.37), (3.46), (3.50) we get $\underline{M} = \frac{\min\{\lambda_{\min}(P),\alpha,\beta\}}{\max\{b,b_1,b_2\}}$, where

$$b = \max\left\{9\left(1+\sum_{i=1}^{2} D_i \sup_{y\in[0,D_i]}\left(\left(\int_{D_i-y}^{D_i} e^{A(D_i-y-\sigma)}B_i(\sigma)d\sigma\right)^2\right),\right.\right. \tag{3.63}$$

$$\max_{i=1,2}\left\{6D_i\left(1+D_i \sup_{0\le x\le y\le D_i} m_i(x,y)^2\right)\right.$$

$$\left.\left.\times \sup_{y\in[0,D_i]}\left(\left(\int_{D_i-y}^{D_i} e^{A(D_i-y-\sigma)}B_i(\sigma)d\sigma\right)^2\right\}\right\}, \tag{3.64}$$

$$b_i = 3\max\left\{1+D_i^2 \sup_{0\le x\le y\le D_i} m_i(x,y)^2, D_i \sup_{0\le x\le D_i} \delta_i(x)^2\right\}, \quad i=1,2. \tag{3.65}$$

Moreover, from (3.27) we conclude that $Z(t)$ is bounded and converges exponentially to zero. From (3.25)–(3.26) one can conclude that $w_i(\cdot,0) \in L^2(0,D_i)$, $i = 1,2$, and thus it follows from (3.28)–(3.31) that $w_i(\cdot,t) \in C(L^2(0,D_i))$, $i = 1,2$. Using the inverse transformations (3.37) and (3.46) we can conclude that $u_i(\cdot,t) \in C(L^2(0,D_i))$, $i = 1,2$. The uniqueness of weak solution is proved using the uniqueness of weak solution to the boundary problems (3.28)–(3.31) (see, e.g., [34]). Thus, the theorem is proved. □

3.1.2 ▪ An Example

We consider the case where $B_2(\sigma)=0$ for all $\sigma\in[0,D_2]$ and $B_1(\sigma)=B\delta(\sigma)+B_0\delta(D_1-\sigma)$, where $\delta(\sigma)$ is the Dirac function. In this case system (3.19)–(3.23) can be represented as

$$\dot X(t) = AX(t)+B_0u(0,t)+Bu(D,t), \tag{3.66}$$

$$\partial_t u(x,t) = \partial_x u(x,t), \tag{3.67}$$

$$u(D,t) = U(t), \tag{3.68}$$

with $u(x,t)=u_1(x,t)$ and $D=D_1$. The transformation of the finite-dimensional state $X(t)$ in relation (3.24) becomes

$$Z(t) = X(t) + \int_0^D e^{-Ay}B_0u(y,t)dy. \tag{3.69}$$

Moreover, the infinite-dimensional transformation of the actuator state $u(x,t)$ in equation (3.25) together with the kernels (3.34)–(3.36) gives ($K=K_1$)

$$w(x,t) = u(x,t) - Ke^{Ax}X(t) - K\int_0^x e^{A(x-y)}B_0u(y,t)dy$$
$$+K\int_x^D e^{A(D+x-y)}Bu(y,t)dy. \tag{3.70}$$

Using (3.69) and (3.70), and since the system has a single input (that is, $q_1(x,y)$ in (3.28) is zero since it is given by (3.36)), we get $\dot Z(t) = AZ(t) + \left(e^{-AD}B_0+B\right)u(D,t)$ and $\partial_t w(x,t) = \partial_x w(x,t)$. With

$$u(D,t) = Ke^{AD}Z(t)$$
$$= Ke^{AD}X(t) + K\int_0^D e^{A(D-y)}B_0u(y,t)dy, \tag{3.71}$$

we get

$$\dot{Z}(t)=\left(A+\left(e^{-AD}B_0+B\right)Ke^{AD}\right)Z(t), \tag{3.72}$$

$$\partial_t w(x,t)=\partial_x w(x,t), \tag{3.73}$$

$$w(D,t)=0. \tag{3.74}$$

Relation (3.37) together with the fact that

$$\begin{aligned} e^{(A+B_eK)(x-y)} &= e^{e^{AD}Acl(x-y)e^{-AD}} \\ &= e^{AD}e^{Acl(x-y)}e^{-AD}, \end{aligned} \tag{3.75}$$

where $B_e=e^{AD}(e^{-AD}B_0+B)$ and $A_{cl}=A+\left(e^{-AD}B_0+B\right)Ke^{AD}$, yields the inverse transformation of (3.70) as

$$\begin{aligned} u(x,t)=&\, w(x,t)+K_e e^{A_{cl}x}Z(t) \\ &-K_e\int_x^D e^{A_{cl}(D+x-y)}\left(e^{-AD}B_0+B\right)w(y,t)dy, \end{aligned} \tag{3.76}$$

with $K_e=Ke^{AD}e^{-A_{cl}D}$. Finally, we express $X(t)$ in terms of $Z(t)$ and $w(x,t)$ as

$$\begin{aligned} X(t)=&\left(I-\int_0^D e^{-Ay}B_0Ke^{AD}e^{Acl(y-D)}dy\right)Z(t)-\int_0^D e^{-Ay}B_0w(y,t)dy \\ &+\int_0^D e^{-Ay}B_0Ke^{AD}e^{-AclD}\int_y^D e^{A_{cl}(D+y-r)} \\ &\times\left(e^{-AD}B_0+B\right)w(r,t)drdy. \end{aligned} \tag{3.77}$$

Now one can use the Lyapunov functional (3.51) to establish exponential stability of the closed-loop system (3.66)–(3.68), (3.71). Instead, we will use the following Lyapunov functional:

$$V(t)=X^T(t)PX(t)+\alpha\int_0^D (1+x)w^2(x,t)dx, \tag{3.78}$$

where α is an arbitrary positive constant. Observe here that the above Lyapunov functional depends on $X(t)$ directly and through $w(x,t)$, while the Lyapunov functional defined in (3.51) depends on $X(t)$ through $Z(t)$ and $w(x,t)$. Hence, in order to prove exponential stability of the closed-loop system (3.66)–(3.68), (3.71), using (3.78), we have to derive a relation for the (X,w) cascade.

To see this one has to solve (3.77) for $Z(t)$ (assuming that the matrix that multiplies $Z(t)$ is invertible) and then plugging the resulting expression into (3.76) and (3.71). Then, having on the right-hand side of (3.76) and (3.71) only $X(t)$ and $w(x,t)$, one can find $u(0,t)$ and $u(D,t)$ only as a function of $X(t)$ and $w(x,t)$. Plugging into (3.66) $u(0,t)$ and $u(D,t)$ we get the (X,w) cascade. One can show using (3.78) that this cascade is exponentially stable.

We prove here this result for a simpler case. We set $B_0=B$. Then (3.66)–(3.68) become $\dot{X}(t)=AX(t)+B(u(0,t)+u(D,t))$, $\partial_t u(x,t)=\partial_x u(x,t)$, $u(D,t)=U(t)$. Consequently, by adding equations (3.71) and (3.76) for $x=0$ we have

$$u(0,t)+u(D,t)=w(0,t)-Ke^{AD}\int_0^D e^{-A_{cl}y}\left(e^{-AD}+I\right)Bw(y,t)dy$$
$$+Ke^{AD}\left(I+e^{-A_{cl}D}\right)Z(t). \tag{3.79}$$

Multiplying (3.77) on the left with $(I+e^{-AD})$, using the facts that $\frac{d\left(e^{-Ay}e^{A_{cl}(y-D)}\right)}{dy}=e^{-Ay}(I+e^{-AD})BKe^{AD}e^{A_{cl}(y-D)}$, $B_0=B$, and changing the order of integration in the last integral in (3.77) we have $(I+e^{-AD})X(t)=\left(I+e^{-AclD}\right)Z(t)-\int_0^D e^{-A_{cl}y}\left(I+e^{-AD}\right)Bw(y,t)dy$. Since A_{cl} is Hurwitz, the matrix $I+e^{-AclD}$ that multiplies $Z(t)$ in the previous relation is invertible. Hence, by solving the previous relation for $Z(t)$ and substituting the resulting expression into (3.79), we get $u(0,t)+u(D,t)=w(0,t)+Ke^{AD}\left(I+e^{-AD}\right)X(t)$. Hence,

$$\dot{X}(t)=\left(A+BK\left(I+e^{AD}\right)\right)X(t)+Bw(0,t), \tag{3.80}$$
$$\partial_t w(x,t)=\partial_x w(x,t), \tag{3.81}$$
$$w(D,t)=0. \tag{3.82}$$

One can now show exponential stability of the closed-loop system (3.80)–(3.82) using (3.78). Exponential stability in the original variables can then be proved by following a pattern of calculations similar to those in Section 3.1.1. In the present case (3.78) has the form

$$V(t)=X^T(t)PX(t)+\alpha\int_{t-D}^{t}(1+\theta+D-t)W^2(\theta)d\theta, \tag{3.83}$$

$$W(\theta)=U(\theta)-Ke^{A(\theta+D-t)}X(t)-K\int_{t-D}^{\theta}e^{A(\theta-\sigma)}BU(\sigma)d\sigma$$
$$+K\int_{\theta}^{t}e^{A(D+\theta-\sigma)}BU(\sigma)d\sigma, \tag{3.84}$$

where $t-D\leq\theta\leq t$. Observe here that in order to choose K in (3.80) such that $A+BK(I+e^{AD})$ is Hurwitz, the pair (A,B) has to be controllable and the matrix $I+e^{AD}$ invertible. Since the matrices A and $I+e^{AD}$ commute, this is equivalent to the controllability condition of the pair $(A,(I+e^{-AD})B)$, which is the controllability condition in [4]. An example where the controllability condition from [4] fails is when the matrix $I+e^{AD}$ is identically zero. This is the case, for example, of a second order oscillator with frequency π and input delay $D=1$.

3.1.3 ▪ Disturbance Rejection

We consider the system

$$\dot{X}(t)=AX(t)+\int_0^D B(x)(u(x,t)+d)dx, \tag{3.85}$$
$$u_t(x,t)=u_x(x,t),\quad x\in[0,D], \tag{3.86}$$
$$u(D,t)=U(t), \tag{3.87}$$

where we assume that d is an unknown nonzero constant and the matrices A and $B(x)$ for all $x\in[0,D]$ are known. We also assume that the state $X(t)$ and the infinite-dimensional actuator state $U(\theta)$ for all $\theta\in[t-D,t]$ are measured.

Assumption 3.2. *The pair* $\left(A, \int_0^D e^{-A(D-x)}B(x)dx\right)$ *is completely controllable and there exist a vector* K *and matrices* P, Q *symmetric and positive definite such that*

$$\left(A+\int_0^D e^{A(D-x)}B(x)dxK\right)^T P + P\left(A+\int_0^D e^{A(D-x)}B(x)dxK\right)=-Q. \tag{3.88}$$

The controller for system (3.85)–(3.87) is

$$U(t)=K\hat{Z}(t)-\hat{d}(t), \tag{3.89}$$

$$\hat{Z}(t)=X(t)+\int_0^D\int_0^x e^{-A(x-y)}B(y)dy\left(u(x,t)+\hat{d}(t)\right)dx. \tag{3.90}$$

The update law for the disturbance is given by

$$\dot{\hat{d}}(t)=\gamma\left(\hat{Z}(t)^T P-\alpha\int_0^D(1+x)w(x,t)Ke^{A_{\mathrm{cl}}(x-D)}dx\right)\Gamma_1, \tag{3.91}$$

where

$$w(x,t)=u(x,t)-Ke^{A_{\mathrm{cl}}(x-D)}\hat{Z}(t)+\hat{d}(t), \tag{3.92}$$

$$A_{\mathrm{cl}}=A+\int_0^D e^{A(D-x)}B(x)dxK, \tag{3.93}$$

$$\Gamma_1=\int_0^D B(x)dx, \tag{3.94}$$

$$\alpha=\lambda_{\min}(Q). \tag{3.95}$$

Defining the matrix

$$\Gamma_2=\int_0^D\int_0^x e^{-A(x-y)}B(y)dydx, \tag{3.96}$$

we obtain the following result.

Theorem 3.3. *Consider the closed-loop systems consisting of the plant* (3.85)–(3.87) *together with the adaptive controller* (3.89)–(3.94). *Let Assumption* 3.2 *be satisfied and choose* γ *such that*

$$\gamma\leq\frac{\frac{1}{8}|\Gamma_1|^{-1}\alpha}{\left(|P|+\alpha(1+D)|K|e^{|A_{\mathrm{cl}}|D}\right)\left(|P\Gamma_2|+\alpha(1+D)\left(1+|K|e^{|A_{\mathrm{cl}}|D}|\Gamma_2|\right)\right)}. \tag{3.97}$$

Then there exists a positive constant R *such that*

$$|X(t)|^2+\|u(t)\|^2+\tilde{d}(t)^2\leq R\left(|X(0)|^2+\|u(0)\|^2+\tilde{d}(0)^2+d^2\right), \tag{3.98}$$

where

$$\|u(t)\|^2=\int_0^D u(x,t)^2dx, \tag{3.99}$$

$$\tilde{d}(t)=d-\hat{d}(t). \tag{3.100}$$

Furthermore,

$$\lim_{t\to\infty} X(t)=0, \tag{3.101}$$
$$\lim_{t\to\infty} U(t)=-d. \tag{3.102}$$

We start the proof of Theorem 3.3 by differentiating (3.90) with respect to time and using (3.85)–(3.87) together with (3.89) to get that

$$\dot{\hat{Z}}(t)=A_{\mathrm{cl}}\hat{Z}(t)+\Gamma_1\tilde{d}(t)+\Gamma_2\dot{\hat{d}}(t), \tag{3.103}$$

where Γ_2 is defined in (3.96), and we used also the fact that $-A\Gamma_2=-\Gamma_1+\int_0^D e^{-A(D-x)}B(x)dx$ (which follows from the fact that $F'(x)=B(x)-AF(x)$, with $F(x)=\int_0^x e^{-A(x-y)}B(y)dy$). Differentiating (3.92) with respect to time and with respect to the spatial variable x and using (3.86)–(3.87), (3.89), and (3.103), we get

$$w_t(x,t)=w_x(x,t)-Ke^{A_{\mathrm{cl}}(x-D)}\Gamma_1\tilde{d}(t)+\left(1-Ke^{A_{\mathrm{cl}}(x-D)}\Gamma_2\right)\dot{\hat{d}}(t), \tag{3.104}$$
$$w(D,t)=0. \tag{3.105}$$

We have the following lemma.

Lemma 3.4. *There exist constants F_u, F_w, F_X and $F_{\hat{Z}}$ such that*

$$\|u(t)\|^2\le F_u\left(\|w(t)\|^2+|\hat{Z}(t)|^2+|\hat{d}(t)|^2\right) \tag{3.106}$$
$$|X(t)|^2\le F_X\left(|\hat{Z}(t)|^2+\|w(t)\|^2\right) \tag{3.107}$$
$$\|w(t)\|^2\le F_w\left(\|u(t)\|^2+|X(t)|^2+|\hat{d}(t)|^2\right) \tag{3.108}$$
$$|\hat{Z}(t)|^2\le F_{\hat{Z}}\left(|X(t)|^2+\|u(t)\|^2+|\hat{d}(t)|^2\right). \tag{3.109}$$

Proof. Using relation (3.92) together with the Young and Cauchy–Schwarz inequalities we get bound (3.106) with

$$F_u=3(1+D)\left(2+|K|^2e^{2|A_{\mathrm{cl}}|D}\right). \tag{3.110}$$

Substituting $u(x,t)+\hat{d}(t)=w(x,t)+Ke^{A_{\mathrm{cl}}(x-D)}\hat{Z}(t)$ into (3.90) and solving for $X(t)$ we get bound (3.107) with

$$F_X=2\left(\left(1+|\Gamma_2||K|e^{|A_{\mathrm{cl}}|D}\right)^2+D^2e^{2|A|D}\left(\int_0^D|B(x)|dx\right)^2\right). \tag{3.111}$$

From (3.90) and the Young and Cauchy–Schwarz inequalities we get bound (3.109) with

$$F_{\hat{Z}}=2\left((1+|\Gamma_2|)^2+D^2e^{2|A|D}\left(\int_0^D|B(x)|dx\right)^2\right). \tag{3.112}$$

From (3.92) and (3.109) we get bound (3.108) with

$$F_w = 3\left(1 + D + |K|^2 e^{2|A_{\mathrm{cl}}|D} F_{\hat{Z}}\right). \qquad \square \tag{3.113}$$

Lemma 3.5. *Let γ and α be as in* (3.97). *Then for the Lyapunov function*

$$V(t) = \hat{Z}(t)^T P \hat{Z}(t) + \alpha \int_0^D (1+x) w(x,t)^2 dx + \frac{1}{\gamma}\tilde{d}(t)^2 \tag{3.114}$$

the following holds:

$$V(t) \leq V(0). \tag{3.115}$$

Proof. Differentiating $V(t)$ with respect to time and using (3.103), (3.104), (3.105) and integrating by parts the w integral, we get

$$\begin{aligned}\dot{V}(t) \leq & -\lambda_{\min}(Q)|\hat{Z}(t)|^2 - \alpha w(0,t)^2 - \alpha \int_0^D w(x,t)^2 dx - \frac{2}{\gamma}\dot{\hat{d}}\tilde{d} \\ & + 2\left(\hat{Z}(t)^T P - \alpha \int_0^D (1+x) w(x,t) K e^{A_{\mathrm{cl}}(x-D)} dx\right)\Gamma_1 \tilde{d} + 2\dot{\hat{d}} \\ & \times \left(\hat{Z}(t)^T P \Gamma_2 + \alpha \int_0^D (1+x) w(x,t)\left(1 - K e^{A_{\mathrm{cl}}(x-D)}\Gamma_2\right) dx\right). \end{aligned} \tag{3.116}$$

With (3.91) and using the fact that $|\dot{\hat{d}}| \leq \gamma\delta_1\left(|\hat{Z}| + \int_0^D |w(x,t)|dx\right)$, where $\delta_1 = |P\Gamma_1| + \alpha(1+D)|K e^{|A_{\mathrm{cl}}|D}\Gamma_1|$, we get

$$\begin{aligned}\dot{V}(t) \leq & -\lambda_{\min}(Q)|\hat{Z}(t)|^2 - \alpha \int_0^D w(x,t)^2 dx - \alpha w(0,t)^2 \\ & + \gamma\delta_1\delta_2\left(|\hat{Z}(t)|^2 + \int_0^D w(x,t)^2 dx\right), \end{aligned} \tag{3.117}$$

where we used the fact that $(p+r)^2 \leq 2(p^2+r^2)$ and $\delta_2 = 4|P\Gamma_2| + 4\alpha(1+D)\left(1 + |K e^{|A_{\mathrm{cl}}|D}\Gamma_2|\right)$. Choosing γ and α as in (3.97)–(3.95) we have that

$$\dot{V}(t) \leq -\frac{\lambda_{\min}(Q)}{2}|\hat{Z}(t)|^2 - \frac{\alpha}{2}\int_0^D w(x,t)^2 dx - \alpha w(0,t)^2. \qquad \square \tag{3.118}$$

We are now ready to derive the stability estimate of Theorem 3.3. Using relations (3.110), (3.111) and (3.114), (3.115) it follows that

$$\begin{aligned}|X(t)|^2 + \|u(t)\|^2 + \tilde{d}(t)^2 & \leq 2(F_X + F_u)\left(\|w(t)\|^2 + |\hat{Z}(t)|^2 + |\tilde{d}(t)|^2 + d^2\right) \\ & \leq 2(F_X + F_u)\left(\frac{1}{\min\left\{\lambda_{\min}(P), \alpha, \frac{1}{\gamma}\right\}} + 1\right) \\ & \quad \times \left(V(0) + d^2\right). \end{aligned} \tag{3.119}$$

Moreover, using (3.114) and the bounds (3.112), (3.113) we get

$$|X(t)|^2 + ||u(t)||^2 + \tilde{d}(t)^2 \leq 2M_1 \left(||u(0)||^2 + |X(0)|^2 + |\tilde{d}(0)|^2 + d^2 \right) \tag{3.120}$$

where $M_1 = \left(\lambda_{\max}(P) + \alpha(1+D) + \frac{1}{\gamma} + 1\right)\left(F_{\hat{Z}} + F_w + 1\right)\left(\frac{1}{\min\{\lambda_{\min}(P), \alpha, \frac{1}{\gamma}\}} + 1\right)(F_X + F_u)$. Thus, by setting $R = 2M_1$ we get the stability result in Theorem 3.3.

We now turn our attention to proving the convergence of $X(t)$ and $U(t)$. We first point out that from (3.115) it follows that $|\hat{Z}(t)|$, $||w(t)||$, and $\left|\hat{d}(t)\right|$ are uniformly bounded. From (3.89) we get that $|U(t)|$ is uniformly bounded for all $t \geq 0$, and hence $|u(0,t)|$ is uniformly bounded for all $t \geq D$. From (3.91) we get the uniform boundness of $|\dot{\hat{d}}(t)|$. Using the fact that $\frac{d \int_0^D w(x,t)^2 dx}{dt} = 2\int_0^D w(x,t) w_t(x,t) dx$, integration by parts, and relation (3.104) we conclude that $\frac{d||w(t)||^2}{dt}$ is uniformly bounded if $|w(0,t)|$ is uniformly bounded. This fact follows from (3.92) and the uniform boundedness of $|u(0,t)|$ for all $t \geq D$. Since from (3.118) it turns out that $|\hat{Z}(t)|$ and $||w(t)||$ are square integrable, and using the uniform boundedness of $\frac{d|\hat{Z}(t)|^2}{dt}$ which follows from (3.103), using Barbalat's lemma (Lemma C.9), we conclude that $\lim_{t\to\infty} |\hat{Z}(t)| = 0$ and that $\lim_{t\to\infty} ||w(t)|| = 0$. Using (3.107) we get the regulation of $|X(t)|$ to zero. We prove now convergence of U. Since $\int_0^\infty \dot{\hat{Z}}(t) dt = \hat{Z}(\infty) - \hat{Z}(0)$ exists and is bounded, and since $\ddot{\hat{Z}}(t)$ is uniformly bounded from (3.103), (3.91), and (3.104), we get that $\dot{\hat{Z}}(t)$ is uniformly continuous, and hence with Barbalat's lemma (Lemma C.9) we get that $\lim_{t\to\infty} \dot{\hat{Z}}(t) = 0$. From (3.91) we get that $\lim_{t\to\infty} \dot{\hat{d}}(t) = 0$, and hence, using (3.103) and the fact that not all components of Γ_1 are zero (since in this case from (3.85) one can observe that d has no influence on the plant, and one can simply choose $\hat{d} = 0$), we get that $\lim_{t\to\infty} \tilde{d}(t) = 0$. Therefore, from (3.89) we get that $\lim_{t\to\infty} U(t) = -d$, which completes the proof.

Simulation of a Scalar Example

We consider the following scalar plant:

$$\dot{X}(t) = \theta X(t) + b_1 \int_0^1 (u(x,t) + d) dx, \tag{3.121}$$

$$u_t(x,t) = u_x(x,t), \tag{3.122}$$

$$u(1,t) = U(t), \tag{3.123}$$

where $\theta = 1$ and $b_1 = 0.5$. We choose the initial condition of the system as $X(0) = 1$ and $U(\theta) = 0$ for all $-D \leq \theta \leq 0$. The parameters of the control law (3.89)–(3.94) are chosen as $\gamma = 0.001$, $A_{\text{cl}} = -2$, and $K = \theta \frac{A_{cl} - \theta}{b_1(1 - e^{-\theta})}$. We assume initially that a constant disturbance of magnitude $d = 0.5$ perturbs the closed-loop system at time $t = 10s$. In Figure 3.1 we show the response of the system and the control effort, and in Figure 3.2 the estimation of the unknown disturbance. The control law compensates for the disturbance and brings the state of the plant to zero.

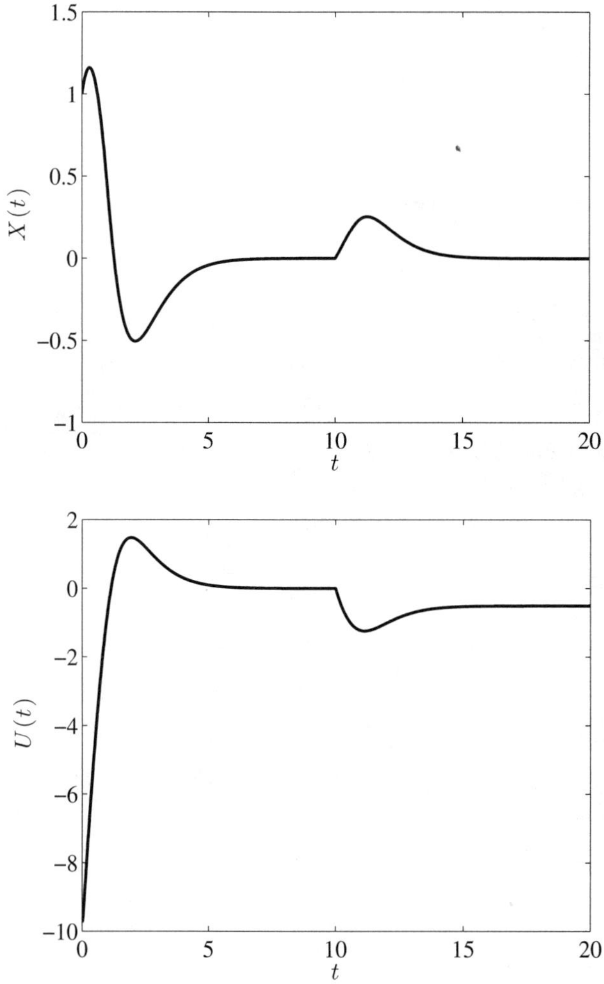

Figure 3.1. *Response of the plant* (3.121)–(3.123) *with the disturbance rejection control algorithm* (3.89)–(3.94) *for initial conditions* $X(0) = 1$ *and* $U(\theta) = 0$ *for all* $-D \leq \theta \leq 0$ *under a step disturbance* $d = 0.5$ *applied at* $t = 10s$.

3.1.4 ▪ Observer Design for Systems with Distributed Sensor Delays

We consider the system

$$\dot{X}(t) = AX(t) + BU(t), \tag{3.124}$$

$$Y_1(t) = \int_0^{D_1} C_1(\sigma)X(t-\sigma)d\sigma, \tag{3.125}$$

$$Y_2(t) = \int_0^{D_2} C_2(\sigma)X(t-\sigma)d\sigma, \tag{3.126}$$

which can be written equivalently as

$$\dot{X}(t) = AX(t) + BU(t), \tag{3.127}$$

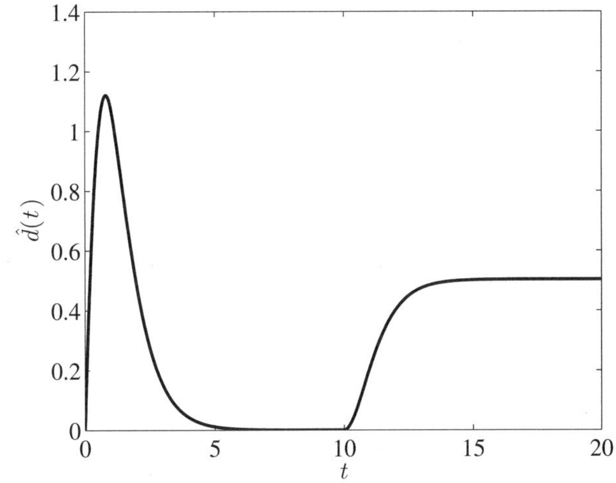

Figure 3.2. *Estimation of the disturbance d for initial conditions $X(0)=1$ and $U(\theta)=0$ for all $-D\leq\theta\leq 0$ under a step disturbance $d=0.5$ applied at $t=10s$.*

$$\partial_t\xi_1(x,t)=\partial_x\xi_1(x,t)+C_1(x)X(t), \tag{3.128}$$
$$\xi_1(D_1,t)=0, \tag{3.129}$$
$$\partial_t\xi_2(z,t)=\partial_z\xi_2(z,t)+C_2(z)X(t), \tag{3.130}$$
$$\xi_2(D_2,t)=0, \tag{3.131}$$
$$Y_1(t)=\xi_1(0,t), \tag{3.132}$$
$$Y_2(t)=\xi_2(0,t). \tag{3.133}$$

Next we state a new observer that compensates for the sensor delays, and we prove exponential convergence of the resulting observer error system.

Theorem 3.6. *Consider the system (3.124)-(3.126) and let the pair $\left(A,\left[\begin{smallmatrix}C_{D_1}\\ C_{D_2}\end{smallmatrix}\right]\right)$ be observable, where*

$$C_{D_i}=\int_0^{D_i}C_i(\sigma)e^{-A\sigma}d\sigma,\quad i=1,2. \tag{3.134}$$

Define the observer

$$\dot{\hat{X}}(t)=A\hat{X}(t)+BU(t)+L_1\left(Y_1(t)-\hat{Y}_1(t)\right)+L_2\left(Y_2(t)-\hat{Y}_2(t)\right), \tag{3.135}$$
$$\begin{aligned}\partial_t\hat{\xi}_1(x,t)=&\ \partial_x\hat{\xi}_1(x,t)+C_1(x)\hat{X}(t)+\gamma_1(x)L_1\left(Y_1(t)-\hat{Y}_1(t)\right)\\&+\gamma_1(x)L_2\left(Y_2(t)-\hat{Y}_2(t)\right),\end{aligned} \tag{3.136}$$
$$\hat{\xi}_1(D_1,t)=0, \tag{3.137}$$
$$\begin{aligned}\partial_t\hat{\xi}_2(z,t)=&\ \partial_z\hat{\xi}_2(z,t)+C_2(z)\hat{X}(t)+\gamma_2(z)L_1\left(Y_1(t)-\hat{Y}_1(t)\right)\\&+\gamma_2(z)L_2\left(Y_2(t)-\hat{Y}_2(t)\right),\end{aligned} \tag{3.138}$$

$$\hat{\xi}_2(D_2,t)=0, \tag{3.139}$$
$$\hat{Y}_1(t)=\hat{\xi}_1(0,t), \tag{3.140}$$
$$\hat{Y}_2(t)=\hat{\xi}_2(0,t), \tag{3.141}$$
$$\gamma_1(x)=\left(C_{D_1}-\int_0^x C_1(y)e^{-Ay}dy\right)e^{Ax}, \tag{3.142}$$
$$\gamma_2(z)=\left(C_{D_2}-\int_0^z C_2(y)e^{-Ay}dy\right)e^{Az}, \tag{3.143}$$

where L_1 and L_2 are chosen such that the matrix $A-L_1C_{D_1}-L_2C_{D_2}$ is Hurwitz. Then for any $(\xi_i(\cdot,0),\hat{\xi}_i(\cdot,0))\in L^2(0,D_i)$, $i=1,2$, the observer error system has a unique solution $\left(X(t)-\hat{X}(t),\xi_1(\cdot,t)-\hat{\xi}_1(\cdot,t),\xi_2(\cdot,t)-\hat{\xi}_2(\cdot,t)\right)\in C([0,\infty),\mathbb{R}^n\times L^2(0,D_1)\times L^2(0,D_2))$ which is exponentially stable in the sense that there exist positive constants $\varkappa$ and λ such that

$$\Xi(t)\le \varkappa\Xi(0)e^{-\lambda t}, \tag{3.144}$$
$$\Xi(t)=\left|X(t)-\hat{X}(t)\right|^2+\int_0^{D_1}\left(\xi_1(x,t)-\hat{\xi}_1(x,t)\right)^2dx$$
$$+\int_0^{D_2}\left(\xi_2(z,t)-\hat{\xi}_2(z,t)\right)^2dz. \tag{3.145}$$

Proof. Introducing the error variables $\tilde{X}(t)=X(t)-\hat{X}(t)$, $\tilde{\xi}_1(x,t)=\xi_1(x,t)-\hat{\xi}_1(x,t)$ and $\tilde{\xi}_2(z,t)=\xi_2(z,t)-\hat{\xi}_2(z,t)$, we obtain

$$\dot{\tilde{X}}(t)=A\tilde{X}(t)-L_1\tilde{\xi}_1(0,t)-L_2\tilde{\xi}(0,t), \tag{3.146}$$
$$\partial_t\tilde{\xi}_1(x,t)=\partial_x\tilde{\xi}_1(x,t)+C_1(x)\tilde{X}(t)-\gamma_1(x)L_1\tilde{\xi}_1(0,t)-\gamma_1(x)L_2\tilde{\xi}_2(0,t), \tag{3.147}$$
$$\tilde{\xi}_1(D_1,t)=0, \tag{3.148}$$
$$\partial_t\tilde{\xi}_1(z,t)=\partial_z\tilde{\xi}_2(z,t)+C_2(z)\tilde{X}(t)-\gamma_2(z)L_1\tilde{\xi}_1(0,t)-\gamma_2(z)L_2\tilde{\xi}_2(0,t), \tag{3.149}$$
$$\tilde{\xi}_2(D_2,t)=0. \tag{3.150}$$

With the transformations

$$\tilde{\zeta}_1(x,t)=\tilde{\xi}_1(x,t)-\gamma_1(x)\tilde{X}(t), \tag{3.151}$$
$$\tilde{\zeta}_2(z,t)=\tilde{\xi}_2(z,t)-\gamma_2(z)\tilde{X}(t), \tag{3.152}$$

and by noting that $\gamma_1(x)$ and $\gamma_2(z)$ in (3.142)–(3.143) are the solutions of the boundary value problems

$$\gamma_1'(x)=\gamma_1(x)A-C_1(x), \tag{3.153}$$
$$\gamma_1(D_1)=0, \tag{3.154}$$
$$\gamma_2'(z)=\gamma_2(z)A-C_2(z), \tag{3.155}$$
$$\gamma_2(D_2)=0, \tag{3.156}$$

we get

$$\dot{\tilde{X}}(t) = \left(A - L_1 C_{D_1} - L_2 C_{D_2}\right)\tilde{X}(t) - L_1\tilde{\zeta}_1(0,t) - L_2\tilde{\zeta}_2(0,t), \tag{3.157}$$

$$\partial_t \tilde{\zeta}_1(x,t) = \partial_x \tilde{\zeta}_1(x,t), \tag{3.158}$$

$$\tilde{\zeta}_1(D_1,t) = 0, \tag{3.159}$$

$$\partial_t \tilde{\zeta}_2(z,t) = \partial_z \tilde{\zeta}_2(z,t), \tag{3.160}$$

$$\tilde{\zeta}_2(D_2,t) = 0. \tag{3.161}$$

To establish exponential stability of the error system, we use the Lyapunov functional

$$V(t) = \tilde{X}^T(t)P\tilde{X}(t) + \alpha\int_0^{D_1}(1+x)\tilde{\zeta}_1^2(x,t)dx + \beta\int_0^{D_2}(1+z)\tilde{\zeta}_2^2(z,t)dz, \tag{3.162}$$

where the positive parameters α, β are to be chosen later and $P = P^T > 0$ satisfies the Lyapunov equation

$$\left(A - L_1 C_{D_1} - L_2 C_{D_2}\right)^T P + P\left(A - L_1 C_{D_1} - L_2 C_{D_2}\right) = -Q \tag{3.163}$$

for some $Q = Q^T > 0$, L_1, and L_2. Taking the time derivative of $V(t)$, and using integration by parts in the integrals in (3.162) and relations (3.157)–(3.161), we get

$$\begin{aligned}\dot{V}(t) \leq{}& -\lambda_{\min}(Q)\left|\tilde{X}(t)\right|^2 - \alpha\int_0^{D_1}\tilde{\zeta}_1^2(x,t)dx - \alpha\zeta_1^2(0,t) - \beta\int_0^{D_2}\tilde{\zeta}_2^2(z,t)dz \\ & - \beta\zeta_2^2(0,t) + \frac{\sqrt{\lambda_{\min}(Q)}\left|\tilde{X}(t)\right|}{\sqrt{2}}\frac{2\sqrt{2}|P||L_1|}{\sqrt{\lambda_{\min}(Q)}}|\zeta_1(0,t)| \\ & + \frac{\sqrt{\lambda_{\min}(Q)}\left|\tilde{X}(t)\right|}{\sqrt{2}}\frac{2\sqrt{2}|P||L_2|}{\sqrt{\lambda_{\min}(Q)}}|\zeta_2(0,t)|.\end{aligned} \tag{3.164}$$

Applying Young's inequality we get

$$\begin{aligned}\dot{V}(t) &\leq -\frac{\lambda_{\min}(Q)}{2}\left|\tilde{X}(t)\right|^2 - \alpha\int_0^{D_1}\tilde{\zeta}_1^2(x,t)dx - \beta\int_0^{D_2}\tilde{\zeta}_2^2(z,t)dz \\ &\leq -\gamma V(t),\end{aligned} \tag{3.165}$$

where

$$\alpha = \frac{4|P|^2|L_1|^2}{\lambda_{\min}(Q)}, \tag{3.166}$$

$$\beta = \frac{4|P|^2|L_2|^2}{\lambda_{\min}(Q)}, \tag{3.167}$$

$$\gamma = \min\left\{\frac{\lambda_{\min}(Q)}{2\lambda_{\max}(P)}, \frac{\alpha}{1+D_1}, \frac{\beta}{1+D_2}\right\}. \tag{3.168}$$

Similarly to the proof of Theorem 3.1, in order to establish exponential stability of the observer estimation error it is sufficient to show that $\underline{M}\Xi(t) \leq V(t) \leq \overline{M}\Xi(t)$ for some positive $\overline{M}$ and $\underline{M}$. Using (3.151)–(3.152) with Young's inequality we get

$$\overline{M} = \max\left\{2, 1+2\sum_{i=1}^{2} D_i \sup_{y\in[0,D_i]} \gamma_i(y)^2\right\} \times \max\{\lambda_{\max}(P), \alpha(1+D_1), \beta(1+D_2)\}, \tag{3.169}$$

$$\underline{M} = \frac{\min\{\lambda_{\min}(P), \alpha, \beta\}}{\max\left\{2, 1+2\sum_{i=1}^{2} D_i \sup_{y\in[0,D_i]} \gamma_i(y)^2\right\}}. \tag{3.170}$$

From (3.151)–(3.152) one can conclude that $\tilde{\zeta}_i(\cdot,0) \in L^2(0,D_i)$, $i=1,2$, and thus it follows from (3.158)–(3.161) that $\tilde{\zeta}_i(\cdot,t) \in C(L^2(0,D_i))$, $i=1,2$. Thus from (3.157) it follows that $\tilde{X}(t)$ is bounded. Using the inverse transformations of (3.151)–(3.152) we can conclude that $\tilde{\xi}_i(\cdot,t) \in C(L^2(0,D_i))$, $i=1,2$. The uniqueness of weak solution is proved using the uniqueness of weak solution to the boundary problems (3.158)–(3.161) (see, e.g., [34]). □

3.2 ▪ Adaptive Control for Systems with Uncertain Plant Parameters

We introduce backstepping-forwarding transformations, of certainty equivalence type, of the finite-dimensional plant state and of the infinite-dimensional actuator state that transform the system into a "target system." By constructing a Lyapunov function for the target system we design update laws for the parameters of the plant, which in the case of the input matrix B is infinite-dimensional. With the help of the available Lyapunov function we prove stability and regulation of the closed-loop system.

3.2.1 ▪ Adaptive Control

We consider the system

$$\dot{X}(t) = AX(t) + \int_0^D B(D-\sigma)U(t-\sigma)d\sigma, \tag{3.171}$$

where $X \in \mathbb{R}^n$ is the state of plant, $U \in \mathbb{R}$ is the input, and $D > 0$ is the delay. For notational simplicity we assume that our system is single-input. However, the results of this section can be extended to the multi-input case, when the delays are the same in each individual input channel. We are concerned here with the case where the matrices $B(x)$, $x \in [0,D]$, and A are unknown and of the form

$$A = A_0 + \sum_{i=1}^{p} \theta_i A_i, \tag{3.172}$$

$$B(x) = B_0(x) + \sum_{i=1}^{p} b_i(x) B_i, \tag{3.173}$$

where the $b_i(x)$, $i=1,2,\ldots,p$, are unknown, scalar continuous functions of x, and θ_i, $i=1,2,\ldots,p$, are unknown constants.

In order to help better understand the structure of system (3.171)–(3.173) we derive its transfer function, namely, $G(s)$, from the input $U(t)$ to the state $X(t)$. We first rewrite system (3.171) in the following equivalent form using transport PDE representation for the actuator state $U(\sigma)$, $\sigma \in [t-D,t]$:

$$\dot{X}(t)=AX(t)+\int_0^D B(x)u(x,t)dx, \tag{3.174}$$

$$u_t(x,t)=u_x(x,t), \quad x\in[0,D], \tag{3.175}$$

$$u(D,t)=U(t). \tag{3.176}$$

Taking the Laplace transform of (3.175), we obtain the following boundary value problem with respect to the spatial variable x:

$$su(s,x)=u'(s,x), \tag{3.177}$$

$$u(s,D)=U(s). \tag{3.178}$$

Solving the above boundary value problem we get

$$u(s,x)=e^{s(x-D)}U(s). \tag{3.179}$$

Taking the Laplace transform of (3.174) and using (3.179) we get

$$G(s)=(sI-A)^{-1}\int_0^D B(x)e^{s(x-D)}dx. \tag{3.180}$$

Our adaptive controller is based on infinite-dimensional update laws for the estimation of the unknown functions $b_i(x)$, $i=1,2,\ldots,p$, for all $x\in[0,D]$ and on finite-dimensional update laws for the estimation of the constant parameters θ_i, $i=1,2,\ldots,p$. We employ the update laws using projector operators (see [156] for the use of projector operators in PDEs). Consequently, we make the following assumption.

Assumption 3.7. *There exist known constants $\underline{\theta}_i$, $\overline{\theta}_i$, ρ_i and known continuous functions $b_i^{\mathrm{n}}(x)$, $i=1,2,\ldots,p$, such that*

$$\int_0^D \left(b_i(x)-b_i^{\mathrm{n}}(x)\right)^2 dx \leq \rho_i, \tag{3.181}$$

$$\underline{\theta}_i \leq \theta_i \leq \overline{\theta}_i, \tag{3.182}$$

for all $i=1,2,\ldots,p$.

Furthermore, we have to make an assumption regarding the controllability of a specific pair of matrices (this fact becomes clear later on) for all $B(x)$ such that $\int_0^D \left(b_i(x)-b_i^{\mathrm{n}}(x)\right)^2 dx \leq \rho_i$, $i=1,2,\ldots,p$, and for all A such that $\underline{\theta}_i \leq \theta_i \leq \overline{\theta}_i$, $i=1,2,\ldots,p$. We thus make the following assumption.

Assumption 3.8. *We assume that the pair $\left(A, \int_0^D e^{-A(D-x)}B(x)dx\right)$ is uniformly completely controllable for all $\underline{\theta}_i \leq \theta_i \leq \overline{\theta}_i$, $i=1,2,\ldots,p$, and for all $\int_0^D \left(b_i(x)-b_i^{\mathrm{n}}(x)\right)^2 dx \leq \rho_i$, $i=1,2,\ldots,p$, and that there exist a vector-valued function $K(\cdot): C^1(\Lambda;R^n)$ and matrices*

$P(\cdot): C^1(\Lambda;\mathbb{R}_+^{n\times n})$, $Q(\cdot): C(\Lambda;\mathbb{R}_+^{n\times n})$, *where*

$$\Lambda=\left\{(\beta,\theta)\in\mathbb{R}^{n+p} \mid \beta=\int_0^D e^{-A(D-x)}B(x)dx \quad \textit{for all} \right.$$

$$\left. \int_0^D \left(b_i(x)-b_i^{\mathrm{n}}(x)\right)^2 dx \le \rho_i \textit{ and } \underline{\theta}_i \le \theta_i \le \overline{\theta}_i,\ i=1,2,\dots,p\right\}, \tag{3.183}$$

symmetric and positive definite such that

$$\left(A+\int_0^D e^{-A(D-x)}B(x)dxK(\beta,\theta)\right)^T P(\beta,\theta)$$
$$+P(\beta,\theta)\left(A+\int_0^D e^{-A(D-x)}B(x)dxK(\beta,\theta)\right)=-Q(\beta,\theta). \tag{3.184}$$

We make next our final assumption that is used in our stability analysis (see also [25]).

Assumption 3.9. *The quantities* $\underline{\lambda}=\inf_{(\beta,\theta)\in\Lambda}\min\{\lambda_{\min}(Q(\beta,\theta)),\lambda_{\min}(P(\beta,\theta))\}$ and $\overline{\lambda}=\sup_{(\beta,\theta)\in\Lambda}\lambda_{\max}(P(\beta,\theta))$ exist and are known.

The controller for system (3.171) is

$$U(t)=K(\hat{\beta},\hat{\theta})\hat{Z}(t), \tag{3.185}$$

where

$$\hat{Z}(t)=X(t)+\int_0^D\int_0^x e^{-\hat{A}(t)(x-y)}\hat{B}(y,t)dyu(x,t)dx. \tag{3.186}$$

The update laws are given by

$$\dot{\hat{b}}_i(x,t)=\gamma_b\mathrm{Proj}\left\{\tau_{b_i},\hat{b}_i,b_i^{\mathrm{n}},\rho_i\right\}(x), \tag{3.187}$$
$$\dot{\hat{\theta}}_i(t)=\gamma_\theta\mathrm{Proj}_{[\underline{\theta}_i,\overline{\theta}_i]}\{\tau_{\theta_i},\hat{\theta}_i\}, \tag{3.188}$$

where the projector operators are defined as

$$\mathrm{Proj}\left\{\tau,\hat{\xi},\xi^{\mathrm{n}},\rho\right\}(x)=\begin{cases}\tau(x)-\left(\hat{\xi}(x)-\xi^{\mathrm{n}}(x)\right)\frac{\langle\hat{\xi}-\xi^{\mathrm{n}},\tau\rangle}{\|\hat{\xi}-\xi^{\mathrm{n}}\|^2} & \text{if } \|\hat{\xi}-\xi^{\mathrm{n}}\|^2=\rho \\ & \text{and } \langle\hat{\xi}-\xi^{\mathrm{n}},\tau\rangle>0, \\ \tau(x) & \text{otherwise,}\end{cases} \tag{3.189}$$

where

$$\|\hat{\xi}-\xi^{\mathrm{n}}\|^2=\int_0^D\left(\hat{\xi}(x)-\xi^{\mathrm{n}}(x)\right)^2dx, \tag{3.190}$$
$$\langle\hat{\xi}-\xi^{\mathrm{n}},\tau\rangle=\int_0^D\left(\hat{\xi}(x)-\xi^{\mathrm{n}}(x)\right)\tau(x)dx \tag{3.191}$$

and

$$\mathrm{Proj}_{[\underline{r},\overline{r}]}\{\tau,\hat{\zeta}\} = \begin{cases} 0 & \text{if } \hat{\zeta} = \underline{r} \text{ and } \tau < 0, \\ 0 & \text{if } \hat{\zeta} = \overline{r} \text{ and } \tau > 0, \\ \tau & \text{otherwise}, \end{cases} \tag{3.192}$$

where

$$\tau_{b_i}(x) = \frac{\hat{Z}(t)^T P\left(\hat{\beta},\hat{\theta}\right) - \int_0^D (1+y) g(y,t) w(y,t) dy}{1+\Xi(t)} B_i u(x,t), \tag{3.193}$$

$$\tau_{\theta_i} = \frac{\int_0^D (1+x) w(x,t) g(x,t) dx - \hat{Z}(t)^T P\left(\hat{\beta},\hat{\theta}\right)}{1+\Xi(t)} \times A_i \left(\int_0^D \int_0^x e^{-\hat{A}(x-y)} \hat{B}(y,t) dy\, u(x,t) dx - \hat{Z}(t) \right), \tag{3.194}$$

$$w(x,t) = u(x,t) - g(x,t)\hat{Z}(t), \tag{3.195}$$

$$\Xi(t) = \hat{Z}(t)^T P\left(\hat{\beta},\hat{\theta}\right)\hat{Z}(t) + \int_0^D (1+x) w(x,t)^2 dx, \tag{3.196}$$

$$g_t(x,t) = -g(x,t) A_{\mathrm{cl}}(\hat{\beta}(t),\hat{\theta}(t)) + g_x(x,t), \tag{3.197}$$

$$g(D,t) = K(\hat{\beta}(t),\hat{\theta}(t)), \tag{3.198}$$

$$A_{\mathrm{cl}}(\hat{\beta}(t),\hat{\theta}(t)) = \hat{A} + \hat{\beta}(t) K(\hat{\beta}(t),\hat{\theta}(t)), \tag{3.199}$$

$$\hat{\beta}(t) = \int_0^D e^{-\hat{A}(D-x)} \hat{B}(x,t) dx. \tag{3.200}$$

We now state our main result.

Theorem 3.10. *Consider the closed-loop systems consisting of the plant* (3.171) *together with the adaptive controller* (3.185)–(3.200), *and let Assumptions* 3.7–3.9 *be satisfied. There exists a constant M such that for any γ_θ and γ_b such that*

$$\gamma_\theta + \gamma_b < \frac{\frac{1}{2}\min\{\underline{\lambda},1\}^2}{M} \tag{3.201}$$

there exist positive constants R and ρ such that

$$\Omega(t) \leq R\left(e^{\rho\Omega(0)} - 1\right), \tag{3.202}$$

where

$$\Omega(t) = |X(t)|^2 + \|u(t)\|^2 + \left\|\tilde{b}(t)\right\|^2 + \left|\tilde{\theta}(t)\right|^2 \tag{3.203}$$

and

$$\|u(t)\|^2 = \int_0^D u^2(x,t) dx, \tag{3.204}$$

$$\left\|\tilde{b}(t)\right\|^2 = \int_0^D \left|\tilde{b}(x,t)\right|^2 dx, \tag{3.205}$$

$$\tilde{b}(x,t) = b(x) - \hat{b}(x,t), \tag{3.206}$$

$$\tilde{\theta}(t) = \theta - \hat{\theta}(t). \tag{3.207}$$

Furthermore,

$$\lim_{t\to\infty} X(t) = 0, \tag{3.208}$$

$$\lim_{t\to\infty} U(t) = 0. \tag{3.209}$$

We start proving the above theorem by first noting that relations (3.186) and (3.195) define two transformations, namely, $\hat{Z}(t)$, and $w(x,t)$, of the form

$$(X(t), u(x,t)) \to \left(\hat{Z}(t), w(x,t)\right). \tag{3.210}$$

With these transformations, system (3.174)–(3.176) is mapped to the target system. Differentiating (3.186) with respect to time we get that

$$\begin{aligned}\dot{\hat{Z}}(t) &= \hat{A}\hat{Z}(t) + \tilde{A}\hat{Z}(t) + \int_0^D e^{-\hat{A}(D-x)}\hat{B}(x,t)dx\,U(t) + \int_0^D \tilde{B}(x,t)u(x,t)dx \\ &\quad + \int_0^D \int_0^x e^{-\hat{A}(x-y)}\dot{\hat{B}}(y,t)dy\,u(x,t)dx - \tilde{A}\int_0^D \int_0^x e^{-\hat{A}(x-y)}\hat{B}(y,t)dy \\ &\quad \times u(x,t)dx - \dot{\hat{A}}\int_0^D \int_0^x (x-y)e^{-\hat{A}(x-y)}\hat{B}(y,t)dy\,u(x,t)dx,\end{aligned} \tag{3.211}$$

where we also used (3.174)–(3.176). Using (3.185) we get

$$\begin{aligned}\dot{\hat{Z}}(t) &= A_{\mathrm{cl}}\left(\hat{\beta},\hat{\theta}\right)\hat{Z}(t) + \tilde{A}\hat{Z}(t) + \int_0^D \tilde{B}(x,t)u(x,t)dx \\ &\quad + \int_0^D \int_0^x e^{-\hat{A}(x-y)}\dot{\hat{B}}(y,t)dy\,u(x,t)dx - \tilde{A}\int_0^D \int_0^x e^{-\hat{A}(x-y)}\hat{B}(y,t)dy \\ &\quad \times u(x,t)dx - \dot{\hat{A}}\int_0^D \int_0^x (x-y)e^{-\hat{A}(x-y)}\hat{B}(y,t)dy\,u(x,t)dx.\end{aligned} \tag{3.212}$$

Moreover, differentiating (3.195) with respect to time and with respect to the spatial variable x we get

$$\begin{aligned}w_t(x,t) &= u_x(x,t) - \left(g_t(x,t) + g(x,t)A_{\mathrm{cl}}\left(\hat{\beta},\hat{\theta}\right)\right)\hat{Z}(t) - g(x,t)\int_0^D \tilde{B}(x,t) \\ &\quad \times u(x,t)dx - g(x,t)\int_0^D \int_0^x e^{-\hat{A}(x-y)}\dot{\hat{B}}(y,t)dy\,u(x,t)dx + g(x,t) \\ &\quad \times \tilde{A}\int_0^D \int_0^x e^{-\hat{A}(x-y)}\hat{B}(y,t)dy\,u(x,t)dx + g(x,t)\dot{\hat{A}} \\ &\quad \times \int_0^D \int_0^x (x-y)e^{-\hat{A}(x-y)}\hat{B}(y,t)dy\,u(x,t)dx + \tilde{A}\hat{Z}(t)\end{aligned} \tag{3.213}$$

and

$$w_x(x,t) = u_x(x,t) - g_x(x,t)\hat{Z}(t), \tag{3.214}$$

respectively. Since $g(x,t)$ satisfies (3.197)–(3.198) we have that

$$\begin{aligned} w_t(x,t) &= w_x(x,t) \\ &\quad - g(x,t)\left(\int_0^D \tilde{B}(x,t)u(x,t)dx + \int_0^D \int_0^x e^{-\hat{A}(x-y)}\dot{\hat{B}}(y,t)dy\,u(x,t)dx\right) \\ &\quad + g(x,t)\left(\tilde{A}\int_0^D \int_0^x e^{-\hat{A}(x-y)}\hat{B}(y,t)dy\,u(x,t)dx\right. \\ &\quad \left. + \dot{\hat{A}}\int_0^D \int_0^x (x-y)e^{-\hat{A}(x-y)}\hat{B}(y,t)dy\,u(x,t)dx - \tilde{A}\hat{Z}(t)\right), \end{aligned} \tag{3.215}$$

$$w(D,t) = 0. \tag{3.216}$$

From relation (3.186) one should notice that $\hat{Z}$ is given explicitly in terms of (X,u). Substituting this explicit relation for $\hat{Z}$ (in terms of (X,u)) into (3.195) one can observe that $w(x,t)$ is also written in terms of (X,u). Hence, the state $(\hat{Z},w)$ defines a transformation of the state (X,u). We define now the inverse of the transformation in (3.195) as

$$u(x,t) = w(x,t) + g(x,t)\hat{Z}(t). \tag{3.217}$$

Consequently we can write (3.215)–(3.216) only in terms of the transformed variables, i.e., in terms of $w(x,t)$ and $\hat{Z}(t)$. Using (3.186) and (3.217) we define the inverse transformation of $\hat{Z}(t)$ as

$$\begin{aligned} X(t) &= \left(I - \int_0^D \int_0^x e^{-\hat{A}(x-y)}\hat{B}(y,t)dy\,g(x,t)dx\right)\hat{Z}(t) \\ &\quad - \int_0^D \int_0^x e^{-\hat{A}(x-y)}\hat{B}(y,t)dy\,w(x,t)dx. \end{aligned} \tag{3.218}$$

Thus now system (3.174)–(3.176) with state (X,u) is mapped into the target system that is composed of relations (3.212), (3.215)–(3.216) and has state $(\hat{Z},w)$. The target system is obtained from the original system through a direct transformation of the form $(X,u) \to (\hat{Z},w)$ which is defined in relations (3.186) and (3.195). With the inverse of this transformation, which is of the form $(\hat{Z},w) \to (X,u)$ and is given in (3.217) and (3.218), one obtains the original system from the target system.

We state now a lemma that is concerned with the uniform boundedness of the original and transformed variables.

Lemma 3.11. *There exist constants M_u, M_w, M_X, and $M_{\hat{Z}}$ such that*

$$\|u(t)\|^2 \le M_u\left(\|w(t)\|^2 + |\hat{Z}(t)|^2\right), \tag{3.219}$$

$$|X(t)|^2 \le M_X\left(|\hat{Z}(t)|^2 + \|w(t)\|^2\right), \tag{3.220}$$

$$\|w(t)\|^2 \le M_w\left(\|u(t)\|^2 + |X(t)|^2\right), \tag{3.221}$$

$$|\hat{Z}(t)|^2 \le M_{\hat{Z}}\left(|X(t)|^2 + \|u(t)\|^2\right). \tag{3.222}$$

Proof. First observe that the signals $A(\hat{\theta})$, $K(\hat{\beta},\hat{\theta})$, and $P(\hat{\beta},\hat{\theta})$ are continuously differentiable with respect to $(\hat{\beta},\hat{\theta})$. Hence, since $(\hat{\beta},\hat{\theta})$ is uniformly bounded, the signals $A(\hat{\theta})$, $K(\hat{\beta},\hat{\theta})$, $P(\hat{\beta},\hat{\theta})$ and their derivatives are also uniformly bounded. Denote by M_A, M_K, M_P the bounds of $A(\hat{\theta})$, $K(\hat{\beta},\hat{\theta})$, $P(\hat{\beta},\hat{\theta})$, respectively, and with M_A', M_K', M_P' the bound of their derivatives. We first prove that g satisfying (3.197)–(3.198) is uniformly bounded; i.e., there exists a $\mu > 0$ such that

$$\int_0^D |g(x,t)|^2 dx \leq \mu^2 \quad \text{for all } t \geq 0. \tag{3.223}$$

Taking the time derivative of

$$V(t) = \int_0^D e^{mx} |g(x,t)|^2 dx, \tag{3.224}$$

where $m > 0$ is arbitrary, and using (3.197)–(3.200) we get that

$$\dot{V}(t) \leq e^{mD} M_K^2 - \left(m - 2\left(M_A + e^{M_A D}\sqrt{DM_2}M_K\right)\right) V(t), \tag{3.225}$$

where we also used integration by parts and the fact that A_{cl} in (3.199) satisfies $|A_{\text{cl}}| \leq M_A + e^{M_A D}\sqrt{DM_2}M_K$, where

$$M_2 = (p+1)\left(D|B_0|^2 + 2\sum_{i=1}^{p} |B_i|^2 \left(\rho_i + \|b_i^n\|^2\right)\right). \tag{3.226}$$

Choosing $m = 2\left(M_A + e^{M_A D}\sqrt{DM_2}M_K\right) + 1$, with Lemma B.4 from Appendix B we get (3.223) with

$$\mu^2 = e^{mD}\left(\int_0^D |g(x,0)|^2 dx + M_K^2\right). \tag{3.227}$$

Note that since from (3.198) $g(D,0) = K(\hat{\beta}(0),\hat{\theta}(0))$, a possible choice for the initial condition of g is $g(x,0) = K(\hat{\beta}(0),\hat{\theta}(0))$ for all $x \in [0,D]$, such that the boundary condition (3.198) is compatible with the initial condition $g(x,0)$. In this case, (3.227) becomes $\mu^2 = e^{mD} M_K^2(1+D)$. From relations (3.186)–(3.200), (3.218) and using the Young and Cauchy–Schwarz inequalities [51], one can show that bounds (3.219)–(3.222) hold with

$$M_u = 2\left(1+\mu^2\right), \tag{3.228}$$

$$M_X = 5\left(De^{2M_A D}|M_2|\mu^2 + 1\right), \tag{3.229}$$

$$M_w = 2\left(1+\mu^2\right)\left(1+M_{\hat{Z}}\right), \tag{3.230}$$

$$M_{\hat{Z}} = 2\left(1 + D^2 e^{2M_A D}|M_2|\right). \qquad \square \tag{3.231}$$

Before we construct a Lyapunov functional for proving stability of the closed-loop system we state the following lemma, which is concerned with an important property of the projector operator defined in (3.189).

Lemma 3.12. *The following holds for* (3.189)

$$-\int_0^D \tilde{\xi}(x)\mathit{Proj}\left\{\tau,\hat{\xi},\xi^{\mathrm{n}},\rho\right\}(x)dx \leq -\int_0^D \tilde{\xi}(x)\tau(x)dx. \tag{3.232}$$

Proof. From (3.189) it follows that

$$-\int_0^D \tilde{\xi}(x)\mathrm{Proj}\left\{\tau,\hat{\xi},\xi^{\mathrm{n}},\rho\right\}(x)dx = -\int_0^D \tilde{\xi}(x)\tau(x)dx + \begin{cases} F & \text{if } \|\hat{\xi}-\xi^{\mathrm{n}}\|^2=\rho \text{ and} \\ & \langle\hat{\xi}-\xi^{\mathrm{n}},\tau\rangle > 0, \\ 0 & \text{otherwise,}\end{cases} \tag{3.233}$$

where

$$F = \langle\hat{\xi}-\xi^{\mathrm{n}},\tilde{\xi}\rangle \frac{\langle\hat{\xi}-\xi^{\mathrm{n}},\tau\rangle}{\|\hat{\xi}-\xi^{n}\|^2}. \tag{3.234}$$

We prove now that the second term in (3.233) is always nonpositive. Assume that $\|\hat{\xi}-\xi^{\mathrm{n}}\|^2=\rho$ and $\langle\hat{\xi}-\xi^{\mathrm{n}},\tau\rangle > 0$. Then it is sufficient to show that

$$\int_0^D \tilde{\xi}(x)\left(\hat{\xi}(x)-\xi^{\mathrm{n}}(x)\right)dx \leq 0. \tag{3.235}$$

It holds that

$$\int_0^D \tilde{\xi}(x)\left(\hat{\xi}(x)-\xi^{\mathrm{n}}(x)\right)dx = \int_0^D (\xi(x)-\xi^{\mathrm{n}}(x))\left(\hat{\xi}(x)-\xi^{\mathrm{n}}(x)\right)dx - \int_0^D \left(\hat{\xi}(x)-\xi^{\mathrm{n}}(x)\right)^2 dx. \tag{3.236}$$

Using the fact that $\|\hat{\xi}-\xi^{\mathrm{n}}\|^2=\rho$ and the Cauchy–Schwarz inequality we get

$$\int_0^D \tilde{\xi}(x)\left(\hat{\xi}(x)-\xi^{\mathrm{n}}(x)\right)dx \leq \sqrt{\int_0^D (\xi(x)-\xi^{\mathrm{n}}(x))^2 dx}\sqrt{\rho}-\rho. \tag{3.237}$$

With (3.181) we get (3.235). □

Remark 3.13. *Instead of assuming in Assumption* 3.7 *(and, respectively, in the controllability condition of Assumption* 3.8*) that the* $b_i(x)$ *satisfy* (3.181), *one can assume that the* $b_i(x)$ *satisfy*

$$R_{\mathrm{low},i}(x) \leq b_i(x) \leq R_{\mathrm{high},i}(x) \tag{3.238}$$

for some known, continuous functions $R_{\mathrm{low},i}(x)$ *and* $R_{\mathrm{high},i}(x)$. *The new projection operator is*

$$\mathrm{Proj}\{\tau,\hat{\xi},R_{\mathrm{low}},R_{\mathrm{high}}\}(x) = \begin{cases} 0 & \text{if } \hat{\xi}(x)=R_{\mathrm{low}}(x) \text{ and } \tau(x)<0, \\ 0 & \text{if } \hat{\xi}(x)=R_{\mathrm{high}}(x) \text{ and } \tau(x)>0, \\ \tau(x) & \text{otherwise.}\end{cases} \tag{3.239}$$

Note that the projection set of the operator (3.239) *is an infinite-dimensional hyper-rectangle, whereas the projection set of* (3.189) *is spherical. We show now that* (3.239) *satisfies*

$$-\int_0^D \tilde{\xi}(x)\mathrm{Proj}\{\tau,\hat{\xi},R_{\mathrm{low}},R_{\mathrm{high}}\}(x)dx \le -\int_0^D \tilde{\xi}(x)\tau(x)dx. \tag{3.240}$$

Using (3.239) *we get*

$$-\int_0^D \tilde{\xi}(x)G(x)dx = -\int_0^D \tilde{\xi}(x)\tau(x)dx + \begin{cases} \int_0^D \tilde{\xi}(x)\tau(x)dx & \begin{array}{c} \text{if } \hat{\xi}(x)=R_{\mathrm{low}}(x) \text{ and } \tau(x)<0 \\ \text{or} \\ \text{if } \hat{\xi}(x)=R_{\mathrm{high}}(x) \text{ and } \tau(x)>0, \end{array} \\ 0 & \text{otherwise,} \end{cases} \tag{3.241}$$

where

$$G(x)=\mathrm{Proj}\{\tau,\hat{\xi},R_{\mathrm{low}},R_{\mathrm{high}}\}(x). \tag{3.242}$$

Assume first that $\hat{\xi}(x)=R_{\mathrm{low}}(x)$ *and* $\tau(x)<0$*. Since* $\hat{\xi}(x)=R_{\mathrm{low}}(x)\le\xi(x)$*, we get that*

$$\tilde{\xi}(x)\tau(x)=\left(\xi(x)-\hat{\xi}(x)\right)\tau(x)\le 0, \tag{3.243}$$

and hence $\int_0^D \tilde{\xi}(x)\tau(x)dx \le 0$*. The case* $\hat{\xi}(x)=R_{\mathrm{high}}(x)$ *and* $\tau(x)>0$ *can be proved analogously.*

Lemma 3.14. *Let* γ_θ *and* γ_b *be as in* (3.201)*. Then, for the Lyapunov function*

$$V(t)=\log(1+\Xi(t))+\frac{1}{\gamma_b}\int_0^D \tilde{b}(x,t)^T\tilde{b}(x,t)dx+\frac{1}{\gamma_\theta}\tilde{\theta}(t)^T\tilde{\theta}(t), \tag{3.244}$$

the following holds:

$$V(t)\le V(0). \tag{3.245}$$

Proof. Differentiating $V(t)$ with respect to time and using (3.172), (3.173) and (3.212), (3.215) we get

$$\begin{aligned}
\dot{V}(t) \le & \frac{1}{1+\Xi(t)}\Bigg(-\underline{\lambda}\left|\hat{Z}(t)\right|^2+2\hat{Z}(t)^T P\left(\hat{\beta},\hat{\theta}\right)\int_0^D\int_0^x e^{-\hat{A}(x-y)}\sum_{i=1}^p B_i\dot{\hat{b}}_i(y,t)dy \\
& \times u(x,t)dx+\left(\left|\dot{\hat{\theta}}(t)\right|+\int_0^D\left|e^{-\hat{A}(D-x)}\sum_{i=1}^p B_i\dot{\hat{b}}_i(x,t)\right|dx\right)M_P'\left|\hat{Z}(t)\right|^2 \\
& -2\hat{Z}(t)^T P\left(\hat{\beta},\hat{\theta}\right)\sum_{i=1}^p\dot{\hat{\theta}}_i(t)A_i\int_0^D\int_0^x(x-y)e^{-\hat{A}(x-y)}\hat{B}(y,t)dy\,u(x,t)dx \\
& -w(0,t)^2-\int_0^D w(x,t)^2dx-2\int_0^D(1+x)w(x,t)g(x,t)dx
\end{aligned}$$

$$\begin{aligned}
&\times\left(\int_0^D\int_0^x e^{-\hat{A}(x-y)}\sum_{i=1}^p B_i\dot{\hat{b}}_i(y,t)dy\,u(x,t)dx-\sum_{i=1}^p\dot{\hat{\theta}}_i(t)A_i\right.\\
&\left.\times\int_0^D\int_0^x (x-y)e^{-\hat{A}(x-y)}\hat{B}(y,t)dy\,u(x,t)dx\right)+2\hat{Z}(t)^T P\left(\hat{\beta},\hat{\theta}\right)\\
&\times\sum_{i=1}^p A_i\hat{Z}(t)\tilde{\theta}_i(t)-2\hat{Z}(t)^T P\left(\hat{\beta},\hat{\theta}\right)\sum_{i=1}^p A_i\int_0^D\int_0^x e^{-\hat{A}(x-y)}\\
&\times\hat{B}(y,t)dy\,u(x,t)dx\,\tilde{\theta}_i+2\hat{Z}(t)^T P\left(\hat{\beta},\hat{\theta}\right)\int_0^D\sum_{i=1}^p B_i\tilde{b}_i(x)u(x,t)dx\\
&-2\int_0^D(1+x)w(x,t)g(x,t)dx\int_0^D\sum_{i=1}^p B_i\tilde{b}_i(x)u(x,t)dx+2\int_0^D(1+x)\\
&\times w(x,t)g(x,t)dx\sum_{i=1}^p A_i\left(\left(\int_0^D\int_0^x e^{-\hat{A}(x-y)}\hat{B}(y,t)dy\,u(x,t)dx-\hat{Z}(t)\right)\tilde{\theta}_i\right)\\
&-\frac{2}{\gamma_b}\int_0^D\sum_{i=1}^p\tilde{b}_i(x,t)\dot{\hat{b}}_i(x,t)dx-\frac{2}{\gamma_\theta}\sum_{i=1}^p\tilde{\theta}_i(t)\dot{\hat{\theta}}_i(t). \qquad (3.246)
\end{aligned}$$

With (3.187)-(3.188), (3.217), and (3.232) we get

$$\begin{aligned}
\dot{V}(t)\leq&\frac{1}{1+\Xi(t)}\left(-\underline{\lambda}\left|\hat{Z}(t)\right|^2+\left(\left|\dot{\hat{\theta}}(t)\right|+\int_0^D\left|e^{-\hat{A}(D-x)}\sum_{i=1}^p B_i\dot{\hat{b}}_i(x,t)\right|dx\right)\right.\\
&\times M_p'\left|\hat{Z}(t)\right|^2+2\hat{Z}(t)^T P\left(\hat{\beta},\hat{\theta}\right)\int_0^D\int_0^x e^{-\hat{A}(x-y)}\sum_{i=1}^p B_i\dot{\hat{b}}_i(y,t)dy\\
&\left(w(x,t)+g(x,t)\hat{Z}(t)\right)dx-2\hat{Z}(t)^T P\left(\hat{\beta},\hat{\theta}\right)\sum_{i=1}^p\dot{\hat{\theta}}_i(t)A_i\int_0^D\int_0^x(x-y)\\
&\times e^{-\hat{A}(x-y)}\hat{B}(y,t)dy\left(w(x,t)+g(x,t)\hat{Z}(t)\right)dx-w(0,t)^2-\int_0^D w(x,t)^2dx\\
&-2\int_0^D(1+x)w(x,t)g(x,t)dx\left(\int_0^D\int_0^x e^{-\hat{A}(x-y)}\sum_{i=1}^p B_i\dot{\hat{b}}_i(y,t)dy\right.\\
&\times\left(w(x,t)+g(x,t)\hat{Z}(t)\right)dx-\sum_{i=1}^p\dot{\hat{\theta}}_i(t)A_i\int_0^D\int_0^x(x-y)e^{-\hat{A}(x-y)}\hat{B}(y,t)dy\\
&\left.\left.\times\left(w(x,t)+g(x,t)\hat{Z}(t)\right)dx\right)\right). \qquad (3.247)
\end{aligned}$$

We now estimate $\int_0^D\left|\dot{\hat{b}}_i(x)\right|dx$. Using (3.187), (3.189), and the Cauchy-Schwarz inequality we get that

$$\int_0^D\left|\dot{\hat{b}}_i(x)\right|dx\leq\gamma_b 2\sqrt{D}\sqrt{\int_0^D\tau_{b_i}(x)^2dx}. \qquad (3.248)$$

We now estimate $\int_0^D \tau_{b_i}(x)^2 dx$. Using (3.193), (3.219), and (3.223) together with the Young and Cauchy–Schwarz inequalities we get

$$\int_0^D \tau_{b_i}(x)^2 dx \le 2M_u \left(M_P^2 |B_i|^2 + (1+D)^2 \mu^2\right) \left(\frac{\left|\hat{Z}(t)\right|^2 + \|w(t)\|^2}{1+\Xi(t)} \right)^2 . \quad (3.249)$$

With (3.248) we arrive at

$$\int_0^D \left|\dot{\hat{b}}_i(x)\right| dx \le \gamma_b M_{\tau_{b_i}} \frac{\left|\hat{Z}(t)\right|^2 + \|w(t)\|^2}{1+\Xi(t)}, \quad (3.250)$$

where

$$M_{\tau_{b_i}} = 2\sqrt{2DM_u \left(M_P^2 |B_i|^2 + (1+D)^2 \mu^2\right)}. \quad (3.251)$$

With similar arguments one can prove that

$$\left|\dot{\theta}_i(t)\right| \le \gamma_\theta M_{\theta_i} \frac{\left|\hat{Z}(t)\right|^2 + \|w(t)\|^2}{1+\Xi(t)}, \quad (3.252)$$

where

$$M_{\theta_i} = 2|A_i| \left(\sqrt{1+D}\mu + M_P\right)\left(De^{M_A D}\sqrt{M_2}(1+\mu)+1\right). \quad (3.253)$$

Using (3.250), (3.252) together with Young's inequality, one can prove that there exists a positive constant M such that the following holds:

$$\begin{aligned} \dot{V}(t) \le & \frac{1}{1+\Xi(t)}\left(-\underline{\lambda}\left|\hat{Z}(t)\right|^2 - w(0,t)^2 - \int_0^D w(x,t)^2 dx \right. \\ & \left. + \frac{(\gamma_\theta + \gamma_b)M}{1+\Xi(t)} \left(\int_0^D w(x,t)^2 dx + \left|\hat{Z}(t)\right|^2 \right)^2 \right), \end{aligned} \quad (3.254)$$

where

$$\begin{aligned} M = & M_P' \left(e^{M_A D} \sum_{i=1}^p |B_i| M_{\tau_{b_i}} + M_{\theta_i} \sum_{i=1}^p \right) + 2M_P D e^{M_A D} \sum_{i=1}^p |B_i| M_{\tau_{b_i}} (1+\mu) \\ & + 2M_P D e^{M_A D} \sum_{i=1}^p |A_i| M_{\theta_i} \sqrt{M_2}(1+\mu) + 2(1+D)\mu e^{M_A D} \sum_{i=1}^p |B_i| M_{\tau_{b_i}} \\ & \times (D+1+\mu) 2D(1+D)\sqrt{M_2}\mu e^{M_A D} \sum_{i=1}^p |A_i| M_{\theta_i} (D+1+\mu). \end{aligned} \quad (3.255)$$

Since from the definition of $\Xi(t)$ in (3.196) we have

$$1+\Xi(t) \ge \min\{\underline{\lambda}, 1\} \left(\left|\hat{Z}(t)\right|^2 + \int_0^D w(x,t)^2 dx \right), \quad (3.256)$$

we conclude that

$$\dot{V}(t) \leq \frac{1}{1+\Xi(t)}\left(-\underline{\lambda}\left|\hat{Z}(t)\right|^2 - w(0,t)^2 - \int_0^D w(x,t)^2 dx \right.$$
$$\left. + (\gamma_\theta + \gamma_b)\frac{M}{\min\{\underline{\lambda}, 1\}}\left(\int_0^D w(x,t)^2 dx + \left|\hat{Z}(t)\right|^2\right)\right). \tag{3.257}$$

Choosing γ_θ and γ_b as in (3.201), the proof of the lemma is complete. □

Lemma 3.15. *There exist constants* $\underline{M}$ *and* $\overline{M}$ *such that*

$$\underline{M}\Xi(t) \leq \Pi(t) \leq \overline{M}\Xi(t), \tag{3.258}$$

where

$$\Pi(t) = |X(t)|^2 + ||u(t)||^2. \tag{3.259}$$

Proof. The proof is immediate, using (3.219)–(3.222) and the definition of $\Xi(t)$ in (3.196). □

We are now ready to derive the stability estimate of Theorem 3.10. Using (3.244) it follows that

$$\Xi(t) \leq \left(e^{V(t)} - 1\right), \tag{3.260}$$

$$\left\|\tilde{b}(t)\right\|^2 + \left|\tilde{\theta}(t)\right|^2 \leq (\gamma_\theta + \gamma_b)V(t) \leq (\gamma_\theta + \gamma_b)\left(e^{V(t)} - 1\right). \tag{3.261}$$

Consequently

$$\Omega(t) \leq \left(\overline{M} + (\gamma_\theta + \gamma_b)\right)\left(e^{V(t)} - 1\right). \tag{3.262}$$

Moreover, from (3.244) we take

$$V(0) \leq \max\left\{\overline{\lambda}, 1\right\}\left(|\hat{Z}(0)|^2 + ||w(0)||^2\right)$$
$$+ \frac{1}{\gamma_b}\left\|\tilde{b}(0)\right\|^2 + \frac{1}{\gamma_\theta}\left|\tilde{\theta}(0)\right|^2. \tag{3.263}$$

Thus, by setting

$$R = \overline{M} + \gamma_\theta + \gamma_b \tag{3.264}$$

$$\rho = \max\left\{\overline{\lambda}, 1, \frac{1}{\gamma_\theta}, \frac{1}{\gamma_b}\right\}, \tag{3.265}$$

we get the stability result in Theorem 3.10.

We now turn our attention to proving the convergence of $X(t)$ and $U(t)$ to zero. We first point out that from (3.245) it follows that $|\hat{Z}(t)|$, $\|w(t)\|$, $\|\hat{b}(t)\|$, and $|\hat{\theta}(t)|$ are uniformly bounded. From (3.185) it follows that $U(t)$ is uniformly bounded. From (3.174) and (3.219)–(3.220) we conclude that $\frac{dX^2(t)}{dt}$ is uniformly bounded. Finally, since from (3.257) it turns out that $|\hat{Z}(t)|$ and $\|w(t)\|$ are square integrable, using (3.220) and Barbalat's lemma (Lemma C.9), we conclude that $\lim_{t\to\infty} X(t) = 0$. We now turn our attention to proving convergence of $U(t)$. With the help of (3.219) and by the square integrability of $|\hat{Z}(t)|$ we conclude using (3.185) that $U(t)$ is square integrable. It only remains to show that $\frac{dU^2(t)}{dt}$ is uniformly bounded. Hence, it is sufficient to show that $\dot{U}(t)$ is uniformly bounded. From (3.185) and with (3.212), since $\|\dot{\hat{b}}(t)\|$, $\|\tilde{b}(t)\|$ and $|\dot{\hat{\theta}}(t)|$, $|\tilde{\theta}(t)|$ are uniformly bounded, we conclude the uniform boundedness of $\frac{dU^2(t)}{dt}$.

3.2.2 ▪ Simulations

The method that is used to discretize (3.267) and (3.122) in space is the finite-difference method. The resulting finite-dimensional ODEs are solved using Euler's method. The integrodifferential equations (3.266) and (3.121) are solved using Euler's method, where the integrals are computed using the left-endpoint rule for numerical integration.

We consider the scalar plant

$$\dot{X}(t) = \theta X(t) + \int_0^1 (1.5 + \sin(5x))\, u(x,t)dx, \tag{3.266}$$

$$u_t(x,t) = u_x(x,t), \tag{3.267}$$

$$u(1,t) = U(t), \tag{3.268}$$

where $A = \theta = 1$ and $B_0(x) = 0$, $B_1 = 1$, $b_1(x) = 1.5 + \sin(5x)$. We choose the initial condition of the system as $X(0) = 1$ and $U(\theta) = 0$ for all $-D \le \theta \le 0$. We choose the parameters for the controller as $K = \frac{A_{cl} - \hat{\theta}(t)}{\int_0^D e^{-\hat{\theta}(t)(D-x)} \hat{b}_1(x,t)dx}$, $A_{cl} = -0.7$, $\gamma_b = 0.003$, $\gamma_\theta = 0.001$, and $\rho_1 = 2$, $b_i^{\mathrm{n}}(x) = 1$ for all $x \in [0,1]$, $\underline{\theta} = 0.4\theta$, $\overline{\theta} = 2\theta$. We choose $\hat{\theta}(0) = \underline{\theta}$ and $\hat{b}_1(x,0) = \sqrt{\rho_1} + 1$ for all $x \in [0,1]$. Note that with this choice of initial estimates for b_1 and θ, A_{cl} is unstable ($A_{cl} = 0.2698$) when $\gamma_b = \gamma_\theta = 0$. In Figure 3.3 we show the response of the system, and in Figure 3.4 the estimates of θ and $b_1(x)$. Finally in Figure 3.5 we show the final profile of the estimate $\hat{b}_1(x,t)$ versus the real $b_1(x)$, as well as the estimation of the norm $\|b_1(t) - b_1^{\mathrm{n}}\|^2$.

3.3 ▪ Notes and References

The predictor feedback design for linear systems with distributed input delay was developed by Manitius and Olbrot [102] and generalized by Artstein [4]. For linear systems with distributed input delay, an alternative control design was proposed in [184]. Yet, this result is applicable to systems having an A matrix with no unstable eigenvalues. Systems with distributed delays appear in population dynamics [4], in networked control systems [147], or in liquid mono-propellant rocket motors [177]; see also the references within these works. The term "forwarding" appeared in the work of Sepulchre, Jankovic, and Kokotovic [148], in which they developed a recursive technique analogous to backstepping, but for systems in the feedforward rather than the feedback form.

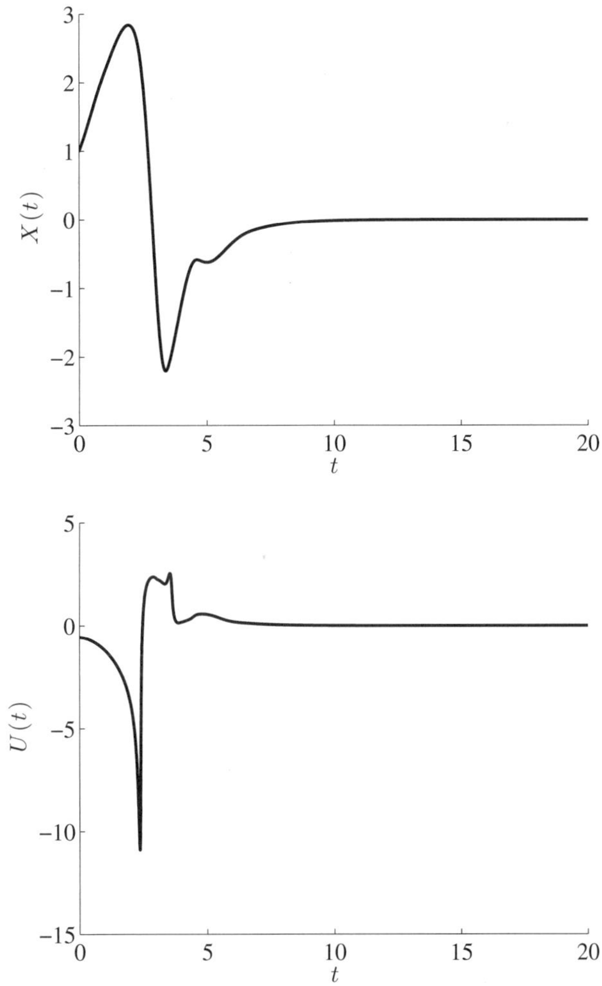

Figure 3.3. *Response of the plant* (3.266)–(3.268) *with the adaptive control algorithm* (3.185)–(3.200) *for initial conditions* $X(0) = 1$, $U(\theta) = 0$ *for all* $-D \leq \theta \leq 0$ *and* $\hat{b}_1(x,0) = \sqrt{\rho_1} + 1$ *for all* $x \in [0,1]$, $\hat{\theta}(0) = \underline{\theta}$.

The benefits of constructing a Lyapunov functional for predictor feedback controllers were highlighted in [85]. Having a Lyapunov functional available, one can derive an inverse-optimal controller, prove robustness of the predictor feedback to a small mismatch in the actuator delay, or study the disturbance attenuation properties of the closed-loop system. An alternative Lyapunov–Krasovskii functional was recently constructed by Mazenc, Niculescu, and Krstic [114].

The usage of an extra input in the developments of Section 3.1.1 might seem to be a relatively easy extension from the single-input case. This fact reinforces our results, which can be easily extended from the single-input case to the multi-input case. Moreover, introducing an extra input highlights the fact that a system can be stabilized by combining, say, two inputs, although the system may not be stabilizable through each input individually. By introducing the backstepping-forwarding transformations as in Section 3.1.1 the

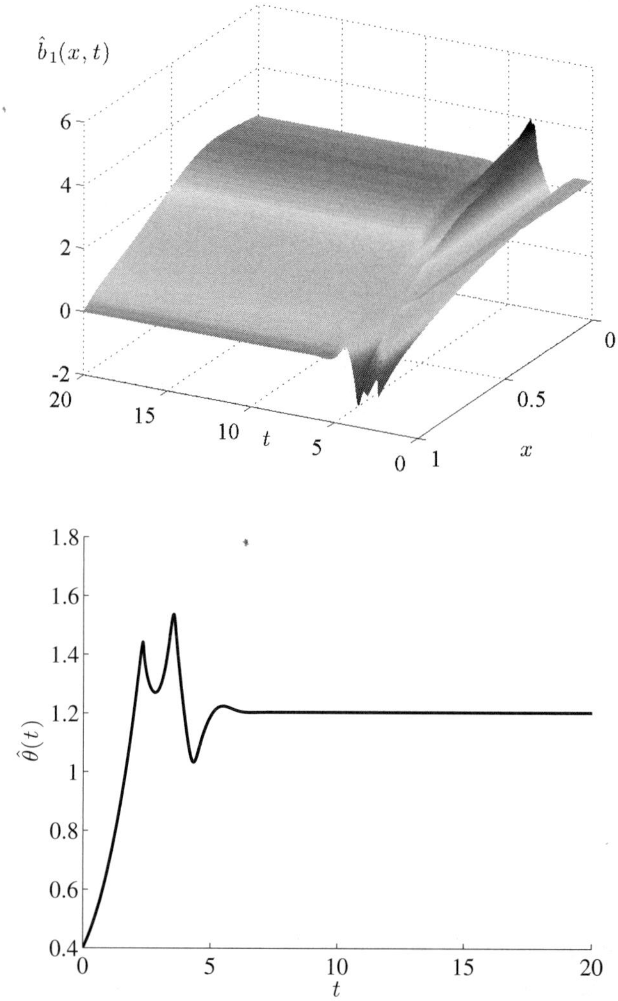

Figure 3.4. *The estimation of the parameters b_1 (top) and θ (bottom) for initial conditions $X(0)=1$, $U(\theta)=0$ for all $-D \leq \theta \leq 0$ and $\hat{b}_1(x,0)=\sqrt{\rho_1}+1$ for all $x \in [0,1]$, $\hat{\theta}(0)=\underline{\theta}$.*

methodology developed in this chapter can be extended to the case of multi-input multi-output linear systems with more complex distributed actuator or sensor dynamics rather than pure delay. In [16] we deal with diffusive and in [17] with wave dynamics. Some of our recent results on the compensation of infinite-dimensional dynamics are summarized in [91].

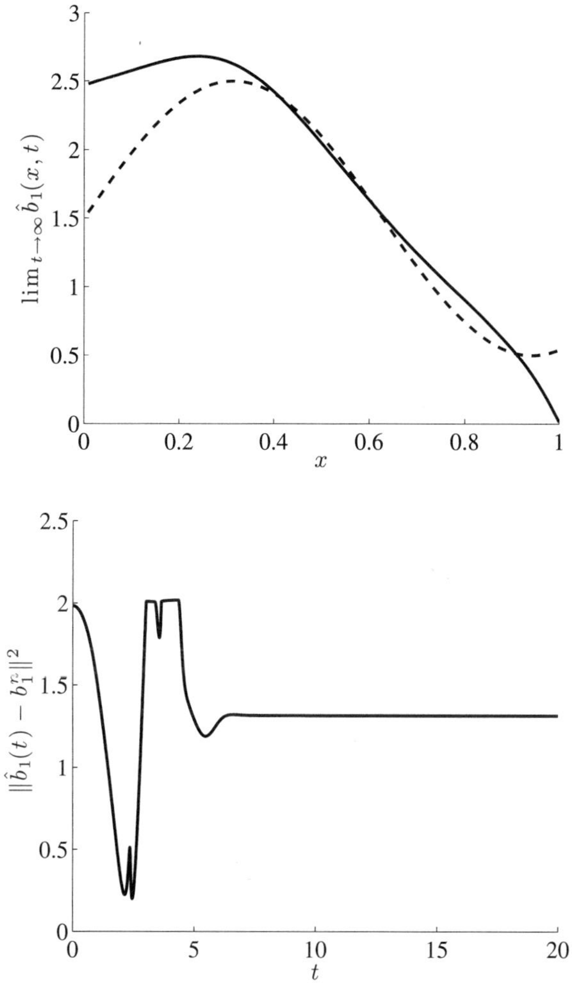

Figure 3.5. *Top: The final profile of the estimate* $\hat{b}_1(x,t)$ *(solid) versus true* $b_1(x)$ *(dashed). Bottom: The estimate* $\int_0^D \left(\hat{b}_1(x,t) - b_1^{\mathrm{n}}(x)\right)^2 dx$.

Chapter 4
Application: Automotive Catalysts

This chapter is devoted to the application of our techniques for delay systems to automotive catalysts. Specifically, we consider the problem of controlling and analyzing the dynamics of the oxygen storage level in three-way catalytic converters. Our model consists of two coupled nonlinear first order hyperbolic PDEs, coupled with an infinite-dimensional (distributed) ODE. The relationship of this model with the models of systems with distributed delays can be explained as follows. By explicitly solving the two PDEs in terms of the state of the infinite-dimensional ODE and plugging these solutions into the equation for the ODE, we get an infinite-dimensional ODE which has, on the right-hand side, integral terms over its spatial domain. These integral terms have some resemblance to the distributed terms in the case of systems with distributed delays.

We develop a model that retains the distributed nature of three-way catalytic converters based on PDE formulation. Based on this model we show that when a square wave air-to-fuel ratio is applied at the inlet of the catalyst, the time-average of the oxygen storage level acquires an equilibrium profile which is asymptotically stable. To suppress various disturbances at the inlet of the catalyst, we design an output feedback control law based on the linearized system around the acquired equilibrium, using the ideal concentration measurements.

Specifically, we retain the native PDE representation of the model. We pursue the development of a simple and at the same time representative PDE-based model in Section 4.1. To a large extent, the model retains the physical parameters of a three-way catalyst, such as storage capacity, mass flow rate, reaction rates, etc. In practice, an oscillating air-to-fuel ratio is typically applied at the inlet of the catalyst. Although it is evident that such an open-loop control strategy is effective, to the best of the authors' knowledge, there has been no attempt to try to explain the effectiveness of such a scheme from the control-theoretic perspective. We analyze the properties of the system under a square wave air-to-fuel ratio at the inlet of the catalyst, although the analysis is not affected by the shape of the input signal, so the same conclusion applies if, for example, a sinusoidal air-to-fuel ratio wave was used instead. We prove in Section 4.2 that the oxygen storage level, averaged over one oscillation period, acquires a unique asymptotically stable equilibrium profile. This in turn implies that the oxygen storage level oscillates with a small magnitude (which depends on the amplitude and the frequency of the imposed air-to-fuel ratio input oscillations) around the equilibrium of the time-averaged oxygen storage level. Typically, in addition to the imposed square wave, unknown disturbances are present at the inlet of the catalyst, due to possible bias and off-stochiometry gain uncertainty of the Universal Exhaust Gas Oxygen (UEGO) sensor. Although the system is asymptotically stable with an open-loop control strategy with the square wave air-to-fuel ratio, this strategy cannot bring the system to the desired set-point in the presence of disturbances. For this reason we design a feedback control strategy which incorporates an integral action for rejecting the input disturbance in Section 4.3. We perform simulation studies which show the performance improvement of the closed-loop system in terms of the catalyst efficiency. We also illustrate a trade-off between the reductants' and the oxygen's efficiencies as the desired equilibrium point of the oxygen storage level varies.

4.1 ▪ Model Development

Although detailed models of three-way catalysts exist that capture accurately the underlying physics and chemistry, they tend to be too complicated for control design. From the control design point of view, the models have to be simple enough to result in a controller implementable in an Electronic Control Unit (ECU) as well as accurate enough in order to capture the important dynamic phenomena. Model simplicity is also important as it allows designers to gain insight into the behavior of the system being modeled.

With this in mind, we start our model derivation by considering the most important reactions that are needed for control purposes. A detailed description of the reactions that are taking place in a three-way catalyst can be found in [61]. However, the total number of reactions could be significantly reduced following [48] and [121]. The aim of a three-way catalyst is to reduce the emissions of oxides of nitrogen NO_x, carbon monoxide CO, and hydrocarbons HC. This is achieved by oxidizing CO and HC, and by reducing NO_x. From the control design point of view, capturing the dynamic behavior of oxygen storage and the air-fuel ratio at the input and the output of the catalyst (essentially, the two air-to-fuel ratios are the measured signals) by the model would be more important than its ability to predict tail pipe emissions. Hence, our model puts the focus on the concentrations of the species that drive the air-to-fuel ratio sensors, i.e., O_2, H_2, and CO. Thus, NO_x, which occurs concurrently and in smaller quantities compared to O_2, is incorporated in the O_2 concentration. Similarly, HC, which occur in small quantities relative to CO, are incorporated in the CO concentration.

While the precious metals also store oxygen, the dominant mechanism of oxygen storage is cerium oxide. The relevant reactions, i.e., oxidation and reduction of cerium oxide, are

$$O_2 + 2Ce_2O_3 \Longleftrightarrow 2Ce_2O_4, \tag{R1}$$
$$H_2O + Ce_2O_3 \Longleftrightarrow H_2 + Ce_2O_4, \tag{R2}$$
$$CO_2 + Ce_2O_3 \Longleftrightarrow CO + Ce_2O_4. \tag{R3}$$

Modeling the three-way catalyst as a one-dimensional adiabatic channel, the main dynamic phenomena in the catalyst are thermodynamic phenomena, which include mass exchange between the channel and the washcoat, as well as gas and solid energy balances [121] (see also [48, Section 2.8.3]). The mass balance equations in the channel are

$$\begin{aligned}\partial_t c_i(z,t) = a\big(T_g(z,t), \dot{m}(z)\big)\big(\partial_z c_i(z,t) + c_i(z,t)\partial_z T_g(z,t)\big)\\ - b_i(c_i(z,t) - x_i(z,t)),\end{aligned} \tag{4.1}$$

where $\partial_t u(z,t)$ denotes the derivative of $u(z,t)$ with respect to time t and $\partial_z u(z,t)$ denotes the derivative of $u(z,t)$ with respect to the spatial variable z. In the washcoat we have

$$\partial_t x_i(z,t) = d\,(b_i(c_i(z,t) - x_i(z,t)) + k\partial_t s_i(z,t)), \tag{4.2}$$

where c_i and x_i denote the species' concentrations in the gas and solid phases, respectively, s_i denotes the reaction rate, a is the gas convection speed which depends on the mass flow $\dot{m}$ and on the temperature of the gas phase T_g, d and k are parameters that depend on the catalyst's characteristics (geometry and specific catalyst active surface), and b_i's are parameters that depend both on the catalyst's characteristics and on the specific species. The energy balances in the gas and solid phases can be formulated as

$$\partial_t T_g(z,t) = g(z)\partial_z T_g(z,t) + h_g\left(T_s(z,t) - T_g(z,t)\right), \tag{4.3}$$

$$\partial_t T_s(z,t) = \mu \partial_{z^2} T_s(z,t) + h_s\left(T_s(z,t) - T_g(z,t)\right) + \sum_i m_i \partial_t s_i, \tag{4.4}$$

where with T_g and T_s we denote the gas and solid temperatures, respectively, $g(z)$, h_g, h_s, μ are parameters that depend on the catalyst's characteristics, and m_i is a parameter that depends both on the catalyst's characteristics and on the specific species. The above equations (4.1)–(4.4) correspond to [121, equations (4), (5), (8), and (9)].

We now work on simplifying the model. Even with the reduced number of reactions and reacting species considered, we are still left with twelve coupled PDEs in the model. So we propose using the time scale separation to reduce the number of equations considered. First, our reaction rates (Eley–Rideal, see later) will depend only on the temperature of the substrate T_s. The dynamics of T_s is slow relative to the dynamics of c_i's and x_i's. Moreover, an ECU control strategy would already have an estimate of T_s. Hence, we shall treat T_s as an external signal rather than model it using (4.4).

Next we note that, due to the parameter d (d is essentially the ratio of the channel diameter and the washcoat thickness) being much larger than 1, the dynamics of the washcoat concentration x_i is much faster than the dynamics of the channel concentration c_i [5]. Using singular perturbation theory [75], we assume steady state concentrations for x_i. Dividing (4.2) with d and setting $\frac{1}{d} = 0$ one can solve (4.2) for $b_i(c_i(z,t) - x_i(z,t))$. Substituting the result to (4.1) we get

$$\partial_t c_i(z,t) = a\partial_z c_i(z,t) + k\partial_t s_i(z,t), \tag{4.5}$$

where $\partial_t s_i = \dot{s}_i$ is the reaction rate for species i and $k = SC/\varepsilon_g$ (SC is the specific catalyst storage capacity in mol/m^3 and ε_g is the volume fraction of the gas phase). To obtain (4.5) we have also used $\partial_z T_g(z,t) = 0$ (i.e., we assume that the rate of change in concentration due to change in gas density is negligible) and dropped the dependence of a on the mass flow and temperature because the gas convection speed is approximately constant in the channel.

The final step is to select the species to track and the appropriate reactions. Note that the selected reactions are given by (R1) to (R3). Next, we assume that the concentration of CO_2 and H_2O in the exhaust gas are dominated by the cylinder combustion and depend little on catalyst reactions. Indeed, they tend to be in the 13%, respectively, 14%, ±1% range. This small variation has negligible impact on the reaction rates below, and hence c_{CO_2}, c_{H_2O} are considered constant in the model. We are left with the 3 active gas components to be tracked (H_2, CO, O_2) and the relative oxygen storage in the catalyst Ψ

$(0 \leq \Psi(z,t) \leq 1, z \in [0,D])$:

$$\partial_t c_{CO} = a\partial_z c_{CO}(z,t) - kk^b_{CO} c_{CO}(z,t)\Psi(z,t), \tag{4.6}$$

$$\begin{aligned} \partial_t c_{H_2} &= a\partial_z c_{CH_2}(z,t) + k\left(k^f_{H_2} c_{H_2O}(z,t)(1-\Psi(z,t))\right. \\ &\quad \left. - k^b_{H_2} c_{H_2}(z,t)\Psi(z,t)\right), \end{aligned} \tag{4.7}$$

$$\partial_t c_{O_2} = a\partial_z c_{O_2}(z,t) - kk^f_{O_2} c_{O_2}(z,t)(1-\Psi(z,t))^2, \tag{4.8}$$

$$\begin{aligned} \dot{\Psi}(z,t) &= 2k^f_{O_2} c_{O_2}(z,t)(1-\Psi(z,t))^2 - k^b_{CO} c_{CO}(z,t)\Psi(z,t) \\ &\quad + k^f_{H_2} c_{H_2O}(z,t)(1-\Psi(z,t)) - k^b_{H_2} c_{H_2}(z,t)\Psi(z,t). \end{aligned} \tag{4.9}$$

The model (4.9) uses the Eley–Rideal mechanism between reactants in the gas and the substrate (i.e., does not model adsorption/desorption). The forward reaction parameters (with superscript f) are functions of substrate temperature and are of the form $A_{R_i} e^{-\frac{E_{R_i}}{RT_s}}$, R being the universal gas constant. The backward reaction parameters are the products of the forward parameters and $e^{\frac{\Delta G_{R_i}}{RT_s}}$. The values of the parameters needed to compute the forward and backward reaction parameters in (4.9) can be found in Tables 2.6 and 2.7 in [48]. Only the H_2 concentration tracks both forward and backward reactions, retaining the ability to model the water-gas shift reaction that impacts catalyst operation and oxygen sensor reading [121]. As mentioned, c_{H_2O} is assumed constant.

In the rest of this chapter we shall consider an even simpler version of this model, in which the concentrations of reductants c_{CO} and c_{H_2} are combined into one quantity c_r. This simplification is partially justified because the qualitative behavior of the concentrations of CO and H_2 with respect to changes in the fuel-to-air ratio are very similar (see [5]). The price to pay is that the water-gas shift effect is gone from the model. To accomplish this simplification we assume the ratio $CO : H_2 = 3 : 1$ in the burned gas (this is typical for feed-gas emissions), which allows us to set $k_{rb} = \frac{x}{1+x} k^b_{CO} + \frac{1}{1+x} k^b_{H_2}$, $x = 3$. With these simplifications we rewrite (4.9) as

$$\partial_t c_r(z,t) = a\partial_z c_r(z,t) - kk_{rb} c_r(z,t)\Psi(z,t), \tag{4.10}$$

$$\partial_t c_o(z,t) = a\partial_z c_o(z,t) - kk_{of} c_o(z,t)(1-\Psi(z,t))^2, \tag{4.11}$$

$$\dot{\Psi}(z,t) = -k_{rb} c_r(z,t)\Psi(z,t) + 2k_{of} c_o(z,t)(1-\Psi(z,t))^2. \tag{4.12}$$

The value of each parameter in the model as well as its physical interpretation are shown in Table 4.1. For the PDE system of equations (4.10)–(4.12) to be well posed one has to impose boundary conditions. We derive these boundary conditions from the concentrations of the species at the inlet of the catalyst. Assuming that the gasoline that enters the catalyst at $z = 0$ is approximately $CH_{1.9}$ (i.e., 1.9 atoms of H for every C), while oxygen concentration in the air is about 21%, and denoting with $\phi(t)$ the normalized fuel-to-air ratio in the inlet of the catalyst, we have the following concentrations at the inlet of the catalyst, i.e., at $z = 0$:

$$0.21O_2 + 0.142CH_{1.9} \Longrightarrow 0.142CO_2 + 0.135H_2O,$$

$$\text{Stochiometric combustion } \phi = 1, \tag{R4}$$

$$0.21O_2 + 0.142\phi CH_{1.9} \Longrightarrow 0.142\phi(t)CO_2 + 0.135\phi H_2O + 0.21(1-\phi)O_2,$$

$$\text{Lean combustion } \phi < 1, \tag{R5}$$

Table 4.1. *Parameters of the catalyst.*

Physical parameter	Model parameter	Value
Gas convection speed $\left(\frac{\text{m}}{\text{s}}\right)$	a	$-\frac{\dot{m}}{A_{cs}\rho}$
Catalyst's cross-sectional area $\left(\text{m}^2\right)$	A_{cs}	πr^2
Radius of the catalyst (m)	r	0.1
Mass flow $\left(\frac{\text{kg}}{\text{s}}\right)$	$\dot{m}$	0.04
Length of the catalyst (m)	D	0.09
Model parameter $\left(\frac{\text{mol}}{\text{m}^3}\right)$	k	$\frac{SC}{\epsilon_g}$
Backward reaction rate constant $\left(\frac{\text{m}^3}{\text{mol}\times\text{s}}\right)$	k_{rb}	$\frac{47.3}{3}$
Storage capacity of the catalyst $\left(\frac{\text{mol}}{\text{m}^3}\right)$	SC	10
Forward reaction rate constant $\left(\frac{\text{m}^3}{\text{mol}\times\text{s}}\right)$	k_{of}	$\frac{3099}{3}$
Average molar mass $\left(\frac{\text{kg}}{\text{mol}}\right)$	$\overline{M}$	0.03
Density of the gas phase $\left(\frac{\text{kg}}{\text{m}^3}\right)$	ρ	0.43
Universal gas constant $\left(\frac{J}{\text{mol}K}\right)$	R	8.31
Volume fraction of the gas phase (—)	ϵ_g	0.8
Temperature of the gas phase (K)	T_{Gas}	730

$$\begin{aligned} &0.21O_2 + 0.142\phi CH_{1.9} \Longrightarrow (-0.173\phi + 0.315)CO_2 \\ &\quad + (0.03\phi + 0.105)H_2O + 0.42(\phi(t) - 1)\left(\frac{3}{4}CO + \frac{1}{4}H_2\right), \end{aligned}$$

Rich combustion $\phi > 1$. (R6)

The derivation assumed equilibrated concentrations of oxidants and reductants because (a) it simplifies the analysis, (b) from the UEGO and Heated Exhaust Gas Oxygen (HEGO) sensor signals one could only extract equilibrated information (see the discussion after equation (4.60)), and (c) the catalyst would act to equilibrate the gas. Using (R4)–(R6) and assuming that a perturbation, namely, $W(t)$, constantly perturbs the control input $V(t)$ at the inlet of the catalyst (i.e., $\phi = V + W$), we get the following boundary conditions at the inlet of the catalyst ($z = 0$):

$$\xi_r(0,t) = 0.21U(t)(\text{sgn}(U(t)) + 1), \tag{4.13}$$
$$\xi_o(0,t) = 0.105U(t)(\text{sgn}(U(t)) - 1), \tag{4.14}$$

where $U(t) = V(t) - 1 + W(t)$. Since these are relative molecular concentrations, we multiply (4.13)–(4.14) with $\frac{\rho}{\overline{M}}$ to get absolute molar concentrations (i.e., to convert from % to $\frac{\text{mol}}{\text{m}^3}$), where ρ is the density of the gas phase and $\overline{M}$ is the average molar mass of the mixture. Setting $x = D - z$ we rewrite (4.10)–(4.12) and (4.13)–(4.14) as

$$\partial_t c_r(x,t) = -a\partial_x c_r(x,t) - kk_{rb}c_r(x,t)\Psi(x,t), \tag{4.15}$$
$$\partial_t c_o(x,t) = -a\partial_x c_o(x,t) - kk_{of}c_o(x,t)(1 - \Psi(x,t))^2, \tag{4.16}$$
$$\dot{\Psi}(x,t) = -k_{rb}c_r(x,t)\Psi(x,t) + 2k_{of}c_o(x,t)(1 - \Psi(x,t))^2, \tag{4.17}$$

$$c_r(D,t) = \frac{0.21\rho}{\overline{M}} U(t)(\mathrm{sgn}(U(t))+1), \tag{4.18}$$

$$c_o(D,t) = \frac{0.105\rho}{\overline{M}} U(t)(\mathrm{sgn}(U(t))-1). \tag{4.19}$$

4.2 ▪ Analysis of the Model under a Square Wave Air-to-Fuel Ratio Input

Many on-board control algorithms use an oscillating air-to-fuel ratio command imposed either by a relay-type feedback from the HEGO sensor [126] or a gain loop schedule. Using the notation of (4.18), (4.19) with $U(t) = V(t)-1$ (W is neglected), the square wave fuel-to-air ratio V is

$$V(t) = 1 + P_1(t) - P_2(t), \tag{4.20}$$

$$P_1(t) = \left\{ \begin{array}{ll} c_1, & iT \le t \le iT + \frac{T}{2} \\ 0, & iT + \frac{T}{2} \le t \le iT + T \end{array} \right\}, \tag{4.21}$$

$$P_2(t) = \left\{ \begin{array}{ll} 0, & iT \le t \le iT + \frac{T}{2} \\ c_2, & iT + \frac{T}{2} \le t \le iT + T \end{array} \right\}, \tag{4.22}$$

where $c_1, c_2 > 0$ and $T > 0$ are the magnitudes and period of oscillations, respectively, and $i = 1,2,\ldots$.

Although the air-to-fuel ratio oscillations are common in practice, the analysis of the qualitative properties of such an algorithm remains unavailable. Looking at relation (4.17) one can see that the oxygen storage level reaches its upper limit of one when the imposed air-to-fuel ratio is slightly lean and its lower limit of zero when the applied air-to-fuel ratio is slightly rich. However, when applying a symmetric square wave air-to-fuel ratio (period $T = 3s$ and magnitude $\pm 1\%$) at the inlet of the catalyst, one gets the image shown in Figure 4.1. From Figure 4.1 one can observe that the spatial mean of the oxygen storage level

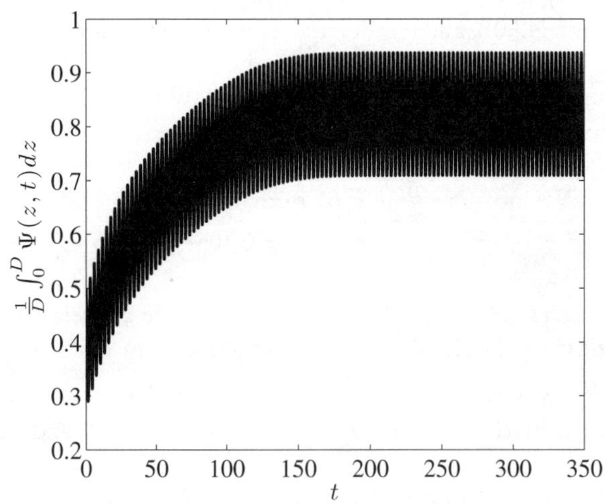

Figure 4.1. *The spatial mean of the oxygen storage level for a symmetric (open-loop) square wave air-to-fuel ratio with magnitude $\pm 1\%$ and period $T = 3s$.*

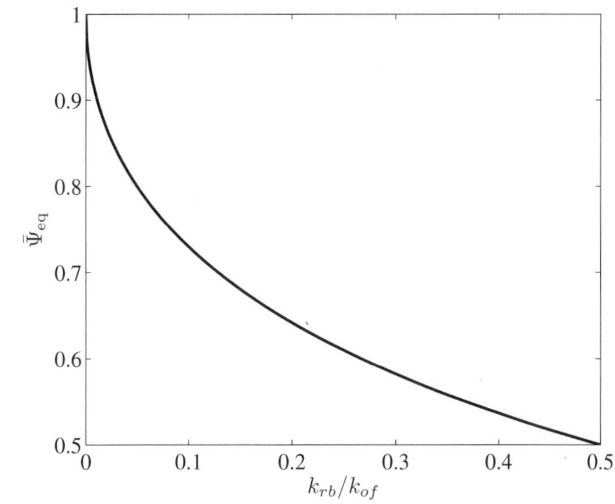

Figure 4.2. *Constant equilibrium profile of the time-averaged oxygen storage level as a function of the ratio* $\frac{k_{rb}}{k_{of}}$.

starting from a point around 0.4 increases and reaches a point around .8. Then it oscillates around this specific point. Therefore, Figure 4.1 suggests that the oxygen storage level admits an attractive periodic pattern around a spatially constant profile. We prove below that under a square wave air-to-fuel ratio (of sufficiently small T), the time-averaged oxygen storage level acquires a unique equilibrium profile which is constant for all $x \in [0, D]$ and asymptotically stable. The equilibrium profile of the oxygen level is constant along the channel of the catalyst because we assume constant T_s along the length of the catalyst. In a more realistic case of T_s varying with x, so would the equilibrium value of $\bar{\Psi}(x)$. Note that one could choose a nonsymmetric pulse in order to get a different equilibrium point. The equilibrium point of the time-averaged oxygen storage level depends on the ratio of the reaction parameters (see relation (4.33)). In Figure 4.2 we show the equilibrium of the time-averaged oxygen storage level as a function of the ratio $\frac{k_{rb}}{k_{of}}$.

For the parameter values in Table 4.1 we get an equilibrium at $\bar{\Psi}_{\text{eq}} = 0.884$. Since the time-averaged oxygen storage level is asymptotically stable, the spatial mean of the oxygen storage level converges to a limit cycle around the point $\bar{\Psi}_{\text{eq}} = 0.884$. This is expected since for sufficiently small T, averaging (Section 10.4 in [75]) guarantees that the spatial mean of the oxygen storage level admits an exponentially stable periodic solution around the equilibrium of the spatial mean of the time-averaged oxygen storage level, with magnitude that decreases as T decreases.

4.2.1 ▪ Proof of Equilibrium Acquisition of the Oxygen Storage Level and Its Asymptotic Stability

Using (4.20) and neglecting for the moment W, relations (4.18), (4.19) can be written as $c_r(D,t) = 2gP_1(t)$, $c_o(D,t) = gP_2(t)$. Since the time-average of $P_1(t)$ and $P_2(t)$ is $0.5c_1$ and $0.5c_2$, respectively, assuming a fast varying square wave air-to-fuel ratio, we consider the time-average of system (4.15)–(4.17) [75, Section 10.4] given as

$$\partial_t \bar{c}_r(x,t) = -a\partial_x \bar{c}_r(x,t) - kk_{rb}\bar{c}_r(x,t)\bar{\Psi}(x,t), \tag{4.23}$$
$$\partial_t \bar{c}_o(x,t) = -a\partial_x \bar{c}_o(x,t) - kk_{of}\bar{c}_o(x,t)\left(1-\bar{\Psi}(x,t)\right)^2, \tag{4.24}$$
$$\dot{\bar{\Psi}}(x,t) = -k_{rb}\bar{c}_r(x,t)\bar{\Psi}(x,t) + 2k_{of}\bar{c}_o(x,t)\left(1-\bar{\Psi}(x,t)\right)^2, \tag{4.25}$$
$$\bar{c}_r(D) = gc_1, \tag{4.26}$$
$$\bar{c}_o(D) = 0.5gc_2. \tag{4.27}$$

We now study the equilibrium of (4.23)–(4.27). At equilibrium, it holds that

$$\partial_x \bar{c}_r(x,t) = a_1\bar{c}_r(x,t)\bar{\Psi}(x,t), \tag{4.28}$$
$$\partial_x \bar{c}_o(x,t) = a_2\bar{c}_o(x,t)\left(1-\bar{\Psi}(x,t)\right)^2. \tag{4.29}$$

Solving (4.28), (4.29), and assuming a symmetric pulse, i.e., $\bar{c}_r(D) = gc$ and $\bar{c}_o(D) = 0.5gc$, we get

$$\bar{c}_r(x) = gce^{-a_1\int_x^D \bar{\Psi}(y)dy}, \tag{4.30}$$
$$\bar{c}_o(x) = \frac{gc}{2}e^{-a_2\int_x^D (1-\bar{\Psi}(y))^2 dy}. \tag{4.31}$$

Integrating (4.25) from x to D we get

$$0 = \frac{a}{k}(\bar{c}_r(D) - \bar{c}_r(x)) - 2\frac{a}{k}(\bar{c}_o(D) - \bar{c}_o(x)), \tag{4.32}$$

where we used (4.28), (4.29) and $aa_1 = -kk_{rb}$, $aa_2 = -kk_{of}$. Hence, $a_1\int_x^D \bar{\Psi}(y)dy = a_2\int_x^D (1-\bar{\Psi}(y))^2 dy$ for all $x \in [0,D]$. Since this holds for $x = D$, it holds for all $x \in [0,D]$ if $a_1\bar{\Psi}(x) = a_2(1-\bar{\Psi}(x))^2$ for all $x \in [0,D]$. Solving this relation and using $a_1 = -\frac{kk_{rb}}{a}$, $a_2 = -\frac{kk_{of}}{a}$ we get

$$\bar{\Psi}_{\text{eq}} = 1 + 0.5a_3 \pm \sqrt{0.25a_3^2 + a_3}, \tag{4.33}$$

where $a_3 = \frac{k_{rb}}{k_{of}}$. Since $0 \leq \bar{\Psi}(x) \leq 1$, we keep the root with the negative sign.

We analyze next the stability properties of the equilibrium. We use singular perturbation theory [75] in the sense that we assume that the dynamics of $\bar{c}_r$ and $\bar{c}_o$ are instantaneous. For an accurate application of the singular perturbation argument in the time-averaged system, one has to assume that the square wave fuel-to-air ratio oscillates much faster than the concentrations' dynamics, which in turn are much faster than the time-averaged oxygen storage dynamics. For our model this implies that one has to have $\frac{1}{T} \gg k \gg 1$ and $-a \gg 1$. Under this assumption, the quasi-steady-states of $\bar{c}_r$ and $\bar{c}_o$ are given by (4.30), (4.31). Substituting the solutions into (4.25) and keeping only first order terms we get

$$\delta\dot{\bar{\Psi}}(x,t) \approx -\Gamma(x)\delta\bar{\Psi}(x,t) + \Gamma(x)a_1\bar{\Psi}_{\text{eq}}\int_x^D \delta\bar{\Psi}(y,t)dy, \tag{4.34}$$

where $\delta\bar{\Psi}(x,t)=\bar{\Psi}(x,t)-\bar{\Psi}_{\rm eq}$ and

$$\Gamma(x)=gc\left(k_{rb}+2k_{of}\left(1-\bar{\Psi}_{\rm eq}\right)\right)e^{a_1(x-D)\bar{\Psi}_{\rm eq}}. \tag{4.35}$$

Defining an operator $\mathscr{A}:L^2[0,D]\mapsto L^2[0,D]$ as

$$\mathscr{A}u=-\Gamma(x)u(x)+\Gamma(x)a_1\bar{\Psi}_{\rm eq}\int_x^D u(y)dy, \tag{4.36}$$

we obtain the following result.

Theorem 4.1. *The spectrum of the operator* $\mathscr{A}$ *is*

$$\sigma(\mathscr{A})=\{\lambda\in\mathbb{R}:\lambda\in[-\Gamma(D),-\Gamma(0)]\}. \tag{4.37}$$

Proof. Let us find the resolvent of $\mathscr{A}$, that is, all $\lambda\in\mathbb{C}$ such that for any given $g(x)\in L^2([0,D])$ there exists an $f(x)\in L^2([0,D])$ such that

$$\lambda f(x)-\Gamma(x)\left(-f(x)+a_1\bar{\Psi}_{\rm eq}\int_x^D f(y)dy\right)=g(x). \tag{4.38}$$

Assume for the moment that $g(x)$ is differentiable; using (4.38) and the facts that $\Gamma(x)>0$ for all $x\in[0,D]$ and $\Gamma'(x)=a_1\bar{\Psi}_{\rm eq}\Gamma(x)$, we can differentiate (4.38), and assuming $\lambda\neq\Gamma(x)$ for all $x\in[0,D]$, we can solve the resulting equation as

$$(\lambda+\Gamma(x))f'(x)+a_1\bar{\Psi}_{\rm eq}(\Gamma(x)-\lambda)f(x)=g'(x)-a_1\bar{\Psi}_{\rm eq}g(x). \tag{4.39}$$

Hence, if $\lambda\neq\Gamma(x)$ for all $x\in[0,D]$, we have that

$$f(x)=\frac{1}{\lambda+\Gamma(x)}g(x)+k\int_x^D e^{k\int_x^y\frac{\Gamma(r)-\lambda}{\Gamma(r)+\lambda}dr}\frac{\Gamma(y)}{\lambda+\Gamma(y)}g(y)dy, \tag{4.40}$$

where $k=a_1\bar{\Psi}_{\rm eq}$. Hence,

$$\sigma(\mathscr{A})\subseteq\left\{\lambda\in\mathbb{C}:\lambda+\Gamma(\xi)=0\text{ for some }\xi\in[0,D]\right\}. \tag{4.41}$$

We show now that actually $\sigma(\mathscr{A})=\left\{\lambda\in\mathbb{C}:\lambda+\Gamma(\xi)=0\text{ for some }\xi\in[0,D]\right\}$. Assume that $-\Gamma(D)<\lambda<-\Gamma(0)$ and that for any $g\in L^2([0,D])$, (4.38) has a solution $f\in L^2([0,D])$. For each fixed λ the set $M_\lambda=\{x\in[0,D]:\lambda+\Gamma(x)=0\}$ is of measure zero, since $\lambda=-\Gamma(x)\Rightarrow x=\frac{1}{k}\log\left(-\frac{\lambda}{\Gamma(0)}\right)$. Define now the sets

$$E_1=\left[\frac{1}{k}\log\left(-\frac{\lambda}{\Gamma(0)}\right)-\epsilon,\frac{1}{k}\log\left(-\frac{\lambda}{\Gamma(0)}\right)+\epsilon\right], \tag{4.42}$$

$$E_2=\left[\frac{1}{k}\log\left(-\frac{\lambda}{\Gamma(0)}\right)-\frac{\epsilon}{2},\frac{1}{k}\log\left(-\frac{\lambda}{\Gamma(0)}\right)+\frac{\epsilon}{2}\right]. \tag{4.43}$$

Using the continuity of the function $\Gamma(x)$ in $[0,D]$ and Proposition 3.7 from [47] we conclude that $\lambda \in (-\Gamma(D), -\Gamma(0))$ belong to $\sigma(\mathscr{A})$. Assume now $\lambda = -\Gamma(0)$ and integrate (4.38) from 0 to D to get $(\lambda + \Gamma(0)) \int_0^D f(x)dx = \int_0^D g(x)dx$, and hence $\int_0^D g(x)dx = 0$. Such a function cannot represent all functions in $L^2([0,D])$, that is, $\mathscr{R}(\lambda\mathscr{I} - \mathscr{A}) \neq L^2([0,D])$, where $\mathscr{R}(\lambda\mathscr{I} - \mathscr{A})$ is the range of the operator $\lambda\mathscr{I} - \mathscr{A}$. Thus $\lambda = -\Gamma(0)$ belongs to the spectrum of $\mathscr{A}$. Consider next $\lambda = -\Gamma(D)$, $g(x) = c \neq 0$. For any f, relation (4.38) cannot be satisfied, since for $x = D$ the left-hand side of (4.38) is zero, whereas $g(D) = c \neq 0$. Thus, constant nonzero functions are not in the range of $\lambda\mathscr{I} - \mathscr{A}$ with $\lambda = -\Gamma(D)$, i.e., $-\Gamma(D) \in \sigma(\mathscr{A})$. □

4.3 ▪ Observer-Based Output Feedback

The fact that the time-averaged system under a square wave air-to-fuel ratio in the input admits a unique equilibrium profile enables us to apply linear control design techniques to the average system. We focus on the oxygen storage level and design an output feedback controller with integral action, such that constant disturbances can be rejected. We modify our open-loop control law as

$$V(t) = 1 + P_1(t) - P_2(t) + u(t), \tag{4.44}$$

where u is yet to be designed. With a symmetric square wave fuel-to-air ratio, i.e., setting $c_1 = c_2 = c$ in (4.21), (4.22) and assuming that for all $t \geq 0$, $|u(t) + W| < c$, with (4.44), relations (4.18), (4.19) can be written as $c_r(D,t) = 2g\left(1 + \frac{u(t)+W}{c}\right)P_1(t)$, $c_o(D,t) = g\left(1 - \frac{u(t)+W}{c}\right)P_2(t)$, where $g = \frac{0.21\rho}{M}$. Assuming a sufficiently small T, we can approximate the solutions of system (4.15)–(4.19) with those of the time-averaged system as

$$\partial_t \bar{c}_r(x,t) = -a\partial_x \bar{c}_r(x,t) - kk_{rb}\bar{c}_r(x,t)\bar{\Psi}(x,t), \tag{4.45}$$

$$\partial_t \bar{c}_o(x,t) = -a\partial_x \bar{c}_o(x,t) - kk_{of}\bar{c}_o(x,t)\left(1 - \bar{\Psi}(x,t)\right)^2, \tag{4.46}$$

$$\dot{\bar{\Psi}}(x,t) = -k_{rb}\bar{c}_r(x,t)\bar{\Psi}(x,t) + 2k_{of}\bar{c}_o(x,t)\left(1 - \bar{\Psi}(x,t)\right)^2, \tag{4.47}$$

$$\bar{c}_r(D,t) = gc + g\,(u(t) + W), \tag{4.48}$$

$$\bar{c}_o(D,t) = 0.5gc - 0.5g\,(u(t) + W). \tag{4.49}$$

We use the singular perturbation argument as in Section 4.2.1. The quasi-steady-states of $\bar{c}_r$ and $\bar{c}_o$ are obtained by solving (4.28) and (4.29), respectively, with boundary conditions given in (4.48), (4.49), as

$$\bar{c}_r(x,t) = ge^{-a_1\int_x^D \bar{\Psi}(y,t)dy}\,(c + u(t) + W), \tag{4.50}$$

$$\bar{c}_o(x,t) = 0.5ge^{-a_2\int_x^D (1-\bar{\Psi}(y,t))^2 dy}\,(c - u(t) - W). \tag{4.51}$$

We now linearize the outputs $\bar{y}_r(t) = \bar{c}_r(0,t)$, $\bar{y}_o(t) = \bar{c}_o(0,t)$ around $y_{r,0} = cge^{-a_1 D\bar{\Psi}_{\text{eq}}}$, $y_{o,0} = \frac{c}{2}ge^{-a_2 D(1-\bar{\Psi}_{\text{eq}})^2}$, respectively, and $u = -W$. The Taylor series expansions yield

$$\bar{y}_{r,l}(t) = y_{r,0} - a_1 cge^{-a_1 D\bar{\Psi}_{\text{eq}}}\delta X(t) + ge^{-a_1 D\bar{\Psi}_{\text{eq}}}(u(t) + W), \tag{4.52}$$

$$\bar{y}_{o,l}(t) = y_{o,0} + a_2 c g (1-\bar{\Psi}_{\mathrm{eq}}) e^{-a_2 D (1-\bar{\Psi}_{\mathrm{eq}})^2} \delta X(t) - \frac{g}{2} e^{-a_2 D (1-\bar{\Psi}_{\mathrm{eq}})^2} (u(t)+W), \tag{4.53}$$

$$\delta X(t) = X(t) - D\bar{\Psi}_{\mathrm{eq}}, \tag{4.54}$$

$$X(t) = \int_0^D \bar{\Psi}(x) dx. \tag{4.55}$$

Integrating (4.47) from 0 to D and using (4.28), (4.29), with $a_1 = -\frac{kk_{rb}}{a}$, $a_2 = -\frac{kk_{of}}{a}$ we get

$$\delta \dot{X}(t) = \frac{a}{k} (\bar{c}_r(D,t) - \bar{c}_r(0,t) - 2(\bar{c}_o(D,t) - \bar{c}_o(0,t))). \tag{4.56}$$

Using relations (4.48), (4.49) and (4.52), (4.53) we arrive at

$$\delta \dot{X}(t) = \frac{cga}{k} e^{-a_1 D \bar{\Psi}_{\mathrm{eq}}} \left(a_1 + 2a_2 \left(1 - \bar{\Psi}_{\mathrm{eq}}\right)\right) \delta X(t) + \frac{2ga}{k} \left(1 - e^{-a_1 D \bar{\Psi}_{\mathrm{eq}}}\right)(u(t)+W), \tag{4.57}$$

where we also use the fact that $a_1 \bar{\Psi}_{\mathrm{eq}} = a_2 (1-\bar{\Psi}_{\mathrm{eq}})^2$ (see relation (4.33)). The observer that we suggest is

$$\delta \dot{\hat{X}}(t) = \frac{cga}{k} e^{-a_1 D \bar{\Psi}_{\mathrm{eq}}} \left(a_1 + 2a_2 \left(1 - \bar{\Psi}_{\mathrm{eq}}\right)\right) \delta \hat{X}(t) + L_1 \left(\bar{y}_r(t) - \bar{y}_{r,\mathrm{es}}(t)\right) + L_2 \left(\bar{y}_o(t) - \bar{y}_{o,\mathrm{es}}(t)\right) + g e^{-a_1 D \bar{\Psi}_{\mathrm{eq}}} (L_1 - 0.5 L_2) u(t), \tag{4.58}$$

where the estimated outputs $\bar{y}_{r,\mathrm{es}}$, $\bar{y}_{o,\mathrm{es}}$ are based on the linearized outputs (4.52), (4.53) and are defined as

$$\bar{y}_{r,\mathrm{es}}(t) = y_{r,0} - a_1 c g e^{-a_1 D \bar{\Psi}_{\mathrm{eq}}} \delta \hat{X}(t) + g e^{-a_1 D \bar{\Psi}_{\mathrm{eq}}} u(t), \tag{4.59}$$

$$\bar{y}_{o,\mathrm{es}}(t) = y_{o,0} + a_2 c g (1-\bar{\Psi}_{\mathrm{eq}}) e^{-a_2 D (1-\bar{\Psi}_{\mathrm{eq}})^2} \delta \hat{X}(t) - \frac{g}{2} e^{-a_2 D (1-\bar{\Psi}_{\mathrm{eq}})^2} u(t). \tag{4.60}$$

Observer (4.58) is not exactly a copy of the plant (4.57) plus output injection. The difference is that the coefficient that multiplies the input u in the observer is different from the corresponding one of the plant. This can be explained as follows. The linearized output injection terms are

$$\bar{y}_r(t) - \bar{y}_{r,\mathrm{es}}(t) = -a_1 c g e^{-a_1 D \bar{\Psi}_{\mathrm{eq}}} \left(\delta X(t) - \delta \hat{X}(t)\right) + g e^{-a_1 D \bar{\Psi}_{\mathrm{eq}}} W, \tag{4.61}$$

$$\bar{y}_o(t) - \bar{y}_{o,\mathrm{es}}(t) = a_2 c g (1-\bar{\Psi}_{\mathrm{eq}}) e^{-a_2 D (1-\bar{\Psi}_{\mathrm{eq}})^2} \left(\delta X(t) - \delta \hat{X}(t)\right) - \frac{g}{2} e^{-a_1 D \bar{\Psi}_{\mathrm{eq}}} W, \tag{4.62}$$

where we used (4.52), (4.53), (4.59), (4.60). The equilibrium point of (4.57) is $\delta X = 0$, $u = -W$. Since the desired equilibrium of (4.58) is $\delta \hat{X} = 0$, $u = -W$, we conclude that the coefficient of u in (4.58) is $g e^{-a_1 D \bar{\Psi}_{\mathrm{eq}}} (L_1 - 0.5 L_2)$.

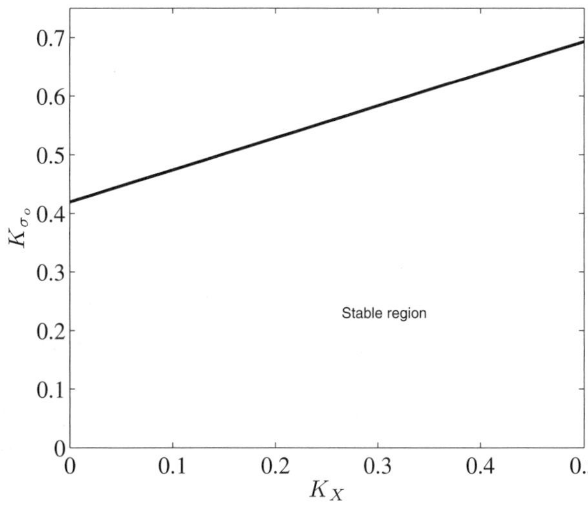

Figure 4.3. *A stable region for the closed-loop system* (4.57), (4.58), (4.64) *with the controller* (4.63) *in the parametric space* (K_X, K_{σ_o}).

We assume that the measured quantities are the output and input concentrations of the combined reductants and the oxygen. This is plausible since these quantities can be calculated using an inverse model of the air-to-fuel ratio sensors precatalyst (UEGO sensor) and postcatalyst (HEGO sensor) [5]. Yet our integral action is one-sided. We choose an integral action based on the reductants' measurement when the equilibrium for the oxygen storage level is low, or on the oxygen's measurement when the equilibrium is high. Define

$$u(t) = K_x \delta \hat{X}(t) + K_{\sigma_o} \bar{\sigma}_o(t), \tag{4.63}$$

$$\frac{d\bar{\sigma}_o(t)}{dt} = \bar{c}_o(0,t) - \frac{c}{2} g e^{-a_2 D(1-\bar{\Psi}_{\text{eq}})^2}. \tag{4.64}$$

Substituting u in (4.63) into (4.57), (4.58), (4.64), the equilibrium $\delta X = \delta \hat{X} = 0$, $\sigma_o = -\frac{W}{K_{\sigma_o}}$ of the closed-loop system consisting of the time-averaged spatial mean of the oxygen storage level (4.57), its estimate (4.58), and the integrator state (4.64) can be rendered asymptotically stable with an appropriate choice of K_X, K_{σ_o}, L_1, and L_2. We choose $L_2 = -2L_1 = 4$ and we remain with three conditions on the parameters K_X and K_{σ_o}. In Figure 4.3 we show one of the possible stable regions in the (K_X, K_{σ_o}) plane. The choice $K_X = 0.25, K_\sigma = 0.3, L_1 = -2, L_2 = 4$ places the eigenvalues at -0.23 and $-0.325 \pm 0.365j$.

We consider the response of the system (4.15)–(4.19) with $U(t) = V(t) + W - 1$ under a symmetric square wave air-to-fuel ratio V of period $T = 1$ s and magnitude $\pm 2\%$ under a constant disturbance $W = 0.2\%$ rich applied at $t = 50$s. In Figure 4.4 we show the time-average (which is computed as $\frac{1}{D}\int_0^D \bar{\Psi}(x)dx = \frac{\int_{t-T}^t \frac{1}{D}\int_0^D \Psi(x,\tau)dx d\tau}{T}$) of the spatial mean of the oxygen storage level when the controller is turned off and when it is turned on (i.e., when u in (4.44) is zero and when u is given by (4.63)). As one can observe, the disturbance pulls the air-to-fuel ratio richer, and hence in the open-loop system the oxygen

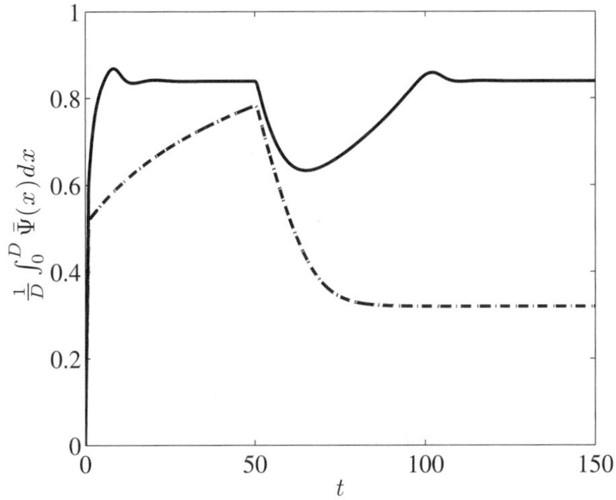

Figure 4.4. *Open-loop (dashed line) and closed-loop (solid line) spatial mean of the time-averaged storage level for a symmetric square wave air-to-fuel ratio with magnitude* $\pm 2\%$ *of period* $T = 1s$*, under a constant disturbance* $W = 0.2\%$ *applied at* $t = 50s$.

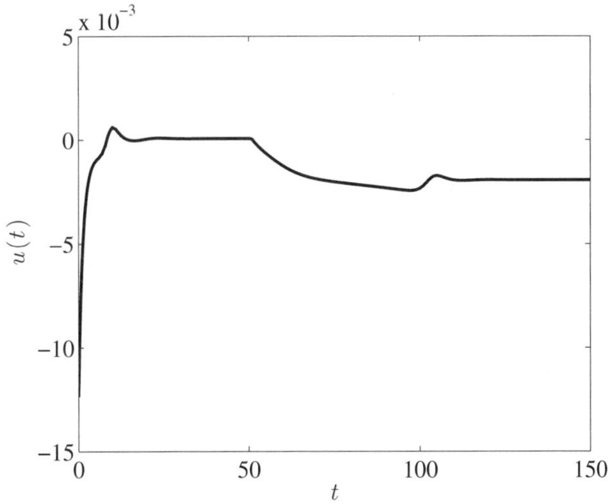

Figure 4.5. *The feedback part u given in* (4.63) *of the controller* (4.44) *for a symmetric square wave air-to-fuel ratio with magnitude* $\pm 2\%$ *and period* $T = 1s$*, under a constant rich disturbance* $W = 0.2\%$ *applied at* $t = 50s$.

storage level oscillates around a lower value than the expected one. However, even with a rich disturbance and open-loop control, the time-averaged oxygen level admits an asymptotically stable equilibrium. In contrast, the closed-loop oxygen storage level reaches an equilibrium close to the expected one. In Figure 4.5 we show the feedback part of V in (4.44). As one can observe, the steady-state value of u is such that it compensates for W (i.e., $-W$). In Figure 4.6 we show the time-average of the concentrations of oxygen and

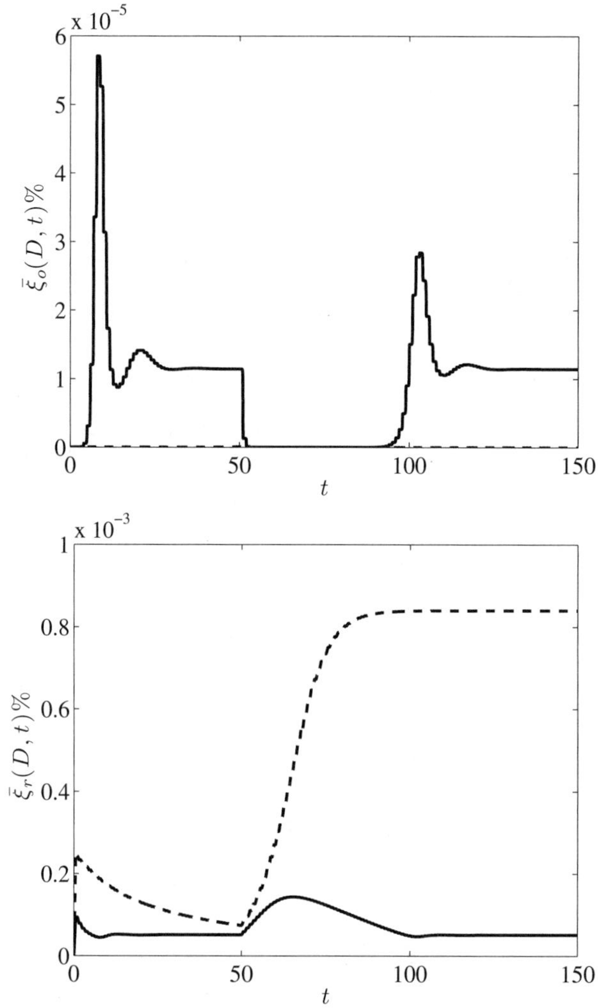

Figure 4.6. *Relative time-averaged concentration of oxygen* $\bar{\xi}_o(D,t) = \frac{\overline{M}}{\rho}\bar{c}_o(0,t)$ *(top) and reductants* $\bar{\xi}_r(D,t) = \frac{\overline{M}}{\rho}\bar{c}_r(0,t)$ *(bottom) at the outlet of the catalyst for the open-loop (dashed lines) and closed-loop (solid lines) systems, under a constant rich disturbance* $W = 0.2\%$ *applied at* $t = 50s$.

reductants (which are computed as $\frac{\overline{M}}{\rho}\bar{c}_i(0,t) = \bar{\xi}_i(D,t) = \frac{1}{T}\int_{t-T}^{t} \xi_i(D,\tau)d\tau,\ i = r, o$) at the outlet of the catalyst, in open- and closed-loop. In the open-loop case the oxygen's relative concentration at the output of the catalyst is negligible, whereas the concentration of reductants is high. This is because the rich disturbance pulls the catalyst towards complete depletion, and hence the concentration of oxygen is low, whereas the concentration of reductants is high. In Figure 4.7 we show the open- and closed-loop response of the oxygen's and reductants' efficiencies under a $W = 0.2$ rich disturbance.

In Figure 4.8 we compare the oxygen's and reductants' efficiencies but now without the effect of the disturbance and for two different regulation points: one at the natural open-loop equilibrium $\bar{\Psi}_{\text{eq}} = 0.884$ and another at the value $\bar{\Psi}_{\text{eq}} = 0.75$ which has to be

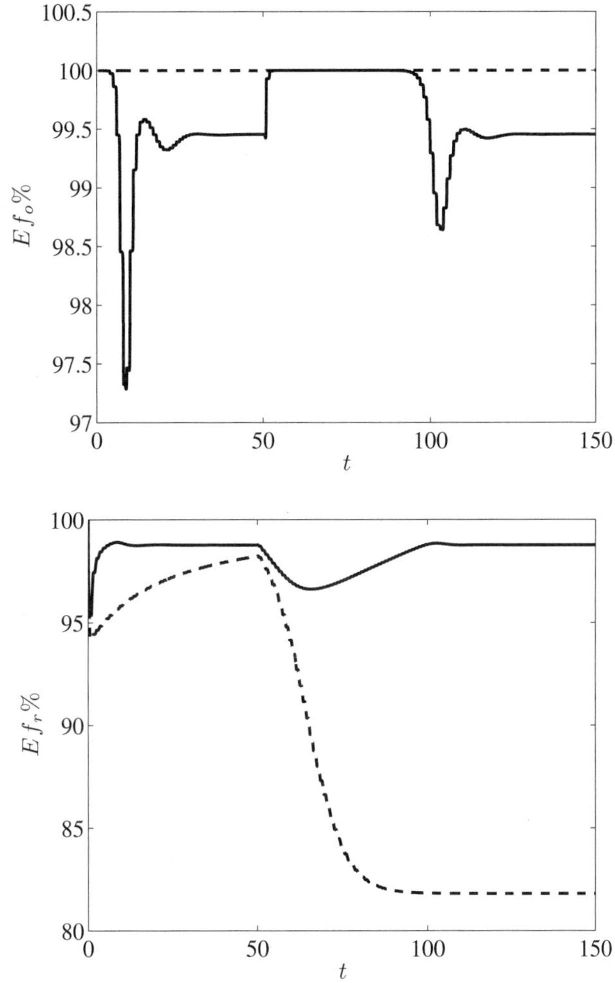

Figure 4.7. *Comparison of the oxygen's (top) and reductants' (bottom) efficiency for the open-loop (dashed lines) and closed-loop (solid lines) systems under a constant rich disturbance $W = 0.2\%$ applied at $t = 50s$.*

imposed by feedback. As one can observe from Figure 4.8, in the case where we regulate the system around an equilibrium lower than $\Psi_{eq} = 0.884$ the oxygen's efficiency is improving, whereas the reductants' efficiency drops. These two figures show that there is a trade-off between the oxygen's and the reductants' efficiencies as the equilibrium profile varies. This is expected because the lower the equilibrium of the oxygen storage, the harder for the reductants to react with the stored oxygen, whereas the easier for oxygen to be stored. Hence, the reductants' efficiency drops, whereas the oxygen's efficiency increases. Analogous conclusions can be made for the efficiencies shown in Figure 4.7 since in each of the two plots the efficiencies for two different equilibrium points of the oxygen storage level are compared, but now the steady-state oxygen storage level is affected by the disturbance and not by the control law.

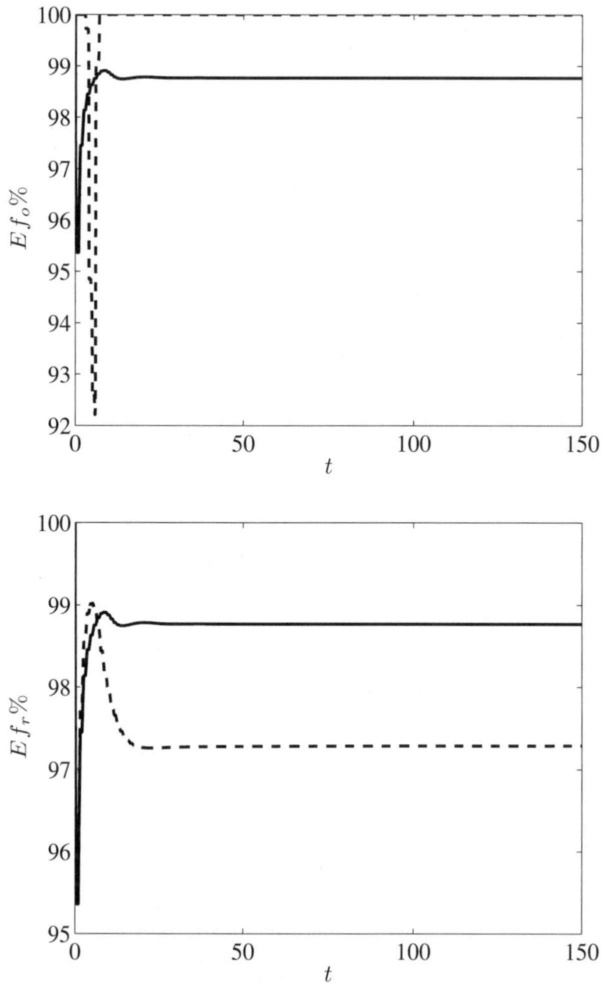

Figure 4.8. *Oxygen's (top) and reductants' (bottom) efficiency trade-off for the closed-loop system for two different set-points* $\Psi_{\text{eq}} = 0.884$ *(solid lines) and* $\Psi_{\text{eq}} = 0.75$ *(dashed lines).*

4.4 ▪ Notes and References

Detailed models capturing the chemical dynamics as well as the thermodynamics in a three-way catalyst can be found in the works by Jobson and coauthors [61], Koltsakis and Stamatelos [81], and Moller and coauthors [121] (see also the book by Guzzella and Onder [48]). These models consist of several nonlinear PDEs and, therefore, present a major challenge for control design. As a consequence, most control design algorithms for the oxygen storage level are based on simpler, lumped-parameter models such as the ones developed by Balenovic and coauthors [6], Fiengo and coauthors [42], Roduner and coauthors [146], and Shafai and coauthors [150]. As an alternative, Auckenthaler [5] considers control design for a spatially distributed model which approximates the catalyst with a sequence of three cells, each represented by a lumped-parameter model.

The reader should notice that in contrast to the requirement of numerically solving the PDE-based model of the catalyst, under the various scenarios considered in this chapter, the proposed controller is only using the lump-parameter (ODE) model for the spatial mean of the catalyst's oxygen storage. This is crucial since our control algorithm has to be simple enough to result in a controller implementable in an ECU.

Chapter 5

Nonlinear Systems with Input Delay

In this chapter we introduce the concepts of *nonlinear* predictor feedback and *nonlinear* infinite-dimensional backstepping transformation. One of the main challenges in developing predictor feedback designs for nonlinear plants is the determination of an implementable form for the future values of the state. Having determined the predictor state, the control law is then obtained by replacing the state in a nominal state feedback law (which stabilizes the delay-free plant) by the state's predictor. In addition, another major obstacle in designing globally stabilizing control laws for nonlinear systems with long input delays is the finite escape phenomenon. The input delay may be so large that the control signal cannot reach the plant before its state escapes to infinity. Therefore, in this chapter we assume that the plant is forward complete, that is, for every initial condition and every bounded input signal the corresponding solution is defined for all times. The last ingredient that one needs for achieving global stabilization is the existence of a (possibly time-varying) feedback law that globally stabilizes the delay-free plant.

We start in Section 5.1 with the design of the nonlinear predictor feedback. In Section 5.2 we introduce the nonlinear infinite-dimensional backstepping transformation, which allows the construction of a Lyapunov functional for the closed-loop system, which we use to prove stability of the closed-loop system in Section 5.3.

5.1 ▪ Nonlinear Predictor Feedback Design for Constant Delay

We consider nonlinear systems with constant input delay, i.e., systems of the form

$$\dot{X}(t) = f(X(t), U(t-D)), \tag{5.1}$$

where $X \in \mathbb{R}^n$, $U \in \mathbb{R}$, $D > 0$ is arbitrary long, and $f : \mathbb{R}^n \times \mathbb{R} \to \mathbb{R}^n$ is a locally Lipschitz mapping. Assume now that the function $\varkappa$ globally stabilizes the delay-free plant; i.e., $\dot{X}(t) = f(X(t), \varkappa(t, X(t)))$ is globally asymptotically stable. The predictor feedback law for plant (5.1) is

$$U(t) = \varkappa(t + D, P(t)), \tag{5.2}$$

where

$$P(t) = X(t) + \int_{t-D}^{t} f(P(\theta), U(\theta))\, d\theta, \tag{5.3}$$

with the initial condition for the integral equation for $P(t)$ given for all $\theta \in [t_0 - D, t_0]$ (t_0 is the initial time which must be given because the closed-loop system is time varying) as

$$P(\theta) = X(t_0) + \int_{t_0 - D}^{\theta} f(P(\sigma), U(\sigma))d\sigma. \tag{5.4}$$

The signal $P(t)$ represents the D-time-unit ahead predictor of X, i.e., $P(t) = X(t+D)$ for all $t \geq t_0$, and it is directly implementable since it incorporates the current state $X(t)$, and the history of $P(\theta)$, $U(\theta)$ for all $t - D \leq \theta \leq t$, which are available for measurement. The computation of $P(t)$ from (5.3) is straightforward with a discretized implementation in which $P(\theta)$ is assigned values based on the right-hand side of (5.3), which involves earlier values of P and the values of the input U. At each time step the integral in (5.3) can be computed using a method of numerical integration (e.g., the trapezoidal rule) with a total number of discrete points N, given by $N(t) = \left\lfloor \frac{D(X(t))}{h} \right\rfloor$, where h is the time-discretization step and $\lfloor a \rfloor$ denotes the integer part of a.

One can see that the signal P defined in (5.3) is the predictor of X as follows. Performing the change of variables $t = \theta + D$ for all $t - D \leq \theta \leq t$ in (5.1) and using the fact that $\frac{d\theta}{dt} = 1$, we get

$$\frac{dX(\theta + D)}{d\theta} = f(X(\theta + D), U(\theta)) \quad \text{for all } t - D \leq \theta \leq t. \tag{5.5}$$

Defining the new signal

$$P(\theta) = X(\theta + D) \quad \text{for all } t - D \leq \theta \leq t \tag{5.6}$$

and solving the resulting ODE in θ for P, starting at $P(t - D) = X(t)$ we arrive at

$$P(\theta) = X(t) + \int_{t-D}^{\theta} f(P(s), U(s))ds \quad \text{for all } t - D \leq \theta \leq t. \tag{5.7}$$

It is a bit difficult to comprehend the mathematical meaning of the relationship (5.7), where $P(\theta)$ appears on both sides of the equation. It is helpful to start from the linear case, $\dot{X}(t) = AX(t) + BU(t - D)$ with a constant delay D. In that case the predictor is given explicitly using the variation of constants formula, with the initial condition $P(t - D) = X(t)$, as $P(t) = e^{AD}X(t) + \int_{t-D}^{t} e^{A(t-\theta)}BU(\theta)d\theta$. For systems that are nonlinear $P(t)$ cannot be written explicitly, for the same reason that a nonlinear ODE cannot be solved explicitly. So we represent $P(\theta)$ implicitly using the nonlinear integral equation (5.7). Therefore, one should view $P(\theta)$ as the output of an operator, parametrized by t, acting on $P(s)$, $U(s)$, $t - D \leq s \leq \theta$, in the same way that the solution $X(t)$ to a nonlinear ODE (i.e., $X(t) = X(t_0) + \int_{t_0}^{t} f(X(s), U(s))ds$) can be viewed as the output of an operator, parametrized by t_0, acting on $X(s)$, $U(s)$, $t_0 \leq s \leq t$.

5.2 ▪ Nonlinear Infinite-Dimensional Backstepping Transformation

5.2.1 ▪ Nonlinear Backstepping Transformation in Standard Delay Notation

Together with the predictor-based control law (5.2), (5.3) we define the infinite-dimensional backstepping transformation of the actuator state given by

$$W(\theta) = U(\theta) - \kappa(\theta + D, P(\theta)) \quad \text{for all } t - D \leq \theta \leq t, \tag{5.8}$$

where $P(\theta)$ is given in (5.7). Using the facts that $U(t)=\kappa(t+D,P(t))$ and that $P(t-D)=X(t)$ for all $t\geq t_0$, transformation (5.8) maps the closed-loop system consisting of the plant (5.1) and the control law (5.2) to the following target system:

$$\dot{X}(t)=f(X(t),\kappa(t,X(t))+W(t-D)), \tag{5.9}$$
$$W(t)=0, \quad \text{for all } t\geq t_0. \tag{5.10}$$

The inverse backstepping transformation is

$$U(\theta)=W(\theta)+\kappa(\theta+D,\Pi(\theta)) \quad \text{for all } t-D\leq\theta\leq t, \tag{5.11}$$

where[1] for all $t-D\leq\theta\leq t$

$$\Pi(\theta)=X(t)+\int_{t-D}^{\theta} f(\Pi(s),\kappa(s+D,\Pi(s))+W(s))ds, \tag{5.12}$$

with initial condition, for all $t_0-D\leq\theta\leq t_0$,

$$\Pi(\theta)=X(t_0)+\int_{t_0-D}^{\theta} f(\Pi(s),\kappa(s+D,\Pi(s))+W(s))ds. \tag{5.13}$$

5.2.2 ▪ Nonlinear Backstepping Transformation in PDE Representation

An alternative representation of system (5.1) can be written by representing the actuator state $U(\theta)$, $\theta\in[t-D,t]$, with a transport PDE:

$$\dot{X}(t)=f(X(t),u(0,t)), \tag{5.14}$$
$$u_t(x,t)=u_x(x,t) \quad \text{for all } x\in[0,D], \tag{5.15}$$
$$u(D,t)=U(t). \tag{5.16}$$

Note that with this representation, $u(x,t)=U(t+x-D)$ for all $x\in[0,D]$. Analogously, the PDE representation of the predictor state given in (5.7) is

$$p(x,t)=X(t)+\int_0^x f(p(y,t),u(y,t))dy \quad \text{for all } x\in[0,D]. \tag{5.17}$$

With this representation

$$P(t)=p(D,t). \tag{5.18}$$

The backstepping transformation (5.8) in the PDE notation is written as

$$w(x,t)=u(x,t)-\kappa(t+x,p(x,t)), \tag{5.19}$$

and together with the control law (5.2), which is written as

$$\begin{aligned} U(t)&=\kappa(t+D,P(t))\\ &=\kappa(t+D,p(D,t)), \end{aligned} \tag{5.20}$$

[1]As was explained in Section 2.1, the quantities P in (5.7) and Π in (5.12) are identical. However, we use two distinct symbols for the same quantity because, in one case, P is expressed in terms of X and U for the direct backstepping transformation, while, in the other case, Π is expressed in terms of X and W for the inverse backstepping transformation.

transforms system (5.14)–(5.16) into the target system (5.9), (5.10), written now as

$$\dot{X}(t) = f(X(t), x(t, X(t)) + w(0,t)), \tag{5.21}$$
$$w_t(x,t) = w_x(x,t) \quad \text{for all } x \in [0,D], \tag{5.22}$$
$$w(D,t) = 0. \tag{5.23}$$

To see this first note that one gets relation (5.21) using (5.19) and the fact that from (5.17) it follows that $p(0,t) = X(t)$. Relation (5.23) follows from (5.19) and the fact that $U(t) = x(t+D, P(t)) = x(t+D, p(D,t))$. For deriving relation (5.22) one has only to show that $p_t(x,t) = p_x(x,t)$ for all $x \in [0,D]$, since already $u_t(x,t) = u_x(x,t)$ for all $x \in [0,D]$. An alternative form of the integral equation (5.17) is as differential equation in x, with appropriate initial condition, given as

$$p_x(x,t) = f(p(x,t), u(x,t)), \tag{5.24}$$
$$p(0,t) = X(t). \tag{5.25}$$

The fact that $p_t(x,t) = p_x(x,t)$ is immediate by noting that $u(x,t)$ is a function of only one variable, $x+t$, and therefore so is $p(x,t)$ based on[2] the ODE (5.24). The inverse backstepping transformation in PDE notation, i.e., the analogue of (5.11), (5.12), is given by

$$u(x,t) = w(x,t) + x(t+x, \pi(x,t)), \tag{5.26}$$

where for all $x \in [0,D]$

$$\pi(x,t) = X(t) + \int_0^x f(\pi(y,t), x(t+y, \pi(y,t)) + w(y,t))\,dy. \tag{5.27}$$

The alternative form of the integral equation (5.27) as an ODE in x is

$$\pi_x(x,t) = f(\pi(x,t), x(t+x, \pi(x,t)) + w(x,t)), \tag{5.28}$$
$$\pi(0,t) = X(t). \tag{5.29}$$

5.3 ▪ Lyapunov-Based Stability Analysis

We provide first the assumptions of this section.

Assumption 5.1. *The plant $\dot{X} = f(X,\omega)$ is strongly forward complete; that is, there exist a smooth positive definite function R and class $\mathcal{K}_\infty$ functions α_1, α_2, and α_3 such that*

$$\alpha_1(|X|) \le R(X) \le \alpha_2(|X|), \tag{5.30}$$
$$\frac{\partial R(X)}{\partial X} f(X,\omega) \le R(X) + \alpha_3(|\omega|) \tag{5.31}$$

for all $X \in \mathbb{R}^n$ and for all $\omega \in \mathbb{R}$.

Assumption 5.1 guarantees that system (5.1) does not exhibit finite escape time; that is, for every initial condition and every locally bounded input signal the corresponding solution is defined for all $t \ge t_0$. The difference between Assumption 5.1 and Theorem

[2]Another way to see this is by noting that $p(x,t) = X(t+x)$ satisfies (5.24) (based on relation (5.1) and the fact that $u(x,t) = U(t+x-D)$) and using a standard existence and uniqueness theorem for the ODE in x given in (5.24).

C.13 is that Theorem C.13 guarantees that $R(\cdot)$ is nonnegative and radially unbounded, whereas we assume in addition that $R(\cdot)$ is positive definite (which can be justified by the fact that we assume $f(0,0)=0$).

Assumption 5.2. *The plant $\dot{X}(t) = f(X(t), \kappa(t,X(t)) + \omega(t))$ is input-to-state stable with respect to ω, and the function $\kappa(t,X)$ is locally Lipschitz, periodic in t, and such that $k(t,0)=0$ for all $t \geq 0$.*

Theorem 5.3. *Consider the closed-loop system consisting of the plant* (5.1) *and the control law* (5.2), (5.3). *Under Assumptions* 5.1 *and* 5.2, *there exists a class $\mathcal{KL}$ function β such that for all initial conditions $X(t_0) \in \mathbb{R}^n$, $U(t_0+\theta); \theta \in [-D,0] \in L_\infty[-D,0]$ the following holds:*

$$\Omega(t) \leq \beta(\Omega(t_0), t - t_0), \tag{5.32}$$

$$\Omega(t) = |X(t)| + \sup_{t-D \leq \theta \leq t} |U(\theta)| \tag{5.33}$$

for all $t \geq t_0 \geq 0$.

The proof of Theorem 5.3 is based on the following two lemmas. The first one is concerned with the stability of the target system (5.21), (5.22).

Lemma 5.4. *There exists a class $\mathcal{KL}$ function β_3 such that*

$$|X(t)| + \sup_{t-D \leq \theta \leq t} |W(\theta)| \leq \beta_3\left(|X(t_0)| + \sup_{t_0-D \leq \theta \leq t_0} |W(\theta)|, t - t_0 \right) \tag{5.34}$$

for all $t \geq t_0 \geq 0$.

Proof. Under Assumption 5.2 and using Theorem C.16 there exist a C^1 function $S : \mathbb{R}_+ \times \mathbb{R}^n \to \mathbb{R}_+$ and class $\mathcal{K}_\infty$ functions α_4, α_5, α_6, α_7 such that

$$\alpha_4(|X(t)|) \leq S(t, X(t)) \leq \alpha_5(|X(t)|), \tag{5.35}$$

$$\dot{S}(t, X(t)) \leq -\alpha_6(|X(t)|) + \alpha_7(|W(t-D)|), \tag{5.36}$$

$$\dot{S}(t, X(t)) = \frac{\partial S(t,X(t))}{\partial t} + \frac{\partial S(t,X(t))}{\partial X} \times f(X(t), \kappa(t, X(t)) + W(t-D)). \tag{5.37}$$

The Lyapunov functional for the target system is then

$$V(t) = S(t, X(t)) + \frac{2}{c} \int_0^{L(t)} \frac{\alpha_7(r)}{r} dr, \tag{5.38}$$

where $\frac{\alpha_7(r)}{r}$ is a class $\mathcal{K}$ function or α_7 has been appropriately majorized so that this is true (with no loss of generality), $c > 0$ is arbitrary, and

$$\begin{aligned} L(t) &= \sup_{t-D \leq \theta \leq t} \left| e^{c(\theta - t + D)} W(\theta) \right| \\ &= \lim_{n \to \infty} \left(\int_{t-D}^{t} e^{2nc(\theta + D - t)} W(\theta)^{2n} d\theta \right)^{\frac{1}{2n}}. \end{aligned} \tag{5.39}$$

We now upper- and lower-bound $L(t)$ in terms of $\sup_{t-D\le\theta\le t}|W(\theta)|$. Since $0\le\theta+D-t\le D$ for all $t-D\le\theta\le t$, we get that

$$L(t)\le e^{cD}\sup_{t-D\le\theta\le t}|W(\theta)|, \tag{5.40}$$

$$L(t)\ge \sup_{t-D\le\theta\le t}|W(\theta)|. \tag{5.41}$$

Taking the time derivative of $L(t)$, with (5.10) we get

$$\begin{aligned}\dot{L}(t)= \lim_{n\to\infty}\frac{1}{2n}\left(\int_{t-D}^{t} e^{2nc(\theta+D-t)}W(\theta)^{2n}d\theta\right)^{\frac{1}{2n}-1}\\ \times\left(-W(t-D)^{2n}-2nc\int_{t-D}^{t} e^{2nc(\theta+D-t)}W(\theta)^{2n}d\theta\right),\end{aligned} \tag{5.42}$$

and hence

$$\dot{L}(t)\le -cL(t). \tag{5.43}$$

With this inequality and (5.36), taking the derivative of (5.38) we get

$$\dot{V}(t)\le -\alpha_6(|X(t)|)+\alpha_7(|W(t-D)|)-2\alpha_7(L(t)). \tag{5.44}$$

With the help of (5.41) we get

$$\dot{V}(t)\le -\alpha_6(|X(t)|)-\alpha_7(L(t)). \tag{5.45}$$

Using (5.35), the definition of $L(t)$ in (5.39), and (5.38) we conclude that there exists a class $\mathcal{K}$ function γ_1 such that $\dot{V}(t)\le-\gamma_1(V(t))$. Using the comparison principle (Lemma B.7 in Appendix B) and Lemma C.6, there exists a class $\mathcal{KL}$ function β_1 such that

$$V(t)\le\beta_1(V(t_0),t-t_0). \tag{5.46}$$

Using (5.35), the definition of $V(t)$ in (5.38), and the properties of class $\mathcal{K}$ functions we arrive at

$$|X(t)|+L(t)\le\beta_2(|X(t_0)|+L(t_0),t-t_0) \tag{5.47}$$

for some class $\mathcal{KL}$ function β_2. Using relations (5.40) and (5.41) we arrive at (5.34) for some class $\mathcal{KL}$ function β_3. □

We relate next the stability of the system in the transformed variables with the stability of the system in the original variables. We have the following lemma.

Lemma 5.5. *There exist class $\mathcal{K}_\infty$ functions α_9 and α_{10} such that*

$$\Omega(t)\le\alpha_{10}(\Xi(t)), \tag{5.48}$$

$$\Xi(t)\le\alpha_9(\Omega(t)) \tag{5.49}$$

for all $t\ge t_0$, where Ω is defined in (5.33) *and*

$$\Xi=|X(t)|+\sup_{t-D\le\theta\le t}|W(\theta)|. \tag{5.50}$$

Proof. From Assumption 5.2, since $\kappa(t,X)$ is periodic in t and locally Lipschitz with $k(t,0)=0$ for all $t\geq 0$, there exists a class $\mathcal{K}_\infty$ function $\hat{\alpha}$ such that

$$|\kappa(t,\xi)|\leq\hat{\alpha}(|\xi|)\quad\text{for all } t\geq 0. \tag{5.51}$$

Using the backstepping transformation (5.8) and relation (5.51) we get that

$$|X(t)|+\sup_{t-D\leq\theta\leq t}|W(\theta)|\leq|X(t)|+\sup_{t-D\leq\theta\leq t}|U(\theta)|+\hat{\alpha}\left(\sup_{t-D\leq\theta\leq t}|P(\theta)|\right). \tag{5.52}$$

Analogously with the inverse backstepping transformation (5.11) and (5.51), we get that

$$|X(t)|+\sup_{t-D\leq\theta\leq t}|U(\theta)|\leq|X(t)|+\sup_{t-D\leq\theta\leq t}|W(\theta)|+\hat{\alpha}\left(\sup_{t-D\leq\theta\leq t}|\Pi(\theta)|\right). \tag{5.53}$$

We relate next the norm of the predictor P with the norm of the system in the (X,U) variables, i.e., with Ω defined in (5.33) and the norm of Π with the norm of the system in the (X,W) variables, i.e., with Ξ defined in (5.50). We first note from (5.31) that

$$\frac{\partial R(P)}{\partial P}f(P,U)\leq R(P)+\alpha_3(|U|). \tag{5.54}$$

Since from (5.7) one can conclude that P satisfies in θ the ODE

$$\frac{dP(\theta)}{d\theta}=f(P(\theta),U(\theta))\quad\text{for all } t-D\leq\theta\leq t, \tag{5.55}$$

we have that

$$\frac{dR(P(\theta))}{d\theta}\leq R(P(\theta))+\alpha_3(|U(\theta)|)\quad\text{for all } t-D\leq\theta\leq t. \tag{5.56}$$

Using the comparison principle (Lemma B.7 in Appendix B) and noting that $P(t-D)=X(t)$ we get for all $\theta\in[t-D,t]$

$$R(P(\theta))\leq e^{\theta-t+D}R(X(t))+\int_{t-D}^{\theta}e^{\theta-s}\alpha_3(|U(s)|)\,ds. \tag{5.57}$$

With the help of (5.30) we arrive at

$$|P(\theta)|\leq\alpha_8\left(|X(t)|+\sup_{t-D\leq\theta\leq t}|U(\theta)|\right)\quad\text{for all } t-D\leq\theta\leq t, \tag{5.58}$$

where the class $\mathcal{K}_\infty$ function α_8 is given as

$$\alpha_8(s)=\alpha_1^{-1}\left(e^D(\alpha_2(s)+\alpha_3(s))\right). \tag{5.59}$$

Therefore, using (5.52) and (5.59) we arrive at

$$|X(t)|+\sup_{t-D\leq\theta\leq t}|W(\theta)|\leq\alpha_9\left(|X(t)|+\sup_{t-D\leq\theta\leq t}|U(\theta)\right), \tag{5.60}$$

where $\alpha_9 \in \mathscr{K}_\infty$ is given by

$$\alpha_9(s) = s + \hat{\alpha}(\alpha_8(s)). \tag{5.61}$$

We proceed analogously for finding bound (5.49). We first observe that Π in (5.12) satisfies the following ODE in θ:

$$\frac{d\Pi(\theta)}{d\theta} = f(\Pi(\theta), \varkappa(\theta + D, \Pi(\theta)) + W(\theta)). \tag{5.62}$$

From Assumption 5.2, [38], and the fact that $\Pi(t-D) = X(t)$, we get that there exist a class $\mathscr{K}\mathscr{L}$ function β_4 and a class $\mathscr{K}$ function γ_2 such that for all $t - D \le \theta \le t$

$$|\Pi(\theta)| \le \beta_4(|X(t)|, \theta - t + D) + \gamma_2\left(\sup_{t-D\le s\le t} |W(s)|\right). \tag{5.63}$$

Therefore,

$$|\Pi(\theta)| \le \gamma_3\left(|X(t)| + \sup_{t-D\le\theta\le t} |W(\theta)|\right) \quad \text{for all } t - D \le \theta \le t, \tag{5.64}$$

where the class $\mathscr{K}$ function γ_3 is defined as

$$\gamma_3(s) = \beta_4(s, 0) + \gamma_2(s). \tag{5.65}$$

Using (5.53) we arrive at

$$|X(t)| + \sup_{t-D\le\theta\le t} |U(\theta)| \le \alpha_{10}\left(|X(t)| + \sup_{t-D\le\theta\le t} |W(\theta)|\right), \tag{5.66}$$

where the class $\mathscr{K}_\infty$ function α_{10} is defined as

$$\alpha_{10}(s) = s + \hat{\alpha}(\gamma_3(s)). \qquad \square \tag{5.67}$$

The proof of Theorem 5.3 is completed by combining (5.34) with (5.60), (5.66) in order to get estimate (5.32) with

$$\beta(s, t) = \alpha_{10}(\beta_3(\alpha_9(s), t)). \tag{5.68}$$

The stability estimate in Theorem 5.3 is given in standard delay notation. One can use the PDE representation of the plant to prove the following corollary.

Corollary 5.6. *Consider the plant* (5.14)–(5.16) *together with the control law* (5.20), (5.17) *and let Assumptions* 5.1 *and* 5.2 *hold. For all initial conditions* $X(t_0) \in \mathbb{R}^n$, $u(\cdot, t_0) \in L^\infty[0, D]$ *the following holds:*

$$|X(t)| + \|u(t)\|_{L^\infty[0,D]} \le \beta\left(|X(t_0)| + \|u(t_0)\|_{L^\infty[0,D]}, t - t_0\right), \tag{5.69}$$

$$\|u(t)\|_{L^\infty[0,D]} = \sup_{x\in[0,D]} |u(x, t)| \tag{5.70}$$

for all $t \ge t_0 \ge 0$.

The proof of the corollary is immediate by noting that $u(x, t) = U(t + x - D)$ for all $x \in [0, D]$ and by using estimate (5.32).

5.4 ▪ Notes and References

For proving our main result, i.e., Theorem 5.3, we assumed strong forward completeness and input-to-state stability (see Assumptions 5.1 and 5.2, respectively). With these assumptions we are able to construct a Lyapunov functional for the closed-loop system under predictor feedback. The availability of a Lyapunov functional enables one, in principle, to study robustness of the predictor feedback to parametric uncertainties, its disturbance attenuation properties, and the inverse-optimal redesign problem. However, global uniform asymptotic stability for the closed-loop system under predictor feedback can be proved by only assuming standard forward completeness and global uniform stabilizability of the delay-free system. The proof can be found in [89].

One of the key ingredients used in the construction of a Lyapunov functional is the construction of the functional L which is given in terms of the transformed variables (X, W) in (5.39). This functional can be also written directly in terms of the original variables (X, U) as

$$L(t) = \sup_{t-D \leq \theta \leq t} \left| e^{c(\theta - t + D)} \left(U(\theta) - \varkappa(\theta + D, P(\theta)) \right) \right|, \tag{5.71}$$

where P is given in terms of (X, U) from (5.7). The two different representations of the functional L, namely, representations (5.39) and (5.71), reveal one of the benefits of the backstepping transformation: If the construction of the functional L in terms of the transformed actuator state W appears to be nontrivial, its form in terms of the original variables (X, U), i.e., relation (5.71), is rather impossible to guess without the backstepping and predictor transformations.

Control designs for nonlinear systems with state delays were developed by Jankovic [56], [57], [58], Karafyllis and coauthors [63], [64], [68], [72], [73], Mazenc and Malisoff [106], and Pepe and coauthors [139], [140], [178]. For nonlinear systems with input delays, control techniques were developed by Karafyllis and coauthors [63], [65], [69] and Mazenc and coauthors [104], [108], [109], [110], [111], [112]. However, only the papers [63], [69] and [107], [109] deal with *long* input delays. Among them, the idea of compensating for a delay was employed in [69], whereas a time-varying distributed delay feedback law was employed in [63]. Papers [107], [109] use a low gain idea rather than compensating for the delay.

Part II

Time-Varying Delays

Chapter 6

Linear Systems with Time-Varying Input Delay

In this chapter we make the first step towards the design of control laws for systems with nonconstant delays. Two major challenges arise in the case of a time-varying delay. The first challenge is how to determine the predictor state. We resolve this challenge by computing the predictor over a nonconstant prediction horizon which we appropriately define. The predictor is then given explicitly by using the variation of constants formula starting with the current state as an initial condition. The second challenge lies in the stability analysis of the closed-loop system under predictor feedback. We resolve this challenge by introducing a backstepping transformation with time-varying kernels, and by appropriately defining a transport PDE representation of the actuator state using a time- and spatial-varying transport speed. This backstepping transformation enables one to construct a Lyapunov functional which proves the stability of the closed-loop system.

We start in the present chapter by considering linear systems with time-varying input delay, probably the simplest form of systems with nonconstant delays. For this class of systems, we design the predictor feedback law which compensates for the time-varying input delay in Section 6.1. We prove stability of the closed-loop system in Section 6.2. The stability analysis is based on the introduction of the backstepping transformation, which allows the construction of a Lyapunov functional for the closed-loop system. In this chapter we use the PDE representation of the actuator state, which reveals an explicit dependence of the transport speed coefficient from both the time and space.

6.1 ▪ Predictor Feedback Design for Time-Varying Input Delay

We consider the system

$$\dot{X}(t) = AX(t) + BU(\phi(t)), \tag{6.1}$$

where $X \in \mathbb{R}^n$ is the state, U is the control input, and $\phi(t)$ is a continuously differentiable function that incorporates the actuator delay. This function will have to satisfy certain conditions that we shall impose in our development, in particular that

$$\phi(t) \le t \quad \text{for all } t \ge 0. \tag{6.2}$$

One can alternatively view the function $\phi(t)$ in the more standard form

$$\phi(t) = t - D(t), \tag{6.3}$$

where $D(t)\geq 0$ is a time-varying delay. However, the formalism involving the function $\phi(t)$ turns out to be more convenient, particularly because the predictor problem requires the inverse function of $\phi(t)$, i.e., $\phi^{-1}(t)$, so we will proceed with the model (6.1). The invertibility of $\phi(\cdot)$ will be ensured by imposing the following assumption.

Assumption 6.1. $\phi:\mathbb{R}_+\to\mathbb{R}$ *is a continuously differentiable function that satisfies*

$$\phi'(t)>0 \quad \textit{for all } t\geq 0, \tag{6.4}$$

and such that

$$\pi_1^*\frac{1}{\sup_{\vartheta\geq\phi^{-1}(0)}\phi'(\vartheta)}>0. \tag{6.5}$$

The meaning of the assumption is that the function $\phi(t)$ is strictly increasing, which, as we shall see, we need in several elements of our analysis.

The main premise of the predictor-based design is that one generates the control input

$$U(\phi(t))=KX(t) \quad \text{for all } \phi(t)\geq 0, \tag{6.6}$$

so that the closed-loop system is

$$\dot{X}(t)=(A+BK)X(t) \quad \text{for all } \phi(t)\geq 0, \tag{6.7}$$

or, alternatively, using the inverse of $\phi(\cdot)$,

$$\dot{X}(t)=(A+BK)X(t) \quad \text{for all } t\geq\phi^{-1}(0). \tag{6.8}$$

The gain vector K is selected so that the system matrix $A+BK$ is Hurwitz.

We now rewrite (6.6) as

$$U(t)=KX(\sigma(t)) \quad \text{for all } t\geq 0, \tag{6.9}$$

where

$$\sigma(t)=\phi^{-1}(t). \tag{6.10}$$

Performing a change of variables in (6.1) as

$$t=\sigma(\theta), \tag{6.11}$$

we get

$$\frac{dX(\sigma(\theta))}{d\theta}=\frac{d\sigma(\theta)}{d\theta}(AX(\sigma(\theta))+BU(\theta)) \quad \text{for all } \phi(t)\leq\theta\leq t. \tag{6.12}$$

Solving explicitly this relation starting at $\theta = \phi(t)$ and defining the predictor state as

$$P(\theta) = X(\sigma(\theta)) \quad \text{for all } \phi(t) \leq \theta \leq t, \tag{6.13}$$

which, with the help of (6.10), also implies that $P(\phi(t)) = X(t)$, we get for all $\phi(t) \leq \theta \leq t$

$$P(\theta) = e^{A(\sigma(\theta)-t)} X(t) + \int_{\phi(t)}^{\theta} \dot{\sigma}(\theta) e^{A(\sigma(\theta)-\sigma(s))} BU(s) ds. \tag{6.14}$$

Using relation (6.10) and the fact that

$$\frac{d}{d\theta} \phi^{-1}(\theta) = \frac{1}{\phi'(\phi^{-1}(\theta))}, \tag{6.15}$$

the predictor of X is given by

$$P(t) = e^{A(\phi^{-1}(t)-t)} X(t) + \int_{\phi(t)}^{t} e^{A(\phi^{-1}(t)-\phi^{-1}(\theta))} B \frac{U(\theta)}{\phi'(\phi^{-1}(\theta))} d\theta. \tag{6.16}$$

Substituting this expression into the control law (6.9), we obtain the predictor feedback

$$U(t) = K \left[e^{A(\phi^{-1}(t)-t)} X(t) + \int_{\phi(t)}^{t} e^{A(\phi^{-1}(t)-\phi^{-1}(\theta))} B \frac{U(\theta)}{\phi'(\phi^{-1}(\theta))} d\theta \right] \quad \text{for all } t \geq 0. \tag{6.17}$$

The division by $\phi'(\phi^{-1}(\theta))$ in this compensator is safe thanks to the assumption (6.4).

We refer to the quantity

$$t - \phi(t) \tag{6.18}$$

as the *delay time* and to the quantity

$$\sigma(t) - t \tag{6.19}$$

as the *prediction horizon*.

Remark 6.2. *To make sure the above discussion is completely clear, we point out that when the system has a constant delay,*

$$\phi(t) = t - D, \tag{6.20}$$

we have

$$\begin{aligned} \sigma(t) &= \phi^{-1}(t) \\ &= t + D \end{aligned} \tag{6.21}$$

and

$$\phi'(\phi^{-1}(\theta)) = 1. \tag{6.22}$$

Hence, controller (6.17) *reduces to* (2.9).

6.2 ▪ Stability Analysis for Time-Varying Delays

In our stability analysis we will use the transport equation representation of the delay and a Lyapunov construction.

First, we introduce the following fairly nonobvious choice for the state of the transport equation:

$$u(x,t) = U(\phi(t + x(\phi^{-1}(t) - t))). \tag{6.23}$$

This choice yields boundary values

$$u(0,t) = U(\phi(t)), \tag{6.24}$$
$$u(1,t) = U(t). \tag{6.25}$$

System (6.1) can now be represented as

$$\dot{X}(t) = AX(t) + Bu(0,t), \tag{6.26}$$
$$u_t(x,t) = \pi(x,t)u_x(x,t), \tag{6.27}$$
$$u(1,t) = U(t), \tag{6.28}$$

where the speed of propagation of the transport equation is given by

$$\pi(x,t) = \frac{1 + x\left(\frac{d(\phi^{-1}(t))}{dt} - 1\right)}{\phi^{-1}(t) - t}. \tag{6.29}$$

To obtain a meaningful stability result, we need the propagation speed function $\pi(x,t)$ to be strictly positive and uniformly bounded from below and from above by finite constants. Guided by the concern for boundedness from above, we examine the denominator $\phi^{-1}(t) - t$. Since we assumed that $\phi(t)$ is strictly increasing (and continuous), so is $\phi^{-1}(t)$. We also recall assumption (6.2). We need to make this inequality strict, since if $\phi(t) = t$, i.e., $\phi^{-1}(t) = t$ for any t, the propagation speed is infinite at that time instant and the transport PDE representation does not make sense for the study of the stability problem. Hence, we make the following assumption.

Assumption 6.3. $\phi : \mathbb{R}_+ \to \mathbb{R}$ *is a continuously differentiable function that satisfies*

$$\phi(t) < t, \quad t \geq 0, \tag{6.30}$$

and such that

$$\pi_0^* = \frac{1}{\sup_{\vartheta \geq \phi^{-1}(0)} (\vartheta - \phi(\vartheta))} > 0. \tag{6.31}$$

Assumption 6.3 can be alternatively stated as

$$\phi^{-1}(t) - t > 0. \tag{6.32}$$

The implication of the assumption on the delay time and prediction horizon is that they are both positive and uniformly bounded.

Now we return to system (6.26)–(6.28), the definition of the transport PDE state (6.23), and the control law (6.17). The control law (6.17) is written in terms of $u(x,t)$ as

$$u(1,t)=K\left[\mathrm{e}^{A\left(\phi^{-1}(t)-t\right)}X(t)\right.$$
$$\left.+\int_0^1 \mathrm{e}^{A(1-y)\left(\phi^{-1}(t)-t\right)}Bu(y,t)\left(\phi^{-1}(t)-t\right)dy\right]. \tag{6.33}$$

We next derive an equivalent representation of the predictor state P using PDE notation. Setting in (6.12) $\theta=\phi(t+x(\sigma(t)-t))$, using the definition $\sigma(t)=\phi^{-1}(t)$, and defining $P(\phi(t+x(\sigma(t)-t)))=p(x,t)$, we get for all $x\in[0,1]$ that

$$\frac{\partial p(x,t)}{\partial x}=(\sigma(t)-t)(Ap(x,t)+Bu(x,t)), \tag{6.34}$$
$$p(0,t)=X(t), \tag{6.35}$$

and hence we also get for all $x\in[0,1]$ that

$$p(x,t)=X(t)+(\sigma(t)-t)\int_0^x (Ap(y,t)+Bu(y,t))\,dy. \tag{6.36}$$

Solving explicitly the boundary value problem in x (6.34), (6.35) we get that

$$p(x,t)=\mathrm{e}^{Ax(\sigma(t)-t)}X(t)+(\sigma(t)-t)\int_0^x \mathrm{e}^{A(x-y)(\sigma(t)-t)}Bu(y,t)\,dy, \tag{6.37}$$

and hence the control law (6.33) can be also written as

$$u(1,t)=Kp(1,t). \tag{6.38}$$

In order to study the exponential stability of the system $(X(t),u(x,t),x\in[0,1])$, we introduce the initial condition

$$u_0(x)=u(x,0)=U\left(\phi\left(\phi^{-1}(0)x\right)\right),\quad x\in[0,1], \tag{6.39}$$

and $X_0=X(0)$.

Now we establish the following stability result.

Theorem 6.4. *Consider the closed-loop system consisting of the plant* (6.26)–(6.28) *and the controller* (6.33), *and let Assumptions* 6.1 *and* 6.3 *hold. There exist a positive constant G and a positive constant g independent of the function $\phi(\cdot)$, such that*

$$|X(t)|^2+\|u(t)\|^2\leq G\mathrm{e}^{-gt}\left(|X_0|^2+\|u_0\|^2\right)\quad \textit{for all } t\geq 0. \tag{6.40}$$

Proof. Consider the transformation of the transport PDE state given by

$$\begin{aligned} w(x,t) &= u(x,t) - Kp(x,t) \\ &= u(x,t) - Ke^{Ax(\phi^{-1}(t)-t)}X(t) \\ &\quad - K\int_0^x e^{A(x-y)(\phi^{-1}(t)-t)}Bu(y,t)(\phi^{-1}(t)-t)dy. \end{aligned} \tag{6.41}$$

Taking the derivatives of $w(x,t)$ with respect to t and x, we get

$$\begin{aligned} w_t(x,t) &= u_t(x,t) - KAx\left(\frac{d(\phi^{-1}(t))}{dt} - 1\right)e^{Ax(\phi^{-1}(t)-t)}X(t) \\ &\quad - Ke^{Ax(\phi^{-1}(t)-t)}(X(t)+Bu(0,t)) - K\int_0^x e^{A(x-y)(\phi^{-1}(t)-t)} \\ &\quad \times \left(A(x-y)(\phi^{-1}(t)-t)+I\right)\left(\frac{d(\phi^{-1}(t))}{dt} - 1\right)Bu(y,t)dy \\ &\quad - K\int_0^x e^{A(x-y)(\phi^{-1}(t)-t)}Bu_t(y,t)(\phi^{-1}(t)-t)dy \\ &= u_t(x,t) - \left(1 + x\left(\frac{d(\phi^{-1}(t))}{dt} - 1\right)\right)K\Big[Ae^{Ax(\phi^{-1}(t)-t)}X(t) \\ &\quad + A\int_0^x e^{A(x-y)(\phi^{-1}(t)-t)}Bu(y,t)(\phi^{-1}(t)-t)dy + Bu(x,t)\Big], \end{aligned} \tag{6.42}$$

where we have used integration by parts, and

$$\begin{aligned} w_x(x,t) &= u_x(x,t) - (\phi^{-1}(t)-t)K\Big[Ae^{Ax(\phi^{-1}(t)-t)}X(t) \\ &\quad + A\int_0^x e^{A(x-y)(\phi^{-1}(t)-t)}Bu(y,t)(\phi^{-1}(t)-t)dy + Bu(x,t)\Big]. \end{aligned} \tag{6.43}$$

With the help of (6.33), we also obtain $w(1,t)=0$; hence, we arrive at the "target system"

$$\dot{X}(t) = (A+BK)X(t) + Bw(0,t), \tag{6.44}$$

$$w_t(x,t) = \pi(x,t)w_x(x,t), \tag{6.45}$$

$$w(1,t) = 0. \tag{6.46}$$

This is a standard cascade configuration,

$$w \to X, \tag{6.47}$$

that we have encountered many times before. We focus first on the Lyapunov analysis of the w-subsystem. We take a Lyapunov function

$$L(t) = \frac{1}{2}\int_0^1 e^{bx}w^2(x,t)dx, \tag{6.48}$$

where b is any positive constant. The time derivative of $L(t)$ is

$$
\begin{aligned}
\dot{L}(t) &= \int_0^1 e^{bx} w(x,t) w_t(x,t) dx \\
&= \int_0^1 e^{bx} w(x,t) \pi(x,t) w_x(x,t) dx \\
&= \frac{1}{2} \int_0^1 e^{bx} \pi(x,t) d w^2(x,t) \\
&= \frac{e^{bx}}{2} \pi(x,t) w^2(x,t) \Big|_0^1 - \frac{1}{2} \int_0^1 (b\pi(x,t) + \pi_x(x,t)) e^{bx} w^2(x,t) dx \\
&= -\frac{\pi(0,t)}{2} w^2(0,t) - \frac{1}{2} \int_0^1 (b\pi(x,t) + \pi_x(x,t)) e^{bx} w^2(x,t) dx .
\end{aligned}
\tag{6.49}
$$

Noting that

$$
\pi(0,t) = \frac{1}{\phi^{-1}(t) - t} \geq \pi_0^*, \tag{6.50}
$$

we get

$$
\dot{L}(t) \leq -\frac{\pi_0^*}{2} w^2(0,t) - \frac{1}{2} \int_0^1 (b\pi(x,t) + \pi_x(x,t)) e^{bx} w^2(x,t) dx . \tag{6.51}
$$

Next, we observe that

$$
\pi_x(x,t) = \frac{\dfrac{d\left(\phi^{-1}(t)\right)}{dt} - 1}{\phi^{-1}(t) - t} \tag{6.52}
$$

is a function of t only. Hence,

$$
\begin{aligned}
b\pi(x,t) + \pi_x(x,t) &= \frac{b\left[1 + x\left(\dfrac{d\left(\phi^{-1}(t)\right)}{dt} - 1\right)\right] + \dfrac{d\left(\phi^{-1}(t)\right)}{dt} - 1}{\phi^{-1}(t) - t} \\
&= \frac{b - 1 + \dfrac{d\left(\phi^{-1}(t)\right)}{dt} + bx\left(\dfrac{d\left(\phi^{-1}(t)\right)}{dt} - 1\right)}{\phi^{-1}(t) - t} .
\end{aligned}
\tag{6.53}
$$

Since this is a linear function of x, it follows that it has a minimum at either $x = 0$ or $x = 1$, so we get

$$
b\pi(x,t) + \pi_x(x,t) \geq \frac{\min\left\{ b - 1 + \dfrac{d\left(\phi^{-1}(t)\right)}{dt},\ (b+1)\dfrac{d\left(\phi^{-1}(t)\right)}{dt} - 1 \right\}}{\phi^{-1}(t) - t} . \tag{6.54}
$$

Next, we note that

$$\begin{aligned}\frac{d\left(\phi^{-1}(t)\right)}{dt} &= \frac{1}{\phi'\left(\phi^{-1}(t)\right)}\\ &\geq \frac{1}{\sup_{\vartheta\geq\phi^{-1}(0)}\phi'(\vartheta)}\\ &= \pi_1^*,\end{aligned} \tag{6.55}$$

which yields

$$b\pi(x,t)+\pi_x(x,t)\geq \frac{\min\left\{b-1+\pi_1^*,\ (b+1)\pi_1^*-1\right\}}{\phi^{-1}(t)-t}. \tag{6.56}$$

Choosing

$$b\geq\left(1-\pi_1^*\right)\max\left\{1,\ \frac{1}{\pi_1^*}\right\}, \tag{6.57}$$

we get

$$b\pi(x,t)+\pi_x(x,t)\geq \pi_0^*\beta^*, \tag{6.58}$$

where

$$\beta^*=\min\left\{b-1+\pi_1^*,\ (b+1)\pi_1^*-1\right\}>0. \tag{6.59}$$

So, returning to $\dot{L}(t)$, we have

$$\dot{L}(t)\leq -\frac{\pi_0^*}{2}w^2(0,t)-\pi_0^*\beta^*L(t). \tag{6.60}$$

Let us now turn our attention to the X-subsystem. We have

$$\frac{d}{dt}\left(X(t)^TPX(t)\right)=-X^T(t)QX(t)+2X^T(t)PBw(0,t), \tag{6.61}$$

where P satisfies a Lyapunov equation

$$P(A+BK)+(A+BK)^TP=-Q. \tag{6.62}$$

With a usual completion of squares, we get

$$\frac{d}{dt}\left(X(t)^TPX(t)\right)\leq -\lambda_{\min}(Q)|X(t)|^2+\frac{2|PB|^2}{\lambda_{\min}(Q)}w^2(0,t). \tag{6.63}$$

Now we take the Lyapunov–Krasovskii functional

$$V(t)=X(t)^TPX(t)+\frac{4|PB|^2}{\pi_0^*\lambda_{\min}(Q)}L(t). \tag{6.64}$$

Its derivative is

$$\dot{V}(t)\leq -\frac{\lambda_{\min}(Q)}{2}|X(t)|^2-\pi_0^*\beta^*\frac{4|PB|^2}{\pi_0^*\lambda_{\min}(Q)}L(t). \tag{6.65}$$

Finally, with the definition of $V(t)$, we get

$$\dot{V}(t) \leq -\mu V(t), \tag{6.66}$$

where

$$\mu = \min\left\{\pi_0^*\beta^*, \frac{\lambda_{\min}(Q)}{2\lambda_{\max}(P)}\right\}. \tag{6.67}$$

Thus, we obtain

$$V(t) \leq \mathrm{e}^{-\mu t} V(0) \quad \text{for all } t \geq 0. \tag{6.68}$$

Let us now denote

$$\Omega(t) = |X(t)|^2 + \int_0^1 w^2(x,t)dx. \tag{6.69}$$

The following relation holds between $V(t)$ and $\Omega(t)$:

$$\psi_1\Omega(t) \leq V(t) \leq \psi_2\Omega(t), \tag{6.70}$$

where

$$\psi_1 = \min\left\{\lambda_{\min}(P), \frac{2|PB|^2}{\pi_0^*\lambda_{\min}(Q)}\right\}, \tag{6.71}$$

$$\psi_2 = \max\left\{\lambda_{\max}(P), \frac{2|PB|^2}{\pi_0^*\lambda_{\min}(Q)}\mathrm{e}^b\right\}. \tag{6.72}$$

It then follows that

$$\Omega(t) \leq \frac{\psi_2}{\psi_1}\mathrm{e}^{-\mu t}\Omega(0) \quad \text{for all } t \geq 0. \tag{6.73}$$

Now we consider the norm

$$\Xi(t) = |X(t)|^2 + \int_0^1 u^2(x,t)dx. \tag{6.74}$$

We recall the backstepping transformation (6.41) and introduce its inverse,

$$\begin{aligned} u(x,t) = {} & w(x,t) + K\mathrm{e}^{(A+BK)x\left(\phi^{-1}(t)-t\right)}X(t) \\ & + K\int_0^x \mathrm{e}^{(A+BK)(x-y)\left(\phi^{-1}(t)-t\right)}Bu(y,t)\left(\phi^{-1}(t)-t\right)dy. \end{aligned} \tag{6.75}$$

It can be shown that

$$\|w(t)\|^2 \leq \alpha_1(t)\|u(t)\|^2 + \alpha_2|X(t)|^2, \tag{6.76}$$

$$\|u(t)\|^2 \leq \beta_1(t)\|w(t)\|^2 + \beta_2|X(t)|^2, \tag{6.77}$$

where

$$\alpha_1(t) = 3\left(1 + \int_0^1 \left(KM\left(x\left(\phi^{-1}(t) - t\right)\right)B\left(\phi^{-1}(t) - t\right)\right)^2 dx\right), \tag{6.78}$$

$$\alpha_2(t) = 3\int_0^1 \left|KM\left(x\left(\phi^{-1}(t) - t\right)\right)\right|^2 dx, \tag{6.79}$$

$$\beta_1(t) = 3\left(1 + \int_0^1 \left(KN\left(x\left(\phi^{-1}(t) - t\right)\right)B\left(\phi^{-1}(t) - t\right)\right)^2 dx\right), \tag{6.80}$$

$$\beta_2(t) = 3\int_0^1 \left|KN\left(x\left(\phi^{-1}(t), -t\right)\right)\right|^2 dx, \tag{6.81}$$

and where

$$M(s) = \mathrm{e}^{As}, \tag{6.82}$$

$$N(s) = \mathrm{e}^{(A+BK)s}. \tag{6.83}$$

Furthermore, we can show that

$$\alpha_1(t) \le \bar{\alpha}_1 = 3\left(1 + |K|^2|B|^2 \frac{\mathrm{e}^{\frac{2|A|}{\pi_0^*}} - 1}{2\pi_0^*|A|}\right), \tag{6.84}$$

$$\alpha_2(t) \le \bar{\alpha}_2 = 3|K|^2\pi_0^* \frac{\mathrm{e}^{\frac{2|A|}{\pi_0^*}} - 1}{2|A|}, \tag{6.85}$$

$$\beta_1(t) \le \bar{\beta}_1 = 3\left(1 + |K|^2|B|^2 \frac{\mathrm{e}^{\frac{2|A+BK|}{\pi_0^*}} - 1}{2\pi_0^*|A+BK|}\right), \tag{6.86}$$

$$\beta_2(t) \le \bar{\beta}_2 = 3|K|^2\pi_0^* \frac{\mathrm{e}^{\frac{2|A+BK|}{\pi_0^*}} - 1}{2|A+BK|}. \tag{6.87}$$

With a few substitutions, we obtain

$$\phi_1 \Xi(t) \le \Omega(t) \le \phi_2 \Xi(t), \tag{6.88}$$

where

$$\phi_1 = \frac{1}{\max\left\{\bar{\beta}_1, \bar{\beta}_2 + 1\right\}}, \tag{6.89}$$

$$\phi_2 = \max\left\{\bar{\alpha}_1, \bar{\alpha}_2 + 1\right\}. \tag{6.90}$$

Finally, we get

$$\Xi(t) \le \frac{\phi_2 \psi_2}{\phi_1 \psi_1} \mathrm{e}^{-\mu t} \Xi(0) \quad \text{for all } t \ge 0, \tag{6.91}$$

which completes the proof of the theorem with

$$G = \frac{\phi_2 \psi_2}{\phi_1 \psi_1} \tag{6.92}$$

and

$$g = \mu . \tag{6.93}$$

By choosing

$$\beta^* \geq \frac{\lambda_{\min}(Q)}{2\pi_0^* \lambda_{\max}(P)}, \tag{6.94}$$

i.e., by picking b positive and such that

$$b \geq \left(1 - \pi_1^* + \frac{\lambda_{\min}(Q)}{2\pi_0^* \lambda_{\max}(P)}\right) \max\left\{1, \frac{1}{\pi_1^*}\right\}, \tag{6.95}$$

we get

$$g = \frac{\lambda_{\min}(Q)}{2\lambda_{\max}(P)}, \tag{6.96}$$

which means that g is independent of the function $\phi(\cdot)$. □

While Theorem 6.4 provides a stability result in terms of the system norm

$$|X(t)|^2 + \int_0^1 u^2(x,t)dx, \tag{6.97}$$

we would like also to get a stability result in terms of the norm

$$|X(t)|^2 + \int_{\phi(t)}^t U^2(\theta)d\theta. \tag{6.98}$$

Toward that end, we first observe that

$$\int_{\phi(t)}^t U^2(\theta)d\theta = \left(\phi^{-1}(t) - t\right)\int_0^1 \phi'\left(t + x\left(\phi^{-1}(t) - t\right)\right)u^2(x,t)dx, \tag{6.99}$$

$$\int_0^2 u_0^2(x)dx = \frac{1}{\phi^{-1}(0)}\int_{\phi(0)}^0 \frac{1}{\phi'\left(\phi^{-1}(\theta)\right)}U^2(\theta)d\theta. \tag{6.100}$$

With these identities, we obtain the following theorem from Theorem 6.4.

Theorem 6.5. *Consider the closed-loop system consisting of the plant* (6.26)–(6.28) *and the controller* (6.33) *and let Assumptions* 6.1 *and* 6.3 *hold. There exist positive constants G and g (the latter being independent of ϕ) such that*

$$|X(t)|^2 + \int_{\phi(t)}^t U^2(\theta)d\theta \leq hGe^{-gt}\left(|X_0|^2 + \int_{\phi(0)}^0 U^2(\theta)d\theta\right) \quad \textit{for all } t \geq 0, \tag{6.101}$$

where G is the same as in the proof of Theorem 6.4 *and*

$$h = \frac{\sup_{\tau \geq 0} \phi'(\tau)}{\pi_0^* \phi^{-1}(0) \inf_{\tau \in [0,\phi^{-1}(0)]} \phi'(\tau)}. \tag{6.102}$$

6.3 ▪ Notes and References

The predictor feedback design in this chapter was constructed by Nihtila [131] (see also [130] for an adaptive design with time-varying input delay). However, no stability analysis was provided. Linear time-varying systems with time-varying input delay were considered in [4] but without a control design or stability analysis. Recently, Zhou and coauthors [185], [186] proposed a finite-dimensional control law for the stabilization of linear systems with long time-varying input delay. However, this result is restricted to systems that are not exponentially unstable (all eigenvalues of the A matrix lie on the imaginary matrix).

The predictor feedback design and analysis in this chapter are performed under the following four conditions on the delay $D(t)$ and its rate $\dot{D}(t)$:

1. $D(t) \geq 0$. This condition guarantees the causality of the system.

2. $D(t) < \infty$. This condition guarantees that all inputs applied to the plant eventually reach the plant.

3. $\dot{D}(t) < 1$. This condition guarantees that the plant never receives input values that are older than the ones it has already received, that is, the input signal's direction never gets reversed.

4. $\dot{D}(t) > -\infty$. This condition guarantees that the delay cannot disappear instantaneously, but only gradually.

As we shall see, these four conditions need to be satisfied also in the case in which the time variation of the delay is a result of its dependence on the state of the plant, i.e., in the case of state-dependent delays. Yet, since in this case the delay is a function of the state, one can satisfy conditions 2–4 by restricting the initial conditions of the plant and the actuator state under a mild regularity assumption on the delay function (unless some a priori verifiable, but restrictive, conditions on the delay function and the plant are satisfied, in which case conditions 2–4 are a priori satisfied, as in the present case, without restricting the initial conditions of the system).

Chapter 7

Robustness of Linear Constant-Delay Predictors to Time-Varying Delay Perturbations

In this chapter we highlight one of the benefits of the availability of a Lyapunov functional. We show robustness of nominal, constant-delay predictor feedbacks to delay perturbations that depend on time, when the delay perturbation and its rate are small in any one of four different metrics. Specifically, in Section 7.1 we show robustness, in the H_1 stability norm of the actuator state, of global exponential stability of the predictor feedback for the cases where either the delay perturbation and its rate have small magnitude, or their L_1 norm is small, or they converge to zero as the time goes to infinity, or, finally, they have a small moving average for large times. In Section 7.2 we present an example of bilateral teleoperation between two robotic systems through a network. The network induces a constant nominal delay which is subject to an unknown time-varying perturbation that has a small moving average after a long period of time.

7.1 ▪ Robustness to Time-Varying Perturbation for Linear Systems

We consider the system

$$\dot{X}(t) = AX(t) + BU\left(t - \hat{D} - \delta(t)\right), \tag{7.1}$$

where for the rest of the section δ is a function only of the time t. For this linear case, the predictor-based controller is given explicitly as

$$U(t) = K\left(e^{A\hat{D}}X(t) + \int_{t-\hat{D}}^{t} e^{A(t-\theta)}BU(\theta)d\theta\right). \tag{7.2}$$

Theorem 7.1. *Consider the closed-loop system consisting of the plant* (7.1) *and control law* (7.2). *There exists a positive δ_1, such that if the perturbation δ satisfies*

$$|\delta(t)| + |\delta'(t)| < \delta_1 \quad \textit{for all } t \geq 0, \tag{7.3}$$

then the closed-loop system is exponentially stable in the sense that there exist positive constants R and λ such that the following holds for all $t \geq 0$:

$$\Pi_{\mathrm{L}}(t) \leq R\Pi_{\mathrm{L}}(0)e^{-\lambda t}, \tag{7.4}$$

$$\Pi_{\mathrm{L}}(t) = |X(t)|^2 + \int_{t-\hat{D}-\max\{0,\delta(t)\}}^{t} U(\theta)^2 d\theta + \int_{t-\hat{D}}^{t} \dot{U}(\theta)^2 d\theta. \tag{7.5}$$

When (7.3) does not hold, one can still derive exponential stability of the closed-loop system by imposing other conditions on the perturbation δ, such us the ones from [75, Chapter 9.3]. However, in this case one has to guarantee in addition that the propagation speed π is still uniformly bounded from above and below and is strictly positive. Therefore, we make the following assumptions which δ has to a priori satisfy.

Assumption 7.2. *The perturbation δ satisfies*

$$\delta'(t) < 1 \quad \textit{for all } t \geq 0 \tag{7.6}$$

and is such that

$$\pi_1^* = \frac{1}{\sup_{\theta \geq \sigma(0)}(1 - \delta'(\theta))} > 0. \tag{7.7}$$

Assumption 7.2 guarantees that the control signal does not change its direction, i.e., the plant always receives older inputs than the ones it has already received (relation (7.6)), and that the delay perturbation cannot be $-\infty$, i.e., the delay perturbation cannot disappear instantaneously (relation (7.7)).

Assumption 7.3. *The perturbation δ satisfies*

$$\hat{D} + \delta(t) > 0 \quad \textit{for all } t \geq 0 \tag{7.8}$$

and is such that

$$\pi_0^* = \frac{1}{\sup_{\theta \geq \sigma(0)}(\hat{D} + \delta(\theta))} > 0. \tag{7.9}$$

Assumption 7.3 guarantees the causality of the system (relation (7.8)) and that the delay perturbation is upper bounded (relation (7.9)).

Theorem 7.4. *Assume that δ satisfy Assumptions 7.2 and 7.3. There exist positive δ_2 and δ_3 such that if the perturbation δ satisfies either*

$$\int_0^\infty \left(\left|\delta'(\theta)\right| + |\delta(\theta)| \right) d\theta \leq \delta_2, \tag{7.10}$$

or

$$|\delta(t)| + |\delta'(t)| \to 0 \quad \textit{when } t \to \infty, \tag{7.11}$$

or

$$\frac{1}{\Delta} \int_t^{t+\Delta} \left(\left|\delta'(\theta)\right| + |\delta(\theta)| \right) d\theta \leq \delta_3 \quad \textit{for all } t \geq T \tag{7.12}$$

for some positive Δ and nonnegative T, then the closed-loop system consisting of the plant (7.1) and control law (7.2) is exponentially stable in the sense that there exist positive constants R and λ such that the following holds for all $t \geq 0$:

$$\Pi_L(t) \leq R \Pi_L(0) e^{-\lambda t}, \tag{7.13}$$

$$\Pi_L(t) = |X(t)|^2 + \int_{t-\hat{D}-\max\{0,\delta(t)\}}^t U(\theta)^2 d\theta + \int_{t-\hat{D}}^t \dot{U}(\theta)^2 d\theta. \tag{7.14}$$

We introduce now the backstepping transformation of the estimation of the unmeasured actuator state $U(\theta)$ for all $t-\hat{D}-\delta(t)\leq\theta\leq t$, namely, $\hat{u}(x,t)$ defined as

$$\hat{D}\hat{u}_t(x,t)=\hat{u}_x(x,t), \tag{7.15}$$
$$\hat{u}(1,t)=U(t), \tag{7.16}$$

that is,

$$\hat{u}(x,t)=U\left(t+\hat{D}(x-1)\right)\quad\text{for all } x\in[0,1]. \tag{7.17}$$

Lemma 7.5. *Consider the backstepping transformation*

$$\hat{w}(x,t)=\hat{u}(x,t)-Ke^{A\hat{D}x}X(t)-\hat{D}K\int_0^x e^{A\hat{D}(x-y)}B\hat{u}(y,t)dy, \tag{7.18}$$

together with its inverse given by

$$\begin{aligned}\hat{u}(x,t)=&\ \hat{w}(x,t)+Ke^{(A+BK)\hat{D}x}X(t)\\&+\hat{D}K\int_0^x e^{(A+BK)\hat{D}(x-y)}B\hat{w}(y,t)dy.\end{aligned} \tag{7.19}$$

System (7.1) *together with the control law* (7.2) *can be represented as*

$$\dot{X}(t)=(A+BK)X(t)+B\hat{w}(0,t)+B\tilde{u}(0,t), \tag{7.20}$$
$$\hat{D}\hat{w}_t(x,t)=\hat{w}_x(x,t)-\hat{D}Ke^{A\hat{D}x}B\tilde{u}(0,t), \tag{7.21}$$
$$\hat{w}(1,t)=0, \tag{7.22}$$

where the observer error

$$\tilde{u}(x,t)=u(x,t)-\hat{u}(x,t) \tag{7.23}$$

satisfies

$$\tilde{u}_t(x,t)=\pi(x,t)\tilde{u}_x(x,t)-\left(1-\hat{D}\pi(x,t)\right)r(x,t), \tag{7.24}$$
$$\tilde{u}(1,t)=0, \tag{7.25}$$

with

$$\pi(x,t)=\frac{1+x(\dot{\sigma}(t)-1)}{\sigma(t)-t}, \tag{7.26}$$

$$\phi(t)=t-\hat{D}-\delta(t), \tag{7.27}$$
$$\begin{aligned}\sigma(t)&=\phi^{-1}(t)\\&=t+\hat{D}+\delta(\sigma(t)),\end{aligned} \tag{7.28}$$
$$\begin{aligned}r(x,t)=&\ \frac{1}{\hat{D}}\hat{w}_x(x,t)+Ke^{(A+BK)\hat{D}x}(A+BK)X(t)+KB\hat{w}(x,t)\\&+\hat{D}K(A+BK)\int_0^x e^{(A+BK)\hat{D}(x-y)}B\hat{w}(y,t)dy.\end{aligned} \tag{7.29}$$

Furthermore,

$$\hat{D}\hat{w}_{xt}(x,t) = \hat{w}_{xx}(x,t) - \hat{D}^2 K e^{A\hat{D}x} AB\tilde{u}(0,t), \tag{7.30}$$

$$\hat{w}_x(1,t) = \hat{D}Ke^{A\hat{D}}B\tilde{u}(0,t). \tag{7.31}$$

Proof. System (7.1) can be rewritten in the form

$$\dot{X}(t) = AX(t) + Bu(0,t), \tag{7.32}$$

$$u_t(x,t) = \pi(x,t)u_x(x,t), \tag{7.33}$$

$$u(1,t) = U(t), \tag{7.34}$$

where π is defined in (7.26). The predictor feedback (7.2) can be written as

$$U(t) = K\left(e^{A\hat{D}}X(t) + \hat{D}\int_0^1 e^{A\hat{D}(1-y)}B\hat{u}(y,t)dy\right). \tag{7.35}$$

With representation (7.32)–(7.34) for system (7.1), the actuator state $u(x,t)$ is

$$u(x,t) = U(\phi(t + x(\sigma(t) - t))) \quad \text{for all } x \in [0,1], \tag{7.36}$$

or, incorporating δ,

$$u(x,t) = U\left(t + x\left(\hat{D} + \delta(\sigma(t))\right) - \hat{D} - \delta\left(t + x\left(\hat{D} + \delta(\sigma(t))\right)\right)\right) \tag{7.37}$$

for all $x \in [0,1]$. Since the perturbation δ satisfies (7.3), it follows from definition (7.26) and relations (7.27), (7.28) that $\sigma(t) - t > 0$ and that $1 - \delta'(t) > 0$ for all $t \geq 0$. Define the quantities

$$\pi_1^{**} = \frac{1}{\sup_{\theta \geq \sigma(0)}(1 - \delta'(\theta))}, \tag{7.38}$$

$$\pi_0^{**} = \frac{1}{\sup_{\theta \geq \sigma(0)}(\hat{D} + \delta(\theta))}. \tag{7.39}$$

From (7.3) it follows that $\sup_{\theta \geq \sigma(0)}(1-\delta'(\theta)) < \infty$ and that $\sup_{\theta \geq \sigma(0)}(\hat{D}+\delta(\theta)) < \infty$, and hence $\pi_1^{**} > 0$, $\pi_0^{**} > 0$. Since $\sigma(t) - t = \hat{D} + \delta(\sigma(t))$ and $\dot{\sigma}(t) = \frac{1}{1-\delta'(\sigma(t))}$, using (7.3) we conclude that π is positive and uniformly bounded from above and below. Hence, π is a meaningful propagation speed. The rest of the proof is based on algebraic manipulations and is omitted. $\square$

Lemma 7.6. *There exist positive constants r_1 and r_2 such that the derivative of the Lyapunov function*

$$V_L(t) = X(t)^T PX(t) + b_1\int_0^1 e^{bx}\tilde{u}(x,t)^2 dx + \hat{D}b_2\int_0^1 (1+x)\hat{w}(x,t)^2 dx$$
$$+ \hat{D}b_2\int_0^1 (1+x)\hat{w}_x(x,t)^2 dx, \tag{7.40}$$

along the solutions of (7.20)–(7.22), (7.24)–(7.25), (7.30)–(7.31) *satisfies*

$$\dot{V}_{\mathrm{L}}(t) \le -r_1 V_{\mathrm{L}}(t) + r_2 \gamma(t) V_{\mathrm{L}}(t), \tag{7.41}$$

$$\gamma(t) = \max\left\{ |\delta(\sigma(t))|, \left|\delta(\sigma(t))\left(1-\delta'(\sigma(t))\right) - \hat{D}\delta'(\sigma(t))\right| \right\}. \tag{7.42}$$

Proof. Taking the derivative of V_{L} along (7.20)–(7.22), (7.24)–(7.25), and (7.30)–(7.31) and using integration by parts, we get that

$$\begin{aligned}
\dot{V}_{\mathrm{L}}(t) \le & -|X(t)|^2 \lambda_{\min}(Q) + 2X(t)^T PB\hat{w}(0,t) + 2X(t)^T PB\tilde{u}(0,t) \\
& - b_1 b \int_0^1 e^{bx} \pi(x,t) \tilde{u}(x,t)^2 dx - b_1 \pi(0,t)\tilde{u}(0,t)^2 - b_1 \int_0^1 e^{bx} \pi_x(x,t) \\
& \times \tilde{u}(x,t)^2 dx - 2b_1 \int_0^1 e^{bx} \tilde{u}(x,t)\left(1-\hat{D}\pi(x,t)\right) r(x,t) dx + b_2 \\
& \times \left(-\hat{w}(0,t)^2 - \int_0^1 \hat{w}(x,t)^2 dx 2\hat{D}^2 |K|^2 e^{2|A|\hat{D}} |B|^2 \tilde{u}(0,t)^2 - \hat{w}_x(0,t)^2 \right) \\
& - b_2 \int_0^1 \hat{w}_x(x,t)^2 dx - 2b_2 \hat{D} \int_0^1 (1+x)\hat{w}(x,t) K e^{A\hat{D}x} B\tilde{u}(0,t) dx \\
& - 2b_2 \int_0^1 (1+x)\hat{w}_x(x,t) \hat{D}^2 K e^{A\hat{D}x} AB\tilde{u}(0,t) dx.
\end{aligned} \tag{7.43}$$

With calculations similar to those in Section 6.2 and with Lemma 7.5 by choosing $b > (1-\pi_1^{**})\max\left\{1, \frac{1}{\pi_1^{**}}\right\}$ we get that

$$b\pi(x,t) + \pi_x(x,t) \ge \pi_0^{**}\beta^*, \tag{7.44}$$

where

$$\beta^* = \min\left\{ b - 1 + \pi_1^{**}, (b+1)\pi_1^{**} - 1 \right\} > 0. \tag{7.45}$$

Since $\pi(x,t)$ is linear in x, it takes its maximum value either at $x = 0$ or at $x = 1$, and hence

$$\begin{aligned}
\left(1-\hat{D}\pi(x,t)\right) & \le \max\left\{ \left|1-\hat{D}\pi(0,t)\right|, \left|1-\hat{D}\pi(1,t)\right| \right\} \\
& \le M_{2,\mathrm{L}}\gamma(t),
\end{aligned} \tag{7.46}$$

where

$$M_{2,\mathrm{L}} = \pi(0,t) + \pi(1,t) \le \frac{1+\sup_{t\ge t_0}\dot{\sigma}(t)}{\inf_{t\ge t_0}(\sigma(t)-t)}, \tag{7.47}$$

and γ is defined in (7.42). We derive next a bound for $r(x,t)$ in terms of $X, \hat{w}$, and $\hat{w}_x$. Using (7.29) together with the Young and Cauchy–Schwarz inequalities we get that

$$||r(t)||^2 \le M_{1,\mathrm{L}}\left(|X(t)|^2 + ||\hat{w}(t)||^2 + ||\hat{w}_x(t)||^2\right), \tag{7.48}$$

$$||r(t)||^2 = \int_0^1 r(x,t)^2 dx, \tag{7.49}$$

$$\begin{aligned}
M_{1,\mathrm{L}} = & 4\hat{D}^{-1} + 4\hat{D}\left|Ke^{(A+BK)\hat{D}}(A+BK)\right|^2 + 4\hat{D} \\
& \times |KB|^2 + 4\hat{D}\left|Ke^{(A+BK)\hat{D}}(A+BK)\hat{D}B\right|^2.
\end{aligned} \tag{7.50}$$

Using (7.44), (7.46), (7.48) and Young's inequality we get

$$\begin{aligned}\dot{V}_{\mathrm{L}}(t) \leq & -\frac{\lambda_{\min}(Q)}{4}|X(t)|^2 - b_1\pi_0^{**}\beta^*\int_0^1 e^{bx}\tilde{u}(x,t)^2dx - \frac{b_2}{2}\int_0^1 \hat{w}(x,t)^2dx \\ & -\frac{b_2}{2}\int_0^1 \hat{w}_x(x,t)^2dx + \left(\frac{8|PB|}{\lambda_{\min}(Q)} - b_2\right)\hat{w}(0,t)^2 \\ & + \left(\frac{8|PB|}{\lambda_{\min}(Q)} + 8b_2\hat{D}^2|K|e^{|A|\hat{D}}|B|^2\left(2+|A|^2\right) - b_1\pi_0^{**}\right)\tilde{u}(0,t)^2 \\ & + b_1M_{2,\mathrm{L}}(1+M_{1,\mathrm{L}})\gamma(t)\Xi(t), \end{aligned} \tag{7.51}$$

where

$$\Xi(t) = |X(t)|^2 + \int_0^1 e^{bx}\tilde{u}(x,t)^2dx + \int_0^1 \hat{w}(x,t)^2dx + \int_0^1 \hat{w}_x(x,t)^2dx. \tag{7.52}$$

Choosing

$$b_1 = \frac{8|PB|}{\lambda_{\min}(Q)} + 8b_2\hat{D}^2|K|e^{|A|\hat{D}}|B|^2\left(2+|A|^2\right), \tag{7.53}$$

$$b_2 = \frac{8|PB|}{\lambda_{\min}(Q)}, \tag{7.54}$$

and using the fact that

$$M_{4,\mathrm{L}}\Xi(t) \leq V_{\mathrm{L}}(t) \leq M_{3,\mathrm{L}}\Xi(t), \tag{7.55}$$

where

$$M_{3,\mathrm{L}} = \lambda_{\max}(P) + b_1 + 2\hat{D}b_2, \tag{7.56}$$

$$M_{4,\mathrm{L}} = \min\left\{\lambda_{\min}(P), b_1, \hat{D}b_2\right\}, \tag{7.57}$$

we get relation (7.41) with

$$r_1 = \frac{\min\left\{\lambda_{\min}(Q), 4b_1\pi_0^{**}\beta^*, 2b_2\right\}}{4M_{3,\mathrm{L}}}, \tag{7.58}$$

$$r_2 = \frac{b_1M_{2,\mathrm{L}}(1+M_{1,\mathrm{L}})}{M_{4,\mathrm{L}}}. \qquad \square \tag{7.59}$$

Lemma 7.7. *There exists a positive δ_1 such that if the perturbation δ satisfies (7.3), then there exists a positive λ such that V in (7.40) satisfies*

$$\dot{V}_{\mathrm{L}}(t) \leq -\lambda V_{\mathrm{L}}(t). \tag{7.60}$$

Proof. Consider first that $\gamma(t) = \left|\frac{1}{\pi(0,t)} - \hat{D}\right|$. Using (7.28), we get

$$\pi(x,t) = \frac{1 + x\frac{\delta'(\sigma(t))}{1-\delta'(\sigma(t))}}{\hat{D} + \delta(\sigma(t))}, \tag{7.61}$$

and hence

$$\gamma(t) = |\delta(\sigma(t))|. \tag{7.62}$$

Therefore, if δ satisfies (7.3) with $\delta_1 r_2 < r_1$, the lemma is proved. Assume next that $\gamma(t) = \left|\frac{1}{\pi(1,t)} - \hat{D}\right|$. Thus,

$$\gamma(t) = \left|\delta(\sigma(t))(1-\delta'(\sigma(t))) - \hat{D}\delta'(\sigma(t))\right|. \tag{7.63}$$

With (7.38), (7.3) we get $\gamma(t) < \delta_1(1+\delta_1+\hat{D})$. Hence, the lemma is proved if $\delta_1 r_2 < \frac{r_1}{d}$, where $d = 1+\delta_1+\hat{D}$. Note that since one can choose $\pi_0^{**}\beta^*$ sufficiently large (by choosing a large b) such that r_1 in (7.58) is independent of δ_1 (or very large), and since from (7.59) together with (7.50), (7.47), (7.57) r_2 is bounded, one can always find a sufficiently small δ_1 such that relation $\delta_1 r_2 < \frac{r_1}{d}$ is satisfied. □

Lemma 7.8. *There exist positive constants $M_{5,\mathrm{L}}$ and $M_{6,\mathrm{L}}$ such that*

$$M_{5,\mathrm{L}} V_{\mathrm{L}}(t) \le \Gamma_{\mathrm{L}}(t) \le M_{6,\mathrm{L}} V_{\mathrm{L}}(t), \tag{7.64}$$

$$\Gamma_{\mathrm{L}}(t) = |X(t)|^2 + \int_0^1 u(x,t)^2 dx + \int_0^1 \hat{u}(x,t)^2 dx + \int_0^1 \hat{u}_x(x,t)^2 dx. \tag{7.65}$$

Proof. Using relations (7.18), (7.19) together with the Young and Cauchy–Schwarz inequalities we get that

$$||\hat{u}(t)||^2 \le N_1\left(|X(t)|^2 + ||\hat{w}(t)||^2\right), \tag{7.66}$$

$$||\hat{u}_x(t)||^2 \le N_2\left(|X(t)|^2 + ||\hat{w}(t)||^2 + ||\hat{w}_x(t)||^2\right), \tag{7.67}$$

$$||\hat{w}(t)||^2 \le N_3\left(|X(t)|^2 + ||\hat{u}(t)||^2\right), \tag{7.68}$$

$$||\hat{w}_x(t)||^2 \le N_4\left(|X(t)|^2 + ||\hat{u}(t)||^2 + ||\hat{u}_x(t)||^2\right), \tag{7.69}$$

where $||\cdot||$ is defined in (7.49) and

$$N_1 = 3 + 3\left|Ke^{(A+BK)\hat{D}}\right|^2 + 3\hat{D}^2\left|Ke^{(A+BK)\hat{D}}B\right|^2, \tag{7.70}$$

$$N_2 = 4 + 4\left|Ke^{(A+BK)\hat{D}}(A+BK)\hat{D}\right|^2 + 4\hat{D}^2|KB|^2 + 4\hat{D}^2\left|Ke^{(A+BK)\hat{D}}\right|^2 |B|^2\left(1+\hat{D}|A+BK|\right), \tag{7.71}$$

$$N_3 = 3 + 3\left|Ke^{A\hat{D}}\right|^2 + 3\hat{D}^2\left|Ke^{A\hat{D}}B\right|^2, \tag{7.72}$$

$$N_4 = 4 + 4\left|Ke^{A\hat{D}}A\hat{D}\right|^2 + 4\hat{D}^2|KB|^2 + 4\hat{D}^2\left|Ke^{A\hat{D}}B\right|^2 + 4\hat{D}^2\left|Ke^{A\hat{D}}A\hat{D}B\right|^2. \tag{7.73}$$

Using (7.55), (7.66)–(7.69) the lemma is proved with

$$M_{5,\mathrm{L}} = \frac{1}{M_{3,\mathrm{L}}}\left(2e^b + N_3 + N_4 + 1\right), \tag{7.74}$$

$$M_{6,\mathrm{L}} = \frac{3 + 3N_1 + N_2}{M_{4,\mathrm{L}}}. \qquad \square \tag{7.75}$$

Lemma 7.9. *There exist positive constants $M_{7,\mathrm{L}}$ and $M_{8,\mathrm{L}}$ such that*

$$M_{7,\mathrm{L}}\Gamma_{\mathrm{L}}(t) \le \Pi_{\mathrm{L}}(t) \le M_{8,\mathrm{L}}\Gamma_{\mathrm{L}}(t). \tag{7.76}$$

Proof. From (7.17) we get $\hat{u}_x(x,t) = \hat{D}U'\left(t + \hat{D}(x-1)\right)$. Applying a change of variables in (7.65), with (7.36) we get

$$\Gamma_{\mathrm{L}}(t) = |X(t)|^2 + \frac{1}{\sigma(t)-t}\int_{t-\hat{D}-\delta(t)}^{t} \frac{1}{\phi'(\sigma(\theta))}U(\theta)^2 d\theta$$
$$+ \frac{1}{\hat{D}}\int_{t-\hat{D}}^{t} U(\theta)^2 d\theta + \int_{t-\hat{D}}^{t} \dot{U}(\theta)^2 d\theta. \tag{7.77}$$

Hence, the lemma is proved with

$$M_{7,\mathrm{L}} = \frac{1}{\max\left\{1, M_{\mathrm{L}} + \frac{1}{\hat{D}}\right\}} \tag{7.78}$$

$$M_{\mathrm{L}} = \frac{\sup_{t-\hat{D}-\delta(t)\le\theta\le t} \frac{1}{1-\delta'(\sigma(\theta))}}{\inf_{t\ge 0}\left(\hat{D} + \delta(\sigma(t))\right)} \tag{7.79}$$

$$M_{8,\mathrm{L}} = \frac{1}{\min\left\{1, \pi_0^{**}\pi_1^{**}, \frac{1}{\hat{D}}\right\}}. \qquad \square \tag{7.80}$$

***Proof of Theorem* 7.1.** Using Lemma 7.7 and the comparison principle (see also Lemma B.4 from Appendix B), we get $V_{\mathrm{L}}(t) \le V_{\mathrm{L}}(0)e^{-\lambda t}$. With Lemmas 7.8 and 7.9 we get (7.4) with $R = \frac{M_{6,\mathrm{L}}M_{8,\mathrm{L}}}{M_{5,\mathrm{L}}M_{7,\mathrm{L}}}$. $\square$

***Proof of Theorem* 7.4.** We consider first the case where δ satisfies (7.10). Under Assumptions 7.2 and 7.3, Lemmas 7.5, 7.6, 7.8, and 7.9 apply to this case as well (with π_0^{**} and π_1^{**} replaced by π_0^* and π_1^*, respectively). The only difference with the proof of Theorem 7.1 is in the proof of Lemma 7.7. Towards that end, we solve (7.41) to get

$$V_{\mathrm{L}}(t) \le e^{-r_1 t + r_2\int_0^t \gamma(\tau)d\tau} V_{\mathrm{L}}(0). \tag{7.81}$$

Consider that $\gamma(t) = \left|\frac{1}{\pi(0,t)} - \hat{D}\right|$. Applying the change of variables $\tau = \phi(\theta)$ in the integral and using the facts that $\sigma(0) > 0$ and $\phi'(t) = 1 - \delta'(t)$ we get

$$\int_0^\infty \gamma(\tau)d\tau \le \frac{1}{\pi_1^*}\int_0^\infty |\delta(\theta)|d\theta \le \frac{\delta_2}{\pi_1^*}. \tag{7.82}$$

Analogously, for the case $\gamma(t)=\left|\frac{1}{\pi(1,t)}-\hat{D}\right|$, we get

$$\int_0^\infty \gamma(\tau)d\tau \leq \frac{\delta_2}{\pi_1^*}\left(\hat{D}+\frac{1}{\pi_1^*}\right). \tag{7.83}$$

Hence, Lemma 7.7 is proved with $r_2\delta_2 < \min\left\{\pi_1^* r_1, \frac{\pi_1^{*2} r_1}{(1+\pi_1^*\hat{D})}\right\}$ and the fact that $\int_0^t \gamma(\tau)d\tau \leq \int_0^\infty \gamma(\tau)d\tau$. Note that in the present case, Lemma 7.7 can be proved directly from relation (7.41) using Lemma B.4.

For the case where δ satisfies (7.11), Lemma 7.7 is proved by combining Lemma 7.6, Lemma B.6, and the fact that $\gamma(t)$ satisfies

$$\gamma(t)\leq\left(1+\frac{1}{\pi_1^*}+\hat{D}\right)\left(|\delta(\sigma(t))|+|\delta'(\sigma(t))|\right). \tag{7.84}$$

Finally, if δ satisfies (7.12), Lemma 7.7 is proved using Lemma 9.5 in [75]. □

7.2 ▪ A Teleoperation-Like Example

In bilateral teleoperation [54], the operator (e.g., a human) controls a robotic system, called the master, at one end of a communication network. The actions of the master are transmitted (through the network) to another robotic system, called the slave, at the other end of the network. The goal of the control algorithm is to make the slave manipulator behave (in a certain sense) as the master manipulator. A teleoperator-like model of two robotic systems, each one having n degrees of freedom, representing the master and the slave manipulators is written as [54]

$$\ddot{x}_m(t)+\dot{x}_m(t)=\tau_m\left(t-\hat{D}-\delta(t)\right), \tag{7.85}$$

$$\ddot{x}_s(t)+\dot{x}_s(t)=\tau_s\left(t-\hat{D}-2\delta(t)\right), \tag{7.86}$$

where x_m, $x_s \in \mathbb{R}^n$ are the degrees of freedom of the robotic systems and the torques τ_m, $\tau_s \in \mathbb{R}^n$ are to be designed such as coordination between the master and the slave is achieved asymptotically, i.e., $x_m - x_s \to 0$ as $t \to \infty$. The constant delay $\hat{D}$ represents the known network-induced delay which is subject to time-varying perturbations that are often present due to congestion, distance, etc. [31]. For simplicity we assume scalar x_m, x_s, τ_m, τ_s and we rewrite (7.85), (7.86) as

$$\dot{X}_1(t)=X_2(t), \tag{7.87}$$

$$\dot{X}_2(t)=-X_2(t)+U_1\left(t-\hat{D}-\delta(t)\right), \tag{7.88}$$

$$\dot{X}_3(t)=X_4(t), \tag{7.89}$$

$$\dot{X}_4(t)=-X_4(t)+U_2\left(t-\hat{D}-2\delta(t)\right), \tag{7.90}$$

where $X_1=x_m$, $X_2=\dot{x}_m$, $X_3=x_s$, $X_4=\dot{x}_s$, $U_1=\tau_m$, and $U_2=\tau_s$. A simple controller is [54] $\tau_m(t)=-K_p(x_m(t)-x_s(t))-B_m\dot{x}_m(t)-K_p(x_m(t)-r)$ and $\tau_s(t)=K_p(x_m(t)-x_s(t))-B_s\dot{x}_s(t)-K_p(x_s(t)-r)$, where r is the set-point for the positions of the

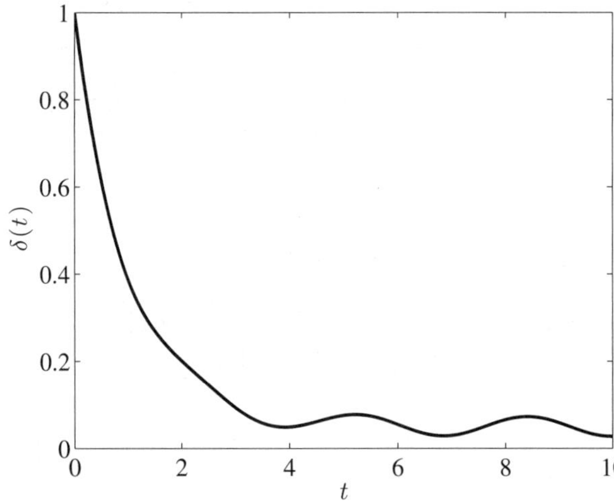

Figure 7.1. *The delay perturbation δ satisfying $\dot{\delta}(t) = -\delta(t) + 0.1\sin(t)^2, \delta(0) = 1$ induced by the network in bilateral teleoperation. The model of the two robotic systems is* (7.85)–(7.86).

manipulators. The predictor-based version of this controller is

$$U_1(t) = -K_p\left(\hat{P}_1(t) - \hat{P}_3(t)\right) - B_m\hat{P}_2(t) - K_p\left(\hat{P}_1(t) - r\right), \tag{7.91}$$

$$U_2(t) = K_p\left(\hat{P}_1(t) - \hat{P}_3(t)\right) - B_s\hat{P}_4(t) - K_p\left(\hat{P}_3(t) - r\right), \tag{7.92}$$

$$\hat{P}_i(t) = X_i(t) + \int_{t-\hat{D}}^{t} \hat{P}_{i+1}(\theta)d\theta, \quad i = 1,3, \tag{7.93}$$

$$\hat{P}_j(t) = X_j(t) + \int_{t-\hat{D}}^{t} \left(-\hat{P}_j(\theta) + U_{\frac{j}{2}}(\theta)\right)d\theta, \quad j = 2,4. \tag{7.94}$$

We choose the desired set-point for x_m and x_s as $r = 2$, the parameters of the controller as $K_p = B_m = B_s = 2$, the known delay as $\hat{D} = 1$, the initial condition of the plant as $x_m(0) = \dot{x}_m = \dot{x}_s = 0$, $x_s(0) = 1$, the initial actuator state as $U(\theta) = 0, -\hat{D} - \delta(0) \leq \theta \leq 0$, and the initial estimation of the actuator state as $U(\theta) = 0, -\hat{D} \leq \theta \leq 0$. We illustrate the robustness properties of the predictor feedback under a time-varying delay perturbation that neither is in L_1, nor converges to zero as the time goes to infinity, nor is small in magnitude. However, after some long period of time its mean is small. This disturbance is shown in Figure 7.1. In Figure 7.2 we show the difference between the positions of the master and the slave for the cases where either there is or there is not a perturbation δ. In both cases, under the nominal predictor feedback the position of the slave tracks the position of the master. In Figure 7.2 we also show the torques applied to the two robotic systems under the perturbation δ. The control efforts are oscillatory as a result of the effect of the oscillatory perturbation.

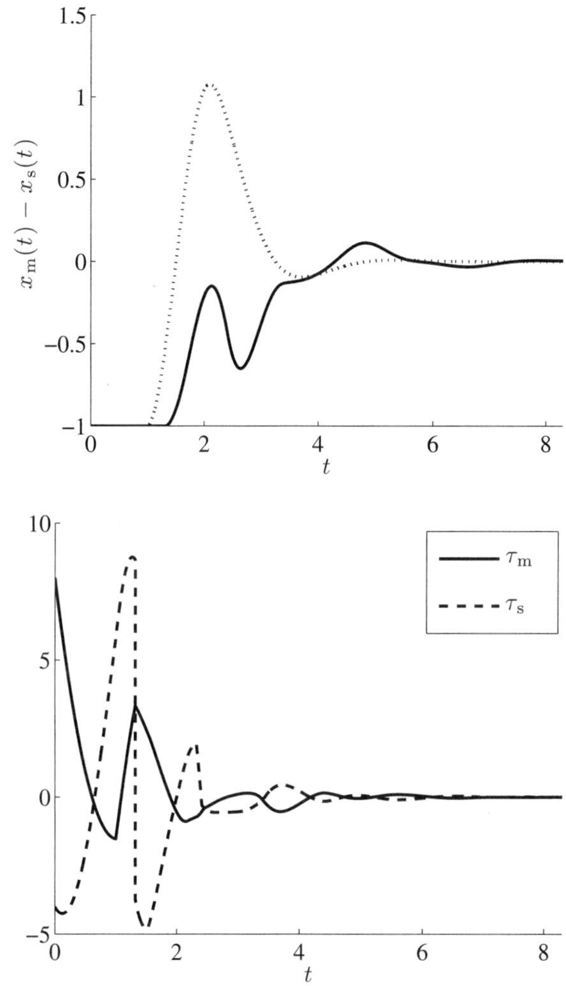

Figure 7.2. *The error (top) between the position of the master and the slave manipulators and the input torques (bottom) for two robotic systems modeled by* (7.85)–(7.86). *The two manipulators are coordinated through a network with the predictor feedback* (7.91)–(7.94) *under an input delay perturbation* $\delta(t) = 0$ *(dotted line) and* δ *satisfying* $\dot{\delta}(t) = -\delta(t) + 0.1\sin(t)^2$, $\delta(0) = 1$ *(solid line), induced by the network. The initial conditions are* $x_m(0) = 0$, $x_s(0) = 1$, $\dot{x}_m(0) = \dot{x}_s(0) = 0$ *and* $\tau_m(\theta) = \tau_s(\theta) = 0$, $-1 - \delta(0) \leq \theta \leq 0$.

7.3 ▪ Notes and References

Robustness results of nominal, finite-dimensional control laws for ODEs to small delays have been developed by Mazenc and coauthors [108] and Teel [162]. Our result is different from the aforementioned works in the fact that the nominal case (the finite-dimensional plant plus the infinite-dimensional actuator state without delay mismatch) and the feedback law are both infinite-dimensional. A robustness result for linear, nominal, constant-delay predictor feedbacks without restriction on the rate of the perturbation was developed by Karafyllis and Krstic [70]. This result is based on a particular form of the small-gain theorem rather than on the construction of a Lyapunov functional.

One might raise the question of robustness to stochastic time-varying delay perturbations. Yet, in our analysis we restrict not only the magnitude of the perturbation δ but also the magnitude of its derivative (which also guarantees the invertibility of $\phi = t - \hat{D} - \delta$), which can be unbounded in the case where δ is white noise, or even when δ is low-pass filtered white noise.

Chapter 8

Nonlinear Systems with Time-Varying Input Delay

We consider in this chapter nonlinear, forward-complete systems with time-varying input delay, and we design the nonlinear predictor feedback law for the compensation of time-varying input delays in Section 8.1. The major challenge that we resolve is the definition of the predictor state since the prediction horizon is not constant. We prove global asymptotic stability of general nonlinear plants under predictor feedback in the presence of a time-varying input delay in Section 8.2. The stability proof is based on a nonlinear infinite-dimensional backstepping transformation with time-varying kernels that we introduce, and which helps us construct a strict Lyapunov functional for the closed-loop system. We present in Section 8.3 an example of unicycle stabilization subject to a time-varying input delay.

8.1 ▪ Nonlinear Predictor Feedback Design for Time-Varying Delay

We consider the following nonlinear system with time-varying input delay:

$$\dot{X}(t) = f(X(t), U(t - D(t))), \tag{8.1}$$

where $X \in \mathbb{R}^n$, $U \in \mathbb{R}$, $t \in \mathbb{R}_+$, and $f : C^1(\mathbb{R}^n \times \mathbb{R}; \mathbb{R}^n)$ with $f(0,0) = 0$. Throughout the chapter we make the following assumptions concerning the plant (8.1).

Assumption 8.1. *The plant $\dot{X} = f(X, \omega)$ is strongly forward complete, that is, there exists a smooth positive definite function R and class $\mathcal{K}_\infty$ functions δ_1, δ_2, and δ_3 such that the following holds:*

$$\delta_1(|X|) \le R(X) \le \delta_2(|X|), \tag{8.2}$$

$$\frac{\partial R(X)}{\partial X} f(X, \omega) \le R(X) + \delta_3(|\omega|) \tag{8.3}$$

for all $X \in \mathbb{R}^n$ and for all $\omega \in \mathbb{R}$.

Assumption 8.1 guarantees that system (8.1) does not exhibit finite escape time, i.e., for every initial condition and every bounded input signal the corresponding solution is defined for all $t \ge 0$, i.e., the maximal interval of existence is $T_{\max} = \infty$. The definition of forward-completeness is the one from Appendix C.2.

Assumption 8.2. *There exists a feedback law $\kappa(t,X) \in C^1\left(\mathbb{R}_+ \times \mathbb{R}^n;\mathbb{R}\right)$, periodic in t with $\kappa(t,0) = 0$ for all $t \geq 0$, such that the plant $\dot{X}(t) = f\left(X(t), \kappa(t,X(t)) + \omega(t)\right)$ is input-to-state stable with respect to ω.*

Naturally, the time-varying delay should be uniformly bounded and positive, and hence there must exist a positive constant m such that $m > D(t) > 0$ for all $t \geq 0$. As it turns out later on, our predictor-based compensator makes use of the inverse of the function

$$\phi(t) = t - D(t), \tag{8.4}$$

namely, $\phi^{-1}(t)$. Hence, we have to impose certain conditions on $\phi'(t)$ that guarantee the invertibility of $\phi(t)$. From now on we refer to the quantity $t - \phi(t) = D(t)$ as the delay time. This is the time interval that indicates how long *ago* the control signal that is currently affecting the plant was actually applied. The main goal of this section is to determine the predictor state, i.e., the quantity P such that $X\left(\phi^{-1}(t)\right) = P(t)$. We further refer to the quantity $\phi^{-1}(t) - t$ as the prediction horizon. This is the time interval which indicates *after* how long an input signal that is currently applied affects the plant. In the constant delay case, the prediction horizon is equal to the delay time, i.e., $t - \phi(t) = D = \phi^{-1}(t) - t$.

We make the following assumptions regarding the function $\phi(t)$.

Assumption 8.3. *The function $\phi(t) = t - D(t)$ satisfies*

$$\phi(t) < t \quad \textit{for all } t \geq 0 \tag{8.5}$$

and

$$\pi_0^* = \frac{1}{\sup_{\theta \geq \phi^{-1}(0)} (\theta - \phi(\theta))} > 0. \tag{8.6}$$

Assumption 8.4. *The derivative of the function $\phi(t)$ satisfies*

$$\phi'(t) > 0 \quad \textit{for all } t \geq 0, \tag{8.7}$$

and it holds that

$$\pi_1^* = \frac{1}{\sup_{\theta \geq \phi^{-1}(0)} \phi'(\theta)} > 0. \tag{8.8}$$

The delay time $D(t)$ has to be positive for all $t \geq 0$ (in order to guarantee causality of the plant), which is guaranteed with condition (8.5) (note that $t - \phi(t) = D(t)$) in Assumption 8.3. Moreover, relation (8.6) guarantees that $D(t)$ is uniformly bounded from above, which implies that the control signal eventually reaches the plant. Relation (8.7) (or equivalently $\frac{dD(t)}{dt} < 1$) in Assumption 8.4 guarantees that the control signal is able to reach the plant and it does not change the direction of propagation of the control signal (the plant keeps receiving control inputs that are never older than the ones it has already received), and the bound (8.8) guarantees that $\frac{dD(t)}{dt}$ is bounded from below (i.e., the delay cannot disappear instantaneously).

The controller for the plant (8.1) that compensates for the time-varying input delay and achieves global asymptotic stability of the closed-loop system is given by

$$U(t)=\kappa\left(\phi^{-1}(t),P(t)\right), \tag{8.9}$$

where $P(t)$ is given from

$$\begin{aligned} P(\theta)=&X(t)+\left(\phi^{-1}(t)-t\right)\int_0^{\frac{\phi^{-1}(\theta)-t}{\phi^{-1}(t)-t}} f\Big(P\left(\phi\left(t+y\left(\phi^{-1}(t)-t\right)\right)\right),\\ &U\left(\phi\left(t+y\left(\phi^{-1}(t)-t\right)\right)\right)\Big)dy\\ =&X(t)+\int_{\phi(t)}^{\theta} f\left(P(\sigma),U(\sigma)\right)\frac{d\sigma}{\phi'\left(\phi^{-1}(\sigma)\right)},\quad \phi(t)\leq\theta\leq t, \end{aligned} \tag{8.10}$$

for $\theta=t$. The initial condition of (8.10) is given for $t=t_0$ ($t_0\geq 0$ is the initial time which must be given because the closed-loop system is time-varying) as

$$P(\theta)=\int_{\phi(t_0)}^{\theta} f\left(P(\sigma),U(\sigma)\right)\frac{d\sigma}{\phi'\left(\phi^{-1}(\sigma)\right)}+X(t_0)\quad \text{for all } \theta\in[\phi(t_0),t_0]. \tag{8.11}$$

For an implementation of the controller (8.9) one has to numerically integrate the finite intervals in (8.10) and (8.11) using one of the numerical quadratures. In our simulations we use the composite left-endpoint rectangle rule. Note that the interval of integration, i.e., the interval $t-\phi(t)=D(t)$, at each time step may not be an integer multiple of the discretization step h. In this case, the total number of points that are used in the computation of the integral is derived by rounding the number $\frac{t-\phi(t)}{h}$ to the nearest integer from below. However, the study of the complexity of the algorithm in an actual implementation is beyond the scope of this book, which concentrates in basic continuous-time designs.

The quantity $P(t)$ given in (8.10) is the $\phi^{-1}(t)-t$-time-units-ahead predictor of $X(t)$, which can be seen as follows: Differentiating relation (8.10) with respect to θ, setting $\theta=t$, and performing a change of variables $\tau=\phi^{-1}(t)$ in the ODE for $X(\tau)$ given in (8.1) (where t is replaced by τ), one observes that $P(t)$ satisfies the same ODE in t as $X(\phi^{-1}(t))$. Since from (8.11) it follows that $P(\phi(t_0))=X(t_0)$, we have that $P(t_0)=X(\phi^{-1}(t_0))$. Hence, indeed $P(t)=X\left(\phi^{-1}(t)\right)$ for all $t\geq t_0$.

The choice of (8.10) becomes clear by considering the case where the input delay is constant, of length, say, D. In this case the predictor that corresponds to (8.10) is

$$\begin{aligned} P_c(t)&=D\int_0^1 f\left(P_c(t+D(y-1)),U(t+D(y-1))\right)dy+X(t)\\ &=\int_{t-D}^{t} f\left(P_c(\theta),U(\theta)\right)d\theta+X(t). \end{aligned} \tag{8.12}$$

The signal $P_c(t)$ in (8.12) is the D-time-units-ahead predictor of $X(t)$ (see also Chapter 5), i.e., $P_c(t)=X(t+D)$ for all $t\geq t_0$. The relationship between (8.10) and (8.12) can be explained as follows: Let us define the predictor time in the case of a time-varying delay as $\phi^{-1}(t)-t$, which is uniformly bounded based on relation (8.8). Analogously, for the case of a constant delay, define the predictor time as $\phi_c^{-1}(t)-t$, which of course equals D.

One can now rewrite (8.12) by substituting D with $\phi_c^{-1}(t)-t$ as

$$\begin{aligned} P_c(t) &= D\int_0^1 f\left(P_c(t+D(y-1)), U(t+D(y-1))\right)dy + X(t) \\ &= \left(\phi_c^{-1}(t)-t\right)\int_0^1 f\left(P_c(t+D(y-1)), U(t+D(y-1))\right)dy + X(t). \end{aligned} \tag{8.13}$$

This is the exact analogue of relation (8.10) for the constant-delay case. Moreover, the choice of the function $\phi\left(t+y\left(\phi^{-1}(t)-t\right)\right)$ inside the integral in equation (8.10) is guided by the choice of an appropriate state, namely, $U\left(\phi\left(t+y\left(\phi^{-1}(t)-t\right)\right)\right)$ for all $y\in[0,1]$ for the infinite-dimensional state of the actuator that also makes $U\left(\phi\left(t+y\left(\phi^{-1}(t)-t\right)\right)\right)$ equal to $U(t)$ in the case where $y=1$ and equal to $U(\phi(t))$ in the case where $y=0$.

8.2 ▪ Stability Analysis

In this section we prove the following result.

Theorem 8.5. *Consider the system* (8.1) *together with the control law* (8.9)–(8.11). *Under Assumptions* 8.1–8.4 *there exists a class $\mathcal{KL}$ function α such that for all initial conditions $X(t_0)\in\mathbb{R}^n$ and $U(t_0+\theta);\theta\in[-D(t_0),0]\in L_\infty[-D(t_0),0]$ the following holds:*

$$|X(t)| + \sup_{\phi(t)\le\theta\le t}|U(\theta)| \le \alpha\left(|X(t_0)| + \sup_{\phi(t_0)\le\theta\le t_0}|U(\theta)|, t-t_0\right) \tag{8.14}$$

for all $t\ge t_0\ge 0$.

We prove the above theorem using a series of technical lemmas. We first introduce an equivalent representation of the plant (8.1) using a transport PDE representation for the actuator state as

$$\dot X(t) = f(X(t), u(0,t)), \tag{8.15}$$

$$u_t(x,t) = \pi(x,t)u_x(x,t), \quad x\in[0,1], \tag{8.16}$$

$$u(1,t) = U(t), \tag{8.17}$$

where

$$\pi(x,t) = \frac{1+x\left(\frac{d(\phi^{-1}(t))}{dt}-1\right)}{\phi^{-1}(t)-t}, \tag{8.18}$$

and $\phi(t)$ is defined in (8.4). The choice of the transport speed $\pi(x,t)$ is guided by the fact that we seek a representation for the infinite-dimensional actuator state $u(x,t)$ such that relations (8.17) and

$$u(0,t) = U(\phi(t)) \tag{8.19}$$

are satisfied. One can verify that $u(x,t)$ is given by

$$u(x,t) = U\left(\phi\left(t+x\left(\phi^{-1}(t)-t\right)\right)\right), \tag{8.20}$$

and consequently both (8.17) and (8.19) are satisfied. We first give an alternative proof of the fact that $P(t)$ is the $\phi^{-1}(t)-t$-time-units-ahead predictor of $X(t)$, based on the PDE representation (8.15)–(8.17).

Lemma 8.6. *The signal $P(t)$ in (8.10) is the $\phi^{-1}(t)-t$-time-units-ahead predictor of $X(t)$, i.e., it holds that*

$$P(t)=X(\phi^{-1}(t)) \quad \text{for all } t\geq t_0. \tag{8.21}$$

Furthermore, an equivalent representation for $P(t)$ is as

$$p(1,t)=\left(\phi^{-1}(t)-t\right)\int_0^1 f(p(y,t),u(y,t))dy+X(t), \tag{8.22}$$

where

$$p(x,t)=P\left(\phi\left(t+x\left(\phi^{-1}(t)-t\right)\right)\right). \tag{8.23}$$

Proof. Consider

$$p(x,t)=\left(\phi^{-1}(t)-t\right)\int_0^x f(p(y,t),u(y,t))dy+X(t), \quad x\in[0,1]. \tag{8.24}$$

Differentiating the above relation with respect to time and using (8.16) we get that

$$\begin{aligned} p_t(x,t)=&\left(\phi^{-1}(t)-t\right)\int_0^x \frac{\partial f(p(y,t),u(y,t))}{\partial p}p_t(y,t)+\left(\phi^{-1}(t)-t\right)\\ &\times\int_0^x \frac{\partial f(p(y,t),u(y,t))}{\partial u}\pi(y,t)u_y(y,t)dy+\left(\frac{d\phi^{-1}(t)}{dt}-1\right)\\ &\times\int_0^x f(p(y,t),u(y,t))dy+f(p(0,t),u(0,t)). \end{aligned} \tag{8.25}$$

Moreover, differentiating (8.24) with respect to the spatial variable x we have

$$\begin{aligned} \pi(x,t)p_x(x,t)=&\left(\phi^{-1}(t)-t\right)\pi(x,t)f(p(x,t),u(x,t))\\ =&\left(\phi^{-1}(t)-t\right)\int_0^x \frac{d\left(\pi(y,t)f(p(y,t),u(y,t))\right)}{dy}dy+\left(\phi^{-1}(t)-t\right)\\ &\times\pi(0,t)f(p(0,t),u(0,t))\\ =&\left(\phi^{-1}(t)-t\right)\int_0^x \pi_y(y,t)f(p(y,t),u(y,t))dy+\left(\phi^{-1}(t)-t\right)\\ &\times\int_0^x \pi(y,t)\frac{\partial f(p(y,t),u(y,t))}{\partial p}p_y(y,t)dy+\left(\phi^{-1}(t)-t\right)\pi(0,t)\\ &\times f(p(0,t),u(0,t))+\left(\phi^{-1}(t)-t\right)\\ &\times\int_0^x \pi(y,t)\frac{\partial f(p(y,t),u(y,t))}{\partial u}u_y(y,t)dy. \end{aligned} \tag{8.26}$$

Comparing (8.25) with (8.26) and using the facts that $\pi(0,t)=\frac{1}{\phi^{-1}(t)-t}$ and $\pi_x(x,t)=\frac{\frac{d\phi^{-1}(t)}{dt}-1}{\phi^{-1}(t)-t}$, which follow from the definition of $\pi(x,t)$ in (8.18), we get that

$$p_t(x,t)-\pi(x,t)p_x(x,t)=\left(\phi^{-1}(t)-t\right)\int_0^x \frac{\partial f(p(y,t),u(y,t))}{\partial p} \times\left(p_t(y,t)-\pi(y,t)p_y(y,t)\right)dy. \tag{8.27}$$

Define now the function $G(x,t)=p_t(x,t)-\pi(x,t)p_x(x,t)$, which satisfies

$$G_x(x,t)=\left(\phi^{-1}(t)-t\right)\frac{\partial f(p(x,t),u(x,t))}{\partial p(x,t)}G(x,t), \tag{8.28}$$

$$G(0,t)=0. \tag{8.29}$$

Hence, $G(x,t)=0$ for all $x\in[0,1]$. Equivalently,

$$p_t(x,t)=\pi(x,t)p_x(x,t) \quad \text{for all } x\in[0,1]. \tag{8.30}$$

Using the above relation and defining $p(1,t)=P(t)$, we get (8.22)–(8.23). Moreover, since from (8.24) it holds that $p(0,t)=X(t)$, using relation (8.23) we get (8.21). □

We next transform our original system given in (8.15)–(8.17) to the target system, which we later prove to be globally asymptotically stable.

Lemma 8.7. *The infinite-dimensional transformation of the actuator state defined by*

$$w(x,t)=u(x,t)-\varkappa(r(x,t),p(x,t)), \tag{8.31}$$

where

$$r(x,t)=t+x\left(\phi^{-1}(t)-t\right), \tag{8.32}$$

together with the control law given in (8.9), *transforms the system* (8.15)–(8.17) *into the target system given by*

$$\dot{X}(t)=f(X(t),\varkappa(t,X(t))+w(0,t)), \tag{8.33}$$

$$w_t(x,t)=\pi(x,t)w_x(x,t), \quad x\in[0,1], \tag{8.34}$$

$$w(1,t)=0. \tag{8.35}$$

Proof. We first point out that $r(x,t)$ in (8.32) satisfies

$$r_t(x,t)=\pi(x,t)r_x(x,t), \tag{8.36}$$

$$r(0,t)=t. \tag{8.37}$$

From (8.31) and using the chain rule together with relations (8.16), (8.30), and (8.36) we get (8.34). Using (8.31) and (8.32) for $x=0$ and $x=1$ together with (8.9), we arrive at (8.33) and (8.35). □

We now define the inverse of the transformation (8.31).

Lemma 8.8. *The inverse of the infinite-dimensional transformation defined in* (8.31) *is given by*

$$u(x,t) = w(x,t) + \varkappa(r(x,t), \xi(x,t)), \tag{8.38}$$

where $\xi(x,t)$ *is defined as*

$$\xi(x,t) = \left(\phi^{-1}(t) - t\right) \int_0^x f(\xi(y,t), \varkappa(r(y,t), \xi(y,t)) + w(y,t))dy + X(t). \tag{8.39}$$

Proof. We first point out that $\xi(x,t)$ is for the closed-loop system (8.33)–(8.35), while $p(x,t)$ is for system (8.15)–(8.17). Although,

$$\xi(x,t) = p(x,t) \quad \text{for all } x \in [0,1], \tag{8.40}$$

$\xi(x,t)$ is driven by the transformed input $w(x,t)$, whereas $p(x,t)$ is driven by the input $u(x,t)$. In other words, the direct transformation is defined as $(X(t), u(x,t)) \mapsto (X(t), w(x,t))$ and is given in (8.31), where $p(x,t)$ is given as a function of $X(t)$ and $u(x,t)$ through relation (8.24). Analogously, the inverse transformation is defined as $(X(t), w(x,t)) \mapsto (X(t), u(x,t))$ and is given in (8.38), where $\xi(x,t)$ is given as a function of $X(t)$ and $w(x,t)$ through relation (8.39). □

We prove next stability of the target system (8.33)–(8.35).

Lemma 8.9. *There exists a class* $\mathcal{KL}$ *function* β *such that*

$$|X(t)| + \sup_{\phi(t) \le \theta \le t} |W(\theta)| \le \beta\left(|X(t_0)| + \sup_{\phi(t_0) \le \theta \le t_0} |W(\theta)|, t - t_0 \right) \tag{8.41}$$

for all $t \ge t_0 \ge 0$.

Proof. Based on Assumption 8.2 and Theorem C.16 we can conclude that there exist a C^1 function $S : \mathbb{R}_+ \times \mathbb{R}^n \to \mathbb{R}_+$and class $\mathcal{K}$ functions γ_1, γ_2, γ_3, and γ_4 such that

$$\gamma_1(|X(t)|) \le S(t, X(t)) \le \gamma_2(|X(t)|), \tag{8.42}$$

$$\dot{S}(t, X(t)) \le -\gamma_3(|X(t)|) + \gamma_4(|w(0,t)|), \tag{8.43}$$

$$\dot{S}(t, X(t)) = \frac{\partial S(t, X(t))}{\partial t} + \frac{\partial S(t, X(t))}{\partial X} f(X(t), \varkappa(t, X(t)) + w(0,t)). \tag{8.44}$$

Consider now the functional

$$\begin{aligned} L(t) &= \sup_{x \in [0,1]} |e^{cx} w(x,t)| \\ &= \lim_{n \to \infty} \left(\int_0^1 e^{2ncx} w^{2n}(x,t) dx \right)^{\frac{1}{2n}}. \end{aligned} \tag{8.45}$$

Following the calculations in Chapter 6, one concludes that

$$\dot{L}(t) \le -c\pi_0^* \min\{1, \pi_1^*\} L(t). \tag{8.46}$$

Consider now the following Lyapunov function for system (8.33)–(8.35), which is positive definite and radially unbounded:

$$V(t)=S(X(t))+\frac{2}{c\pi_0^*\min\{1,\pi_1^*\}}\int_0^{L(t)}\frac{\gamma_4(r)}{r}dr. \tag{8.47}$$

Taking the time derivative of $V(t)$ along the solutions of the target system (8.33)–(8.35) and using (8.43) and (8.46) we have that

$$\dot V(t)\leq-\gamma_3(|X(t)|)+\gamma_4(|w(0,t)|)-2\gamma_4(L(t)). \tag{8.48}$$

Using the fact that $\gamma_4(|w(0,t)|)\leq\gamma_4\left(\sup_{x\in[0,1]}|e^{cx}w(x,t)|\right)=\gamma_4(L(t))$, we get $\dot V(t)\leq-\gamma_3(|X(t)|)-\gamma_4(L(t))$. Using (8.42), the definition of $L(t)$ in (8.45), and (8.47) we conclude that there exists a class $\mathcal{K}$ function γ_5 such that

$$\dot V(t)\leq-\gamma_5(V(t)). \tag{8.49}$$

Consequently, using the comparison principle (Lemma B.7 in Appendix B) and Lemma C.6 we can conclude that there exists a class $\mathcal{KL}$ function β_1 such that

$$V(t)\leq\beta_1(V(t_0),t-t_0). \tag{8.50}$$

Using (8.42), the definition of $V(t)$ in (8.47), and the properties of a class $\mathcal{K}$ function we arrive at

$$|X(t)|+L(t)\leq\beta_2(|X(t_0)|+L(t_0),t-t_0) \tag{8.51}$$

for some class $\mathcal{KL}$ function β_2. Moreover, from (8.45) we get

$$\sup_{x\in[0,1]}|w(x,t)|\leq L(t)\leq e^c\sup_{x\in[0,1]}|w(x,t)|. \tag{8.52}$$

Define now the solution of the transport PDE (8.34)–(8.35) as

$$w(x,t)=W\left(\phi\left(t+x\left(\phi^{-1}(t)-t\right)\right)\right). \tag{8.53}$$

Then combining (8.52) and (8.53) we get

$$\sup_{\theta\in[\phi(t),t]}|W(\theta)|\leq L(t), \tag{8.54}$$

$$L(t_0)\leq e^c\sup_{\theta\in[\phi(t_0),t_0]}|W(\theta)|. \tag{8.55}$$

Combining (8.51) and (8.54)–(8.55), the lemma is proved for some class $\mathcal{KL}$ function β. □

The following two lemmas allow one to establish stability in the original variables.

Lemma 8.10. *There exists a class $\mathcal{K}_\infty$ function ρ_1 such that*

$$\sup_{x\in[0,1]}|p(x,t)|\leq\rho_1\left(|X(t)|+\sup_{x\in[0,1]}|u(x,t)|\right),\quad t\geq t_0. \tag{8.56}$$

Proof. Consider the system

$$\dot{Y}(\tau) = f(Y(\tau), \omega(\tau)), \tag{8.57}$$

with μ being a positive constant. With a time rescaling, namely, $t = \frac{\tau}{\mu}$, we get

$$\frac{dY(t\mu)}{dt} = \mu f(Y(t\mu), \omega(t\mu)). \tag{8.58}$$

Under Assumption 8.1 we get

$$\frac{\partial R(Y'(t))}{\partial Y'(t)} \mu f(Y'(t), \omega'(t)) \le \mu\Big(R(Y'(t)) + \delta_3\big(\big|\omega'(t)\big|\big)\Big), \tag{8.59}$$

where

$$Y'(t) = Y(t\mu), \tag{8.60}$$
$$\omega'(t) = \omega(t\mu). \tag{8.61}$$

By differentiating relation (8.24) with respect to the spatial variable x, we get the following ODE in x:

$$p_x(x,t) = \left(\phi^{-1}(t) - t\right) f(p(x,t), u(x,t)), \tag{8.62}$$

with initial conditions

$$p(0,t) = X(t). \tag{8.63}$$

Viewing $\phi^{-1}(t) - t$ in (8.62) as a parameter (rather as the running variable) and comparing (8.62) with (8.58), with the help of (8.59) we get

$$\begin{aligned}\frac{\partial R(p(x,t))}{\partial p(x,t)} \left(\phi^{-1}(t) - t\right) f(p(x,t), u(x,t)) \le \left(\phi^{-1}(t) - t\right)& \\ \times \left(R(p(x,t)) + \delta_3(|u(x,t)|)\right).&\end{aligned} \tag{8.64}$$

Using the comparison principle (Lemma B.7 in Appendix B) and relation (8.63) we have that

$$\begin{aligned} R(p(x,t)) &\le e^{x\left(\phi^{-1}(t)-t\right)} R(X(t)) + \left(\phi^{-1}(t) - t\right) \int_0^x e^{\left(\phi^{-1}(t)-t\right)(x-y)} \delta_3(|u(y,t)|)\, dy \\ &\le e^{\frac{1}{\pi_0^*}} \left(R(X(t)) + \sup_{x\in[0,1]} \delta_3(|u(x,t)|)\, dy \right), \end{aligned} \tag{8.65}$$

where we used bound (8.6). Using (8.2) and the properties of class $\mathcal{K}_\infty$ functions we get the statement of the lemma. □

Lemma 8.11. *There exists a class $\mathcal{K}_\infty$ function ρ_2 such that*

$$\sup_{x\in[0,1]} |\xi(x,t)| \le \rho_2\left(|X(t)| + \sup_{x\in[0,1]} |w(x,t)| \right), \quad t \ge t_0. \tag{8.66}$$

Proof. Differentiating (8.39) with respect to x we get the following ODE in x:

$$\xi_x(x,t)=(\phi^{-1}(t)-t)f(\xi(x,t),\kappa(r(x,t),\xi(x,t))+w(x,t)), \quad x\in[0,1], \tag{8.67}$$

with initial conditions

$$\xi(0,t)=X(t). \tag{8.68}$$

With a rescaling of the spatial variable x, say y as

$$y=t+x(\phi^{-1}(t)-t), \tag{8.69}$$

and by defining

$$\zeta(y,t)=\xi\left(\frac{y-t}{\phi^{-1}(t)-t},t\right), \tag{8.70}$$

$$\omega(y,t)=w\left(\frac{y-t}{\phi^{-1}(t)-t},t\right), \tag{8.71}$$

with the help of (8.32) we rewrite (8.67)–(8.68) in the new spatial variable y as

$$\zeta_y(y,t)=f(\zeta(y,t),\kappa(y,\zeta(y,t))+\omega(y,t)), \quad y\in[t,\phi^{-1}(t)], \tag{8.72}$$

$$\zeta(t,t)=X(t). \tag{8.73}$$

Under Assumption 8.2 (see Appendix C.14), there exist a class $\mathcal{KL}$ function $\beta_3(\cdot,y)$ and a class $\mathcal{K}$ function γ_6 such that

$$|\zeta(y,t)|\leq\beta_3(|X(t)|,y-t)+\gamma_6\left(\sup_{y\in[t,\phi^{-1}(t)]}|\omega(y,t)|\right), \tag{8.74}$$

where we also used (8.68). Using (8.69)–(8.71) we have for all $x\in[0,1]$ that

$$|\xi(x,t)|\leq\beta_3\left(|X(t)|,x\left(\phi^{-1}(t)-t\right)\right)+\gamma_6\left(\sup_{x\in[0,1]}|w(x,t)|\right). \tag{8.75}$$

By noting that β_3 is a decreasing function of the second argument and by taking the supremum of both sides we arrive at

$$\sup_{x\in[0,1]}|\xi(x,t)|\leq\beta_3(|X(t)|,0)+\gamma_6\left(\sup_{x\in[0,1]}|w(x,t)|\right). \tag{8.76}$$

With standard properties of class $\mathcal{K}_\infty$ and $\mathcal{KL}$ functions, we get the bound of the lemma. □

***Proof of Theorem* 8.5.** Under Assumption 8.2, since $x(t,X)$ is a C^1 function, periodic in t with $k(t,0)=0$ for all $t \geq 0$, there exists a class $\mathcal{K}_\infty$ function $\hat{\rho}$ such that

$$|x(t,\xi)| \leq \hat{\rho}(|\xi|) \quad \text{for all } t \geq 0. \tag{8.77}$$

Using relation (8.31) we have

$$\sup_{x\in[0,1]} |w(x,t)| \leq \sup_{x\in[0,1]} (|u(x,t)| + \hat{\rho}(|p(x,t)|)), \tag{8.78}$$

$$\sup_{x\in[0,1]} |u(x,t)| \leq \sup_{x\in[0,1]} (|w(x,t)| + \hat{\rho}(|\xi(x,t)|)). \tag{8.79}$$

With Lemmas 8.10 and 8.11 we have that

$$\sup_{x\in[0,1]} |w(x,t)| \leq \sup_{x\in[0,1]} |u(x,t)| + \hat{\rho}\circ\rho_1\left(|X(t)| + \sup_{x\in[0,1]} |u(x,t)|\right), \tag{8.80}$$

$$\sup_{x\in[0,1]} |u(x,t)| \leq \sup_{x\in[0,1]} |w(x,t)| + \hat{\rho}\circ\rho_2\left(|X(t)| + \sup_{x\in[0,1]} |w(x,t)|\right). \tag{8.81}$$

Using (8.20) and (8.53) one can conclude that there exist class $\mathcal{K}_\infty$ functions $\hat{\alpha}$ and $\hat{\beta}$ such that

$$|X(t)| + \sup_{\phi(t)\leq\theta\leq t} |W(\theta)| \leq \hat{\alpha}\left(|X(t)| + \sup_{\phi(t)\leq\theta\leq t} |U(\theta)|\right), \tag{8.82}$$

$$|X(t)| + \sup_{\phi(t)\leq\theta\leq t} |U(\theta)| \leq \hat{\beta}\left(|X(t)| + \sup_{\phi(t)\leq\theta\leq t} |W(\theta)|\right). \tag{8.83}$$

From Lemma 8.9 and (8.83) we get

$$|X(t)| + \sup_{\phi(t)\leq\theta\leq t} |U(\theta)| \leq \hat{\beta}\circ\beta\left(|X(t_0)| + \sup_{\phi(t_0)\leq\theta\leq t_0} |W(\theta)|, t-t_0\right). \tag{8.84}$$

Setting $t=t_0$ in (8.82), Theorem 8.5 is proved with $\alpha = \hat{\beta}\circ\beta\circ\hat{\alpha}$. □

8.3 ▪ Stabilization of the Nonholonomic Unicycle with Time-Varying Delay

We consider the problem of stabilizing a mobile robot modeled as

$$\dot{x}(t) = v(\phi(t))\cos(\theta(t)), \tag{8.85}$$

$$\dot{y}(t) = v(\phi(t))\sin(\theta(t)), \tag{8.86}$$

$$\dot{\theta}(t) = \omega(\phi(t)), \tag{8.87}$$

where $(x(t),y(t))$ is the position of the robot, $\theta(t)$ is the heading, $v(t)$ is the speed, and $\omega(t)$ is the turning rate, subject to the input delay given as

$$\phi(t) = t - \frac{1+t}{1+2t}. \tag{8.88}$$

Consequently

$$\phi'(t) = 1 + \frac{1}{(1+2t)^2}, \tag{8.89}$$

$$\phi^{-1}(t) = t + \frac{t+1}{\sqrt{(t+1)^2+1}+t}. \tag{8.90}$$

From expressions (8.88)–(8.90) one can see that the function $\phi(t)$ in (8.88) satisfies both Assumptions 8.3 and 8.4. When $D = 0$ (i.e., $\phi(t) = t$) a time-varying stabilizing controller for this system is proposed in [142] as

$$\omega(t) = -5P(t)^2 \cos\left(3\phi^{-1}(t)\right) - P(t)Q(t)\left(1 + 25\cos^2\left(3\phi^{-1}(t)\right)\right) - \Theta(t), \tag{8.91}$$

$$v(t) = -P(t) + 5Q(t)\left(\sin\left(3\phi^{-1}(t)\right) - \cos\left(3\phi^{-1}(t)\right)\right) + Q(t)\omega(t), \tag{8.92}$$

$$P(t) = X(t)\cos(\Theta(t)) + Y(t)\sin(\Theta(t)), \tag{8.93}$$

$$Q(t) = X(t)\sin(\Theta(t)) - Y(t)\cos(\Theta(t)), \tag{8.94}$$

with

$$X = x, \quad Y = y, \quad \Theta = \theta, \quad \phi^{-1}(t) = t. \tag{8.95}$$

The predictor-based version of (8.91)–(8.94) is given with

$$X(t) = x(t) + \int_{\phi(t)}^{t} \frac{v(s)\cos(\Theta(s))\,ds}{\phi'\left(\phi^{-1}(s)\right)}, \tag{8.96}$$

$$Y(t) = y(t) + \int_{\phi(t)}^{t} \frac{v(s)\sin(\Theta(s))\,ds}{\phi'\left(\phi^{-1}(s)\right)}, \tag{8.97}$$

$$\Theta(t) = \theta(t) + \int_{\phi(t)}^{t} \frac{\omega(s)ds}{\phi'\left(\phi^{-1}(s)\right)}. \tag{8.98}$$

The initial conditions are chosen as $x(0) = y(0) = \theta(0) = 1$ and $\omega(s) = v(s) = 0$ for all $\phi(0) \leq s \leq 0$. From the given initial conditions one can verify that the controller "kicks in" at the time instant $t_0 = \phi^{-1}(0) = \frac{1}{\sqrt{2}}$. In Figure 8.1 we show the trajectory of the robot in the xy plane and the heading, whereas in Figure 8.2 we show the response of the controls $v(t)$ and $\omega(t)$. In the case of the uncompensated controller (8.91)–(8.94), (8.95), the system is unstable.

8.4 ▪ Notes and References

A control design for nonlinear systems with time-varying input delays is proposed by Karafyllis in [63]. Nonlinear systems with time-varying delays are present in numerous real-world applications. Successful application of various control design methods to networked control systems is constrained by the presence of time-varying input delays [32], [53], [123], [173]. Time-varying delays appear also in supply networks [153], [160] and irrigation channels [95]. Last but not least, the reaction time of a driver varies over time [134], [152], and hence can be modeled as a time-varying input delay.

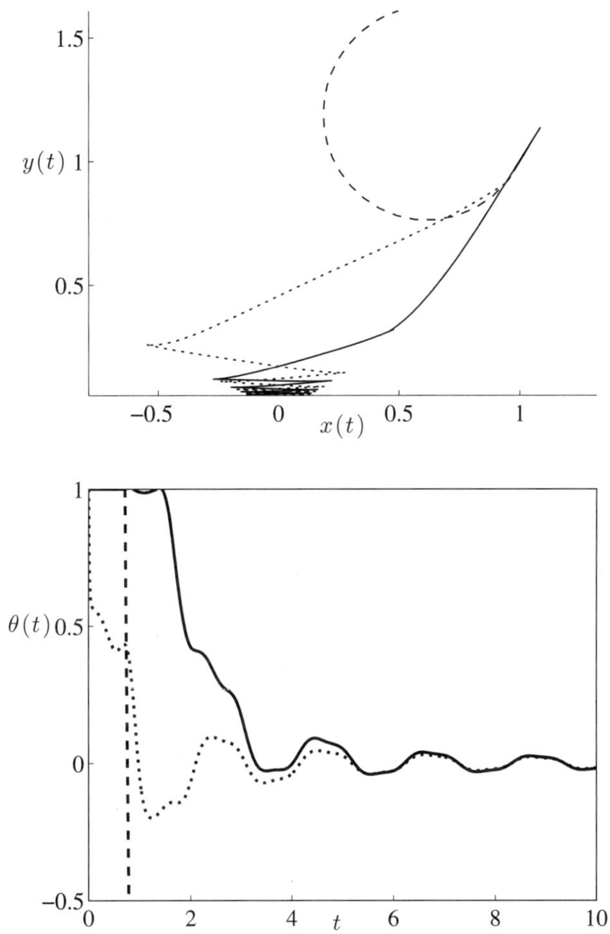

Figure 8.1. *The trajectory of the robot (top) and the heading $\theta(t)$ (bottom) with the compensated controller* (8.91)–(8.94), (8.96)–(8.98) *(solid line), the uncompensated controller* (8.91)–(8.94), (8.95) *(dashed line) and the controller* (8.91)–(8.94), (8.95) *for the delay-free system (dotted line) with initial conditions $x(0) = y(0) = \theta(0) = 1$ and $\omega(s) = v(s) = 0$ for all $\phi(0) \leq s \leq 0$.*

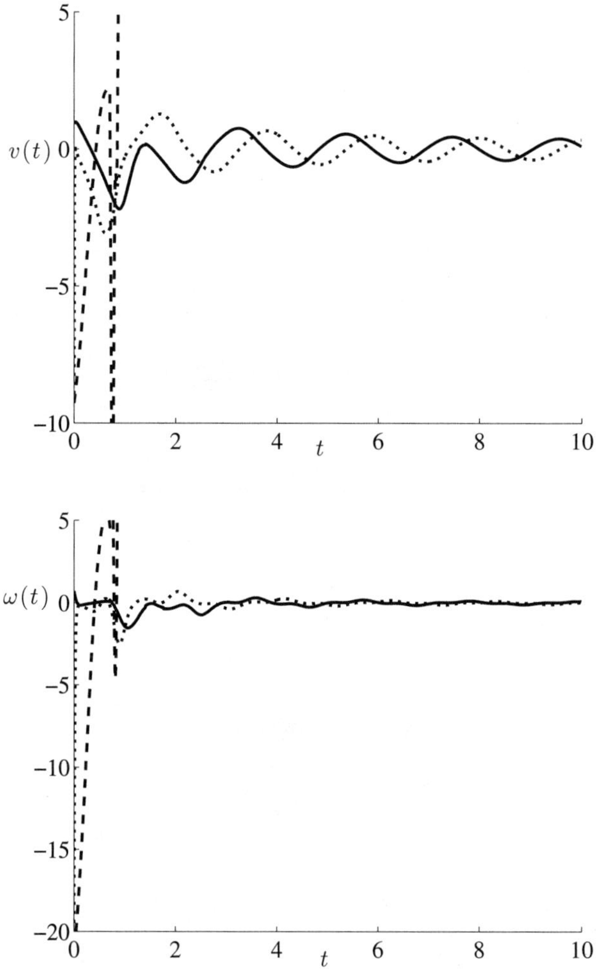

Figure 8.2. *The control efforts $v(t)$ (top) and $\omega(t)$ (bottom) with the controller* (8.91)–(8.94), (8.96)–(8.98) *(solid line), the controller* (8.91)–(8.94), (8.95) *(dashed line) and the controller* (8.91)–(8.94), (8.95) *for the delay-free system (dotted line) with initial conditions $x(0) = y(0) = \theta(0) = 1$ and $\omega(s) = v(s) = 0$ for all $\phi(0) \le s \le 0$.*

Chapter 9

Nonlinear Systems with Simultaneous Time-Varying Delays on the Input and the State

In this chapter we "backstep" one time-varying integrator and design a globally stabilizing controller for nonlinear strict-feedback systems with time-varying delays on the virtual inputs. One of the challenges of designing the predictor feedback law for this class of systems is that the control signal reaches different states of the system after different time intervals. Hence, the predictor states for different states (in addition to being time varying) have to be determined for different prediction horizons. Moreover, the technical details of the stability analysis of the closed-loop system under the predictor feedback design are much more involved than the case of only input delay. The reason is that in the present case one has to assume forward completeness of an infinite-dimensional plant rather than of a finite-dimensional plant as in the case of plants with only input delay (even when there is no input delay, the pant is infinite-dimensional since there is a state delay). Finally, the predictor feedback design does not follow immediately from a nominal design for the delay-free plant, but one has to design the predictor feedback law directly on the delay system. This is achieved by constructing two infinite-dimensional backstepping transformations: one for the delayed internal state and one for the actuator state.

We design the nonlinear predictor feedback law in Section 9.1, and we prove stability of the closed-loop system with the introduction of an infinite-dimensional backstepping procedure in Section 9.2. We illustrate our design with a second order strict-feedback system with time-varying state delay in Section 9.3.

9.1 ▪ Predictor Feedback Design for Systems with Both Input and State Delays

We consider the system

$$\dot{X}_1(t) = f_1(X_1(t)) + X_2(\phi_1(t)), \tag{9.1}$$

$$\dot{X}_2(t) = f_2(X_1(t), X_2(t)) + U(\phi_2(t)), \tag{9.2}$$

where we assume for notational simplicity that X_1 and $X_2 \in \mathbb{R}$, $U \in \mathbb{R}$, and $t \in \mathbb{R}_+$. Moreover, it is assumed that the functions ϕ_1 and ϕ_2 satisfy the following assumptions.

Assumption 9.1. *The functions*

$$\phi_1(t) = t - D_1(t), \tag{9.3}$$

$$\phi_2(t) = t - D_2(t) \tag{9.4}$$

satisfy

$$\phi_1(t) < t \quad \textit{for all } t \geq 0, \tag{9.5}$$
$$\phi_2(t) < t \quad \textit{for all } t \geq 0 \tag{9.6}$$

and

$$\pi^*_{0_{\phi_1}} = \frac{1}{\sup_{\theta \geq \phi_1^{-1}(0)} (\theta - \phi_1(\theta))} > 0, \tag{9.7}$$
$$\pi^*_{0_{\phi_2}} = \frac{1}{\sup_{\theta \geq \phi_2^{-1}(0)} (\theta - \phi_2(\theta))} > 0. \tag{9.8}$$

Assumption 9.2. *The derivative of the functions $\phi_1(t)$ and $\phi_2(t)$ satisfy*

$$\phi_1'(t) > 0 \quad \textit{for all } t \geq 0, \tag{9.9}$$
$$\phi_2'(t) > 0 \quad \textit{for all } t \geq 0, \tag{9.10}$$

and it holds that

$$\pi^*_{1_{\phi_1}} = \frac{1}{\sup_{\theta \geq \phi_1^{-1}(0)} \phi_1'(\theta)} > 0, \tag{9.11}$$
$$\pi^*_{1_{\phi_2}} = \frac{1}{\sup_{\theta \geq \phi_2^{-1}(0)} \phi_2'(\theta)} > 0. \tag{9.12}$$

Assumptions 9.1 and 9.2 also guarantee that for the function

$$\psi(t) = \phi_2(\phi_1(t)), \tag{9.13}$$

it holds that $\psi(t) < t$ and $\psi'(t) > 0$ for all $t \geq 0$, and moreover that

$$\pi^*_{1_{\psi}} = \frac{1}{\sup_{\theta \geq \psi^{-1}(0)} \psi'(\theta)} > 0, \tag{9.14}$$
$$\pi^*_{0_{\psi}} = \frac{1}{\sup_{\theta \geq \psi^{-1}(0)} (\theta - \psi(\theta))} > 0. \tag{9.15}$$

We assume that $f_1 : C^1(\mathbb{R};\mathbb{R})$, $f_2 : \mathbb{R}^2 \mapsto \mathbb{R}$ is locally Lipschitz with respect to its arguments, and it holds that $f_1(0) = 0$, $f_2(0,0) = 0$. For system (9.1)–(9.2) we design a predictor-based feedback for stabilizing the origin. System (9.1)–(9.2) can possibly escape to infinity before the controller reaches it. We thus make the following assumption.

Assumption 9.3. *System* (9.1)–(9.2) *is forward-complete.*

The predictor-based controller for system (9.1)–(9.2) is

$$\begin{aligned} U(t) = &-f_2\left(P_1\left(\psi\left(\phi_2^{-1}(t)\right)\right), P_2(t)\right) - c_2\Big(P_2(t) + c_1 P_1(t) + f_1(P_1(t))\Big) \\ &-\left(c_1 + \frac{\partial f_1(P_1)}{\partial P_1}\right)(f_1(P_1(t)) + P_2(t)) R\left(\phi_2^{-1}(t)\right), \end{aligned} \tag{9.16}$$

where c_1, c_2 are positive constants that satisfy $c_1\pi^*_{1_{\phi_1}} \neq 2c_2$ (a choice made to simplify one step in the analysis) and

$$P_1(t) = X_1(t) + (\psi^{-1}(t) - t)\int_0^1 \Big(f_1(P_1(\psi(t + y(\psi^{-1}(t) - t))))$$
$$+ P_2(\psi(t + (\psi^{-1}(t) - t)y))\Big)dy$$
$$= X_1(t) + \int_{\psi(t)}^{t} (f_1(P_1(\theta)) + P_2(\theta))\frac{d\theta}{\psi'(\psi^{-1}(\theta))}, \tag{9.17}$$
$$P_2(t) = X_2(t) + (\phi_2^{-1}(t) - t)\int_0^1 \Big(f_2\Big(P_1(\psi(t + y(\phi_2^{-1}(t) - t))),$$
$$P_2(\phi_2(t + y(\phi_2^{-1}(t) - t)))\Big) + U(\phi_2(t + y(\phi_2^{-1}(t) - t)))\Big)dy$$
$$= X_2(t) + \int_{\phi_2(t)}^{t} \frac{(f_2(P_1(\psi(\phi_2^{-1}(\theta))), P_2(\theta)) + U(\theta))d\theta}{\phi_2'(\phi_2^{-1}(\theta))}, \tag{9.18}$$
$$R(t) = \frac{d\phi_1^{-1}(t)}{dt}, \tag{9.19}$$

with initial conditions

$$P_1(\theta) = X_1(0) + \int_{\psi(0)}^{\theta} (f_1(P_1(\sigma)) + P_2(\sigma))\frac{d\sigma}{\psi'(\psi^{-1}(\sigma))},$$
$$\theta \in [\psi(0), 0], \tag{9.20}$$
$$P_2(\theta) = X_2(0) + \int_{\phi_2(0)}^{\theta} \frac{(f_2(P_1(\psi(\phi_2^{-1}(\sigma))), P_2(\sigma)) + U(\sigma))d\sigma}{\phi_2'(\phi_2^{-1}(\sigma))},$$
$$\theta \in [\phi_2(0), 0]. \tag{9.21}$$

To make clear the various prediction intervals that appear in the definitions (2.74) and (9.18) for the predictor states of X_1 and X_2, respectively, let us assume that $D_1(t) = D_1$ and $D_2(t) = D_2$ are constant. In that case $\psi(t) = t - D_1 - D_2$ and $\psi(\phi_2^{-1}(t)) = t - D_1$, and hence the prediction horizons for the states X_1 and X_2 are $D_1 + D_2$ and D_2, respectively. This is what one expects since the control signal reaches X_2 after D_2 time units, whereas it reaches X_1 through a delayed integrator and after $D_1 + D_2$ time units.

9.2 ▪ Stability Analysis

We now state the following theorem that is concerned with the stability properties of the closed-loop system that is composed of the plant (9.1)–(9.2) with the controller given in (9.16).

Theorem 9.4. *Consider the plant* (9.1)–(9.2) *together with the controller* (9.16)–(9.21). *Under Assumptions* 9.1–9.3 *there exists a class* $\mathcal{KL}$ *function* β_4 *such that for all initial conditions* $X_1(0) \in \mathbb{R}$, $X_2(\cdot) \in C[-D_1(0), 0]$, *and* $U(\cdot) \in L_\infty[-D_2(0), 0]$ *the following holds for all* $t \geq 0$:

$$\Omega(t) \leq \beta_4(\Omega(0), t), \tag{9.22}$$

where

$$\Omega(t)=|X_1(t)|+\sup_{\phi_1(t)\leq\theta\leq t}|X_2(\theta)|+\sup_{\phi_2(t)\leq\theta\leq t}|U(\theta)|. \tag{9.23}$$

We prove this theorem using a series of technical lemmas.

Lemma 9.5. *The signals $P_1(t)$ and $P_2(t)$ defined in* (9.17) *and in* (9.18) *are the $\psi^{-1}(t)-t$ and $\phi_2^{-1}(t)-t$-time-units-ahead predictors of the $X_1(t)$ and $X_2(t)$, respectively. Moreover, an equivalent representation for* (9.17)*–*(9.18) *is given by*

$$p_1(1,t)=X_1(t)+(\psi^{-1}(t)-t)\times\int_0^1\left(f_1(p_1(y,t))+p_2\left(\frac{\phi_1(t+(\psi^{-1}(t)-t)y)-t}{\phi_2^{-1}(t)-t},t\right)\right)dy, \tag{9.24}$$

$$p_2(1,t)=X_2(t)+(\phi_2^{-1}(t)-t)\times\int_0^1\left(f_2\left(p_1\left(\frac{\phi_2^{-1}(t)-t}{\psi^{-1}-t}y,t\right),p_2(y,t)\right)+u(y,t)\right)dy, \tag{9.25}$$

where

$$p_1(x,t)=P_1(\psi(t+x(\psi^{-1}(t)-t))), \tag{9.26}$$

$$p_2(x,t)=P_2(\phi_2(t+x(\phi_2^{-1}(t)-t))), \tag{9.27}$$

and x varies in $[0,1]$.

Proof. Consider the equivalent representation of system (9.1)–(9.2) using transport PDEs for the delayed state $X_2(t)$ and the controller $U(t)$ as

$$\dot X_1(t)=f_1(X_1(t))+\xi_2(0,t), \tag{9.28}$$

$$\xi_{2_t}(x,t)=\pi_1(x,t)\xi_{2_x}(x,t),\quad x\in[0,1], \tag{9.29}$$

$$\xi_2(1,t)=X_2(t), \tag{9.30}$$

$$\dot X_2(t)=f_2(X_1(t),X_2(t))+u(0,t), \tag{9.31}$$

$$u_t(x,t)=\pi_2(x,t)u_x(x,t),\quad x\in[0,1], \tag{9.32}$$

$$u(1,t)=U(t), \tag{9.33}$$

with

$$\pi_1(x,t)=\frac{1+x\left(\frac{d(\phi_1^{-1}(t))}{dt}-1\right)}{\phi_1^{-1}(t)-t}, \tag{9.34}$$

$$\pi_2(x,t)=\frac{1+x\left(\frac{d(\phi_2^{-1}(t))}{dt}-1\right)}{\phi_2^{-1}(t)-t}. \tag{9.35}$$

Consider the following ODEs in x (to clarify that these are ODEs in x, one should view the time t as a parameter rather than as the running variable of the ODE):

$$p_{1_x}(x,t)=(\psi^{-1}(t)-t)\times\left(f_1(p_1(x,t))+p_2\left(\frac{\phi_1(t+(\psi^{-1}(t)-t)x)-t}{\phi_2^{-1}(t)-t},t\right)\right), \tag{9.36}$$

$$p_{2_x}(x,t)=(\phi_2^{-1}(t)-t)\left(f_2\left(p_1\left(\frac{\phi_2^{-1}(t)-t}{\psi^{-1}-t}x,t\right),p_2(x,t)\right)+u(x,t)\right), \tag{9.37}$$

where x varies in $[0,1]$. The initial conditions for the above system of ODEs are given by

$$p_i(0,t)=X_i(t),\quad i=1,2, \tag{9.38}$$

and

$$p_2(\theta_2,t)=X_2\left(t+\theta_2\left(\phi_2^{-1}(t)-t\right)\right),\quad \theta_2\in\left[\frac{\phi_1(t)-t}{\phi_2^{-1}(t)-t},0\right]. \tag{9.39}$$

Assume for the moment that the following hold true:

$$p_{1_t}(x,t)=\pi_3(x,t)p_{1_x}(x,t),\quad x\in[0,1], \tag{9.40}$$

$$p_{2_t}(x,t)=\pi_2(x,t)p_{2_x}(x,t),\quad x\in[0,1], \tag{9.41}$$

where

$$\pi_3(x,t)=\frac{1+x\left(\frac{d(\psi^{-1}(t))}{dt}-1\right)}{\psi^{-1}(t)-t}. \tag{9.42}$$

Then, by taking into account (9.38), we have that

$$p_1(x,t)=X_1(t+x(\psi^{-1}(t)-t)),\quad x\in[0,1], \tag{9.43}$$

$$p_2(x,t)=X_2(t+x(\phi_2^{-1}(t)-t)),\quad x\in[0,1]. \tag{9.44}$$

By defining

$$p_1(1,t)=P_1(t), \tag{9.45}$$

$$p_2(1,t)=P_2(t), \tag{9.46}$$

we get (9.26). By integrating from 0 to x (9.36)–(9.37) we get

$$p_1(x,t)=X_1(t)+(\psi^{-1}(t)-t)\times\int_0^x\left(f_1(p_1(y,t))+p_2\left(\frac{\phi_1(t+(\psi^{-1}(t)-t)y)-t}{\phi_2^{-1}(t)-t},t\right)\right)dy, \tag{9.47}$$

$$p_2(x,t)=X_2(t)+(\phi_2^{-1}(t)-t)\times\int_0^x\left(f_2\left(p_1\left(\frac{\phi_2^{-1}(t)-t}{\psi^{-1}-t}y,t\right),p_2(y,t)\right)+u(y,t)\right)dy. \tag{9.48}$$

By setting in each $p_i(x,t)$, $x=1$, and using (9.26) we get (9.24)–(9.25).

To see that (9.40) holds, it is sufficient to prove that (9.43)–(9.44) is the unique solution of the ODEs in x given by (9.36)–(9.37) with the initial conditions (9.38)–(9.39). In this case relations (9.40)–(9.41) hold. Thus, it remains to prove that (9.43)–(9.44) is the unique solution of the initial value problem (9.36)–(9.39). Towards that end, we substitute (9.43)–(9.44) into (9.36)–(9.37),

$$X_1'(t+x(\psi^{-1}(t)-t))=f_1(X_1(t+x(\psi^{-1}(t)-t)))+X_2(\phi_1(t+(\psi^{-1}(t)-t)x)), \tag{9.49}$$

$$X_2'(t+x(\phi_2^{-1}(t)-t))=f_2(X_1(t+(\phi_2^{-1}(t)-t)x),X_2(t+x(\phi_2^{-1}(t)-t)))+U(\phi_2(t+x(\phi_2^{-1}(t)-t))), \tag{9.50}$$

where the prime symbol denotes the derivative with respect to the argument of a function. Taking into account (9.1)–(9.2) we conclude that indeed (9.43)–(9.44) is a solution of the ODEs in x given by (9.36)–(9.37). Furthermore, (9.43)–(9.44) satisfy the initial conditions (9.38)–(9.39). Since $X_2(s),\phi_1(0)\le s\le 0$, is continuous and based on Assumption 9.3 and [49, Section 2.2], the state $X_2\left(t+\theta_2\left(\phi_2^{-1}(t)-t\right)\right)$ is continuous for all $\theta_2\in\left[\frac{\phi_1(t)-t}{\phi_2^{-1}(t)-t},0\right]$ and $t\ge 0$. Using [49] we conclude that (9.43)–(9.44) is the unique solution of the ODEs in x given by (9.36)–(9.37) with the initial conditions (9.38)–(9.39). Thus, (9.40)–(9.41) hold. □

It is important here to observe that the total delay from the input $U(t)$ to the state $X_1(t)$ is $t-\psi(t)$ and to the state $X_2(t)$ is $t-\phi_2(t)$. This explains the fact that our predictor intervals are different for each state and specifically must be $\psi^{-1}(t)-t$ for $X_1(t)$ and $\phi_2^{-1}(t)-t$ for $X_2(t)$. Our controller design is based on a recursive procedure that transforms system (9.28)–(9.33) into a target system which is globally asymptotically stable with the controller (9.16)–(9.21). Then, using the invertibility of this transformation, we prove global asymptotic stability of the original system. We now state this transformation, along with its inverse.

Lemma 9.6. *The state transformation defined by*

$$Z_1(t)=X_1(t), \tag{9.51}$$

$$Z_2(t)=X_2(t)+f_1\left(p_1\left(\frac{\mu(t)}{\lambda(t)},t\right)\right)+c_1p_1\left(\frac{\mu(t)}{\lambda(t)},t\right), \tag{9.52}$$

where

$$\mu(t)=\phi_1^{-1}(t)-t, \tag{9.53}$$

$$\lambda(t)=\psi^{-1}(t)-t, \tag{9.54}$$

along with the transformation of the actuator state

$$
\begin{aligned}
w(x,t) = u(x,t) + f_2\left(p_1\left(\frac{\rho(t)}{\lambda(t)}x, t\right), p_2(x,t)\right) + c_2\Bigg(p_2(x,t) \\
+ f_1\left(p_1\left(\frac{\mu(x\rho(t)+t)+x\rho(t)}{\lambda(t)}, t\right)\right) + c_1 p_1\left(\frac{\mu(x\rho(t)+t)+x\rho(t)}{\lambda(t)}, t\right)\Bigg) \\
+ \left(\frac{\partial f_1\left(p_1\left(\frac{\mu(x\rho(t)+t)+x\rho(t)}{\lambda(t)}, t\right)\right)}{\partial p_1\left(\frac{\mu(x\rho(t)+t)+x\rho(t)}{\lambda(t)}, t\right)} + c_1\right) R\left(t + x\left(\phi_2^{-1}(t) - t\right)\right) \\
\times \left(f_1\left(p_1\left(\frac{\mu(x\rho(t)+t)+x\rho(t)}{\lambda(t)}, t\right)\right) + p_2(x,t)\right),
\end{aligned}
\tag{9.55}
$$

where

$$
\rho(t) = \phi_2^{-1}(t) - t, \tag{9.56}
$$

transforms the system (9.1)–(9.2) *to the target system with the control law given by* (9.16). *The target system is given by*

$$
\dot{Z}_1(t) = -c_1 Z_1(t) + Z_2(\phi_1(t)), \tag{9.57}
$$

$$
\dot{Z}_2(t) = -c_2 Z_2(t) + W(\phi_2(t)), \tag{9.58}
$$

where

$$
W(\theta) = 0, \quad \theta \geq 0. \tag{9.59}
$$

Proof. Before we start our recursive procedure, we rewrite the target system using transport PDEs:

$$
\dot{Z}_1(t) = -c_1 Z_1(t) + \zeta_2(0,t), \tag{9.60}
$$

$$
\zeta_{2_t}(x,t) = \pi_1(x,t)\zeta_{2_x}(x,t), \tag{9.61}
$$

$$
\zeta_2(1,t) = Z_2(t), \tag{9.62}
$$

$$
\dot{Z}_2(t) = -c_2 Z_2(t) + w(0,t), \tag{9.63}
$$

$$
w_t(x,t) = \pi_2(x,t) w_x(x,t), \tag{9.64}
$$

$$
w(1,t) = 0. \tag{9.65}
$$

Note that

$$
\zeta_2(x,t) = Z_2\left(\phi_1\left(t + x\left(\phi_1^{-1}(t) - t\right)\right)\right), \quad x \in [0,1]. \tag{9.66}
$$

Define

$$
\zeta_2(x,t) = \xi_2(x,t) + f_1\left(p_1\left(\frac{\mu(t)}{\lambda(t)}x, t\right)\right) + c_1 p_1\left(\frac{\mu(t)}{\lambda(t)}x, t\right). \tag{9.67}
$$

Then using (9.28) and (9.38) we get

$$
\dot{X}_1(t) = -c_1 X_1(t) + \zeta_2(0,t). \tag{9.68}
$$

Using relation (9.40) we have that

$$p_{1_t}(f(x,t),t) = \pi_3(f(x,t),t)\, p_{1_f}(f(x,t),t), \tag{9.69}$$

with

$$f(x,t) = \frac{\mu(t)}{\lambda(t)} x. \tag{9.70}$$

From (9.67) and by using relation (9.69) together with (9.29) we get

$$\begin{aligned} \zeta_{2_t}(x,t) = {} & \pi_1(x,t)\xi_{2_x}(x,t) + \left(\frac{\partial f_1(p_1(f(x,t),t))}{\partial p_1(f(x,t),t)} + c_1 \right) \\ & \times p_{1_f}(f(x,t),t)(f_t(x,t) + \pi_3(f(x,t),t)) \end{aligned} \tag{9.71}$$

and

$$\begin{aligned} \pi_1(x,t)\zeta_{2_x}(x,t) = {} & \pi_1(x,t)\xi_{2_x}(x,t) + \left(\frac{\partial f_1(p_1(f(x,t),t))}{\partial p_1(f(x,t),t)} + c_1 \right) \\ & \times p_{1_f}(f(x,t),t)\, \pi_1(x,t) f_x(x,t). \end{aligned} \tag{9.72}$$

Using (9.42), (9.53)–(9.54) and the definition of $f(x,t)$ in (9.70) we have that

$$\begin{aligned} f_t(x,t) + \pi_3(f(x,t),t) - \pi_1(x,t) f_x(x,t) = {} & \frac{1+\mu'(t)x}{\lambda(t)} - \frac{1+x\mu'(t)}{\mu(t)} \frac{\mu(t)}{\lambda(t)} \\ = {} & 0. \end{aligned} \tag{9.73}$$

Consequently, (9.61) holds. From (9.62) and by using (9.61) we have that

$$\begin{aligned} \dot{Z}_2(t) = {} & f_2(X_1(t), X_2(t)) + u(0,t) + \left(\frac{\partial f_1(p_1(f(1,t),t))}{\partial p_1(f(1,t),t)} + c_1 \right) \\ & \times p_{1_f}(f(1,t),t)\, \pi_1(1,t) f_x(1,t). \end{aligned} \tag{9.74}$$

With the help of (9.34), (9.36), (9.53)–(9.54), and (9.70) we rewrite the previous relation as

$$\begin{aligned} \dot{Z}_2(t) = {} & f_2(X_1(t), X_2(t)) + u(0,t) + \left(\frac{\partial f_1(p_1(f(1,t),t))}{\partial p_1(f(1,t),t)} + c_1 \right) \\ & \times (f_1(p_1(f(1,t),t)) + X_2(t)) \frac{d\phi_1^{-1}(t)}{dt}. \end{aligned} \tag{9.75}$$

Define now for notational convenience

$$g_1(x,t) = \frac{\rho(t)}{\lambda(t)} x, \tag{9.76}$$

$$g_2(x,t) = \frac{\mu(x\rho(t)+t) + x\rho(t)}{\lambda(t)}. \tag{9.77}$$

By noting that $g_1(x,t)$, $g_2(x,t)$, and $R\left(t+x\left(\phi_2^{-1}(t)-t\right)\right)$ satisfy the boundary value problems

$$g_{1_t}(x,t)+\pi_3\left(g_1(x,t),t\right)=\pi_2(x,t)g_{1_x}(x,t), \tag{9.78}$$
$$g_1(0,t)=0, \tag{9.79}$$
$$g_{2_t}(x,t)+\pi_3\left(g_2(x,t),t\right)=\pi_2(x,t)g_{2_x}(x,t), \tag{9.80}$$
$$g_2(0,t)=\frac{\mu(t)}{\lambda(t)}, \tag{9.81}$$
$$R_t\left(t+x\left(\phi_2^{-1}(t)-t\right)\right)=\pi_2(x,t)R_x\left(t+x\left(\phi_2^{-1}(t)-t\right)\right), \tag{9.82}$$

and using (9.55) we get (9.64). Substituting (9.55) for $x=0$ into (9.75) we get (9.63). Using the facts that $g_2(1,t)=\frac{\mu(\rho(t)+t)+\rho(t)}{\lambda(t)}=\frac{\mu\left(\phi_2^{-1}(t)\right)+\phi_2^{-1}(t)-t}{\phi_1^{-1}\circ\phi_2^{-1}(t)-t}=\frac{\phi_1^{-1}\circ\phi_2^{-1}(t)-t}{\phi_1^{-1}\circ\phi_2^{-1}(t)-t}=1$ and $p_1\left(\frac{\rho(t)}{\lambda(t)},t\right)=P_1\left(\phi_1(t)\right)$, with the controller (9.16) we get (9.65). Assuming an initial condition for (9.55) as $w(x,0)=w_0(x)$, and defining a new variable W as $w_0(x)=W\left(\phi_2\left(x\left(\phi_2^{-1}(0)\right)\right)\right)$ for all $x\in[0,1]$, we get that

$$w(x,t)=\begin{cases} W\left(\phi_2\left(t+x\left(\phi_2^{-1}(t)-t\right)\right)\right), & 0\le t+x\left(\phi_2^{-1}(t)-t\right)\le\phi_2^{-1}(0) \\ 0, & t+x\left(\phi_2^{-1}(t)-t\right)\ge\phi_2^{-1}(0)\end{cases}. \tag{9.83}$$

Defining $\theta=t+x\left(\phi_2^{-1}(t)-t\right)$ one gets (9.59). □

The functions $f(x,t)$, $g_1(x,t)$, and $g_2(x,t)$ are increasing with respect to x, and they take values within $[0,1]$. Hence, whenever they appear as the first argument in p_1 or p_2 they guarantee that the first argument of p_1 or p_2 varies within the interval $[0,1]$, which is consistent with the definitions (9.26)–(9.27). We now state the inverse of the transformation (9.51)–(9.55).

Lemma 9.7. *The inverse transformation of* (9.51)–(9.55) *is defined as*

$$X_1(t)=Z_1(t), \tag{9.84}$$
$$X_2(t)=Z_2(t)-\left(f_1\left(\eta_1\left(\frac{\mu(t)}{\lambda(t)},t\right)\right)+c_1\eta_1\left(\frac{\mu(t)}{\lambda(t)},t\right)\right), \tag{9.85}$$
$$\begin{aligned} u(x,t)=&\,w(x,t)-f_2\left(\eta_1\left(\frac{\rho(t)}{\lambda(t)}x,t\right),\eta_2(x,t)-f_1\left(\eta_1\left(g_2(x,t),t\right)\right)\right.\\ &\left.-c_1\eta_1\left(g_2(x,t),t\right)\right)-c_2\eta_2(x,t)-\left(\frac{\partial f_1\left(\eta_1\left(g_2(x,t),t\right)\right)}{\partial\eta_1}+c_1\right)\\ &\times\left(-c_1\eta_1\left(g_2(x,t),t\right)+\eta_2(x,t)\right)R\left(t+x\left(\phi_2^{-1}(t)-t\right)\right), \end{aligned} \tag{9.86}$$

where g_2 is given in (9.77) *and $\eta_1(x,t)$ and $\eta_2(x,t)$ are the predictors of the transformed states $Z_1(t)$ and $Z_2(t)$, and hence they satisfy*

$$\eta_1(x,t) = Z_1(t) + (\psi^{-1}(t) - t)\int_0^x \Bigg(-c_1\eta_1(y,t) + \eta_2\left(\frac{\phi_1\left(t + (\psi^{-1}(t)-t)x\right) - t}{\phi_2^{-1}(t) - t}, t\right)\Bigg)dy, \tag{9.87}$$

$$\eta_2(x,t) = Z_2(t) + (\phi_2^{-1}(t) - t)\int_0^x (-c_2\eta_2(y,t) + w(y,t))dy, \tag{9.88}$$

where x varies in $[0,1]$.

Proof. Using (9.67) we have that

$$\xi_2(x,t) = \zeta_2(x,t) - f_1\left(\eta_1\left(\frac{\mu(t)}{\lambda(t)}x, t\right)\right) + c_1\eta_1\left(\frac{\mu(t)}{\lambda(t)}x, t\right). \tag{9.89}$$

With (9.84) we get (9.28). Consider further

$$\dot{X}_2(t) = -c_2 Z_2(t) + w(0,t) - \left(\frac{\partial f_1(\eta_1(f(1,t),t))}{\partial \eta_1(f(1,t),t)} + c_1\right) \times \left(-c_1\eta_1\left(\frac{\mu(t)}{\lambda(t)}, t\right) + Z_2(t)\right)R(t). \tag{9.90}$$

Then, with (9.86) for $x = 0$, we get (9.31). $\square$

We now prove stability of the target system.

Lemma 9.8. *The target system* (9.57)–(9.59) *is globally exponentially stable in the sense that there exist positive constants G and g such that*

$$\Xi(t) \le G\Xi(0)e^{-gt}, \tag{9.91}$$

where

$$\Xi(t) = |Z_1(t)| + \sup_{\phi_1(t)\le\theta\le t}|Z_2(\theta)| + \sup_{\phi_2(t)\le\theta\le t}|W(\theta)|. \tag{9.92}$$

Proof. Solving explicitly (9.57)–(9.58) and using (9.59) we have for all $t \ge 0$ that

$$|Z_1(t)| \le \left(\Gamma_z + \frac{\Gamma_w}{\sqrt{c_1|c_1\pi^*_{1_{\phi_1}} - 2c_2|\inf_{\theta\ge\phi_1^{-1}(0)}\phi_1'(\theta)}}\right) \times \left(e^{-c_1\min\left\{1, \frac{\pi^*_{1_{\phi_1}}}{2}\right\}t} + e^{-c_2\max\{\phi_1(t),0\}}\right), \tag{9.93}$$

$$\sup_{\phi_1(t)\le\theta\le t}|Z_2(\theta)| \le 2\Gamma_w e^{-c_2\max\{\phi_1(t),0\}}, \tag{9.94}$$

where

$$\Gamma_z = |Z_1(0)| + \phi_1^{-1}(0)e^{c_1\phi_1^{-1}(0)}\sup_{\phi_1(0)\le\theta\le 0}|Z_2(\theta)|, \tag{9.95}$$

$$\Gamma_w = |Z_2(0)| + \phi_2^{-1}(0)e^{c_2\phi_2^{-1}(0)}\sup_{\phi_2(0)\le\theta\le 0}|W(\theta)|. \tag{9.96}$$

With the help of (8.45)–(8.46) and (8.54)–(8.55) we have

$$\sup_{\phi_2(t)\le\theta\le t} |W(\theta)| \le e^{-c\pi^*_{0_{\phi_2}} \min\left\{1,\pi^*_{1_{\phi_2}}\right\} t} e^{c} \sup_{\phi_2(0)\le\theta\le 0} |W(\theta)|. \tag{9.97}$$

Combining (9.93)–(9.97) and using the facts that $\max\{\phi_1(t),0\} \ge t - D_1(t)$ and that

$$\begin{aligned} \sup_{t\ge 0} D_1(t) &= \sup_{t\ge 0} t-\phi_1(t) \\ &\le \sup_{0\le t\le \phi_1^{-1}(0)} t-\phi_1(t) + \sup_{t\ge\phi_1^{-1}(0)} t-\phi_1(t), \end{aligned} \tag{9.98}$$

the proof is complete with

$$\begin{aligned} G &= 3\left(1+b+e^{c}+\phi_1^{-1}(0)e^{c_1\phi_1^{-1}(0)} + b\phi_2^{-1}(0)e^{c_2\phi_2^{-1}(0)}\right) \\ &\quad \times e^{g\left(\phi_1^{-1}(0)-\phi_1(0)+\frac{1}{\pi^*_{0_{\phi_1}}}\right)}, \end{aligned} \tag{9.99}$$

$$b = 2 + \frac{1}{\sqrt{c_1|c_1\pi^*_{1_{\phi_1}} - 2c_2|\inf_{\theta\ge\phi_1^{-1}(0)}\phi_1'(\theta)}}, \tag{9.100}$$

$$g = \min\left\{c_1, \frac{c_1\pi^*_{1_{\phi_1}}}{2}, c_2, c\pi^*_{0_{\phi_2}} \min\left\{1,\pi^*_{1_{\phi_2}}\right\}\right\}. \qquad \square \tag{9.101}$$

We have to relate now the stability of the target system with the stability of the system in the original variables.

Lemma 9.9. *There exists a class $\mathcal{K}_\infty$ function $\hat{\alpha}_1$ such that the following holds for all $t\ge 0$ and all $x\in[0,1]$:*

$$|p_1(x,t)| + |p_2(x,t)| \le \hat{\alpha}_1(\Omega(t)), \tag{9.102}$$

where Ω is defined in (9.23).

Proof. By performing a change of variables as $x = \frac{y}{\phi^{-1}(t)-t}$ in (9.36) and $x = \frac{y}{\phi_2^{-1}(t)-t}$ in (9.37), we rewrite the ODE (in x) system (9.36)–(9.38) as

$$p'_{1_y}(y,t) = f_1(p'_1(y,t)) + p'_2(\phi_1(t+y)-t,t), \quad y\in[0,\phi^{-1}(t)-t], \tag{9.103}$$

$$p'_{2_y}(y,t) = f_2(p'_1(y,t),p'_2(y,t)) + u'(y,t), \quad y\in[0,\phi_2^{-1}(t)-t], \tag{9.104}$$

where

$$p'_1(y,t) = p_1\left(\frac{y}{\phi^{-1}(t)-t},t\right), \quad y\in[0,\phi^{-1}(t)-t], \tag{9.105}$$

$$p'_2(y,t) = p_2\left(\frac{y}{\phi_2^{-1}(t)-t},t\right), \quad y\in[0,\phi_2^{-1}(t)-t], \tag{9.106}$$

$$u'(y,t) = u\left(\frac{y}{\phi_2^{-1}(t)-t},t\right), \quad y\in[0,\phi_2^{-1}(t)-t]. \tag{9.107}$$

Note that we view the function $\phi_1(t+y)-t$, where y is the independent variable, as the function $\phi_1(y)$ in (9.1) (note that $\phi_1(t+y)-t$ satisfies both Assumptions 9.1–9.2 for all $t \geq 0$ and $y \in [0, \psi^{-1}(t)-t]$). Under Assumption 9.3 and using Lemma 3.5 in [62] together with the fact that $f_1(0) = f_2(0,0) = 0$, we conclude that there exist a class $\mathcal{K}$ function μ and a class $\mathcal{K}_\infty$ function ν such that for all $t \geq 0$ the following holds:

$$|X_1(t)| + \sup_{\phi_1(t)\leq\theta\leq t} |X_2(\theta)| \leq \mu(t)\nu\Big(|X_1(0)| + \sup_{\phi_1(0)\leq\theta\leq 0} |X_2(\theta)| + \sup_{\theta\in[0,t]} |U(\phi_2(\theta))|\Big). \tag{9.108}$$

Comparing the ODE (in y) system (9.103)–(9.104) with (9.1)–(9.2) we get for all $y \in [0, \phi_2^{-1}(t)-t]$ that

$$|p_1'(y,t)| + |p_2'(y,t)| \leq \mu\left(\frac{1}{\pi^*_{0_{\phi_2}}}\right)\nu\Big(|p_1'(0,t)| + \sup_{\phi_1(t)-t\leq\theta\leq 0} |p_2'(\theta,t)| + \sup_{\theta\in[0,y]} \left|u'(\theta,t)\right|\Big). \tag{9.109}$$

Using definition (9.23) and (9.43)–(9.44) we get for all $y \in [0, \phi_2^{-1}(t)-t]$

$$|p_1'(y,t)| + |p_2'(y,t)| \leq \mu\left(\frac{1}{\pi^*_{0_{\phi_2}}}\right)\nu(\Omega(t)). \tag{9.110}$$

Define now the following function:

$$\omega(Y,r) = \sup\Big\{|p_1'(h)| : \phi_2^{-1}(t)-t \leq h \leq Y \leq \psi^{-1}(t)-t, \quad |p_1'(\phi_2^{-1}(t)-t)| + \sup_{\phi_2^{-1}(t)-t\leq\theta\leq\psi^{-1}(t)-t} |p_2'(\phi_1(t+\theta)-t)| \leq r, \text{where } p_2'(\phi_1(t+y)-t), \; p_1'(y) \text{ satisfy (9.103), (9.104) for all } \phi_2^{-1}(t)-t \leq y \leq \psi^{-1}(t)-t\Big\}. \tag{9.111}$$

From the forward-completeness assumption we know that $\omega(Y,r)$ is finite. Moreover, one concludes that for all $\phi_2^{-1}(t)-t \leq y \leq \psi^{-1}(t)-t$ the following holds

$$|p_1'(y)| \leq \omega\Big(y, |p_1'(\phi_2^{-1}(t)-t)| + \sup_{\phi_2^{-1}(t)-t\leq\theta\leq\psi^{-1}(t)-t} |p_2'(\phi_1(t+\theta)-t)|\Big). \tag{9.112}$$

Since from the definition of ω we conclude that for each fixed $\phi_2^{-1}(t)-t \leq y \leq \psi^{-1}(t)-t$ the mapping $\omega(y,\cdot)$ is increasing and for each fixed r the mapping $\omega(\cdot,r)$ is increasing, we get for all $\phi_2^{-1}(t)-t \le y \le \psi^{-1}(t)-t$ that

$$|p_1'(y)| \le \omega\left(\frac{1}{\pi^*_{0_\psi}}, |p_1'(\phi_2^{-1}(t)-t)| \right.$$
$$\left. + \sup_{\phi_2^{-1}(t)-t\le\theta\le\psi^{-1}(t)-t} |p_2'(\phi_1(t+\theta)-t)|\right). \tag{9.113}$$

Using the fact that $f_1(0)=0$ we conclude that $\omega(y,0)=0$ for all y, and hence there exists a class $\mathcal{K}_\infty$ function α^* such that

$$|p_1'(y)| \le \alpha^*\Bigg(|p_1'(\phi_2^{-1}(t)-t)|$$
$$+ \sup_{\phi_2^{-1}(t)-t\le\theta\le\psi^{-1}(t)-t} |p_2'(\phi_1(t+\theta)-t)|\Bigg), \tag{9.114}$$

where we absorb the finite constant $\pi^*_{0_\psi}$ into α^*. Therefore, using the fact that $\phi_1(t)-t \le \phi_1(\phi_2^{-1}(t))-t \le \phi_1(\theta+t)-t \le \phi_2^{-1}(t)-t$ for all $\phi_2^{-1}(t)-t \le \theta \le \psi^{-1}(t)-t$ and (9.110), the proof is complete. □

Lemma 9.10. *There exists a class $\mathcal{K}_\infty$ function $\hat{\alpha}_4$ such that the following holds for all $t \ge 0$:*

$$|\eta_1(x,t)|+|\eta_2(x,t)| \le \hat{\alpha}_4(\Xi(t)) \quad \text{for all } x\in[0,1], \tag{9.115}$$

where Ξ is defined in (9.92).

Proof. Relations (9.87)–(9.88) can be solved explicitly as

$$\eta_1(x,t) = Z_1(t)e^{-c_1(\psi^{-1}(t)-t)x} + (\psi^{-1}(t)-t)\int_0^x e^{-c_1(\psi^{-1}(t)-t)(x-y)}$$
$$\times \eta_2\left(\frac{\phi_1(t+(\psi^{-1}(t)-t)y)-t}{\phi_2^{-1}(t)-t}, t\right)dy, \tag{9.116}$$
$$\eta_2(x,t) = Z_2(t)e^{-c_1(\phi_2^{-1}(t)-t)x} + (\phi_2^{-1}(t)-t)$$
$$\times \int_0^x e^{-c_2(\phi_2^{-1}(t)-t)(x-y)} w(y,t)dy. \tag{9.117}$$

From relation (9.117), and using the fact that $0 \le \phi_2^{-1}(t)-t \le \frac{1}{\pi^*_{0_{\phi_2}}}$, we get

$$|\eta_2(x,t)| \le |Z_2(t)| + \frac{1}{\pi^*_{0_{\phi_2}}} \sup_{x\in[0,1]} |w(x,t)| \quad \text{for all } x\in[0,1]. \tag{9.118}$$

By changing variables in the integral in (9.116) and by using the fact that $\eta_2(x,t) = Z_2(t + x(\phi_2^{-1}(t)-t))$ we write

$$|\eta_1(x,t)| \le |Z_1(t)| + \int_0^{\phi_1^{-1}(t)-t} |Z_2(\phi_1(t+y))|dy$$
$$+ \int_{\inf_{t\ge0}\{\phi_1^{-1}(t)-t\}}^{\psi^{-1}(t)-t} \left|\eta_2\left(\frac{\phi_1(t+y)-t}{\phi_2^{-1}(t)-t}, t\right)\right| dy. \tag{9.119}$$

Using (9.118) and the uniform boundness of the functions $\phi_1^{-1}(t)-t$ and $\psi^{-1}(t)-t$ for all $t\geq 0$, the proof of the lemma is complete. □

***Proof of Theorem* 9.4.** Since $f_1(X_1)$, $f_2(X_1,X_2)$, and $f_1'(X_1)$ are continuous, there exist class $\mathcal{K}_\infty$ functions ρ_1, ρ_2, and $\hat{\delta}$ such that

$$f_1(X_1)\leq \rho_1(|X_1|), \tag{9.120}$$

$$f_2(X_1,X_2)\leq \rho_2(|X_1|+|X_2|), \tag{9.121}$$

$$\frac{\partial f_1(X_1)}{\partial X_1}\leq \left.\frac{\partial f_1(X_1)}{\partial X_1}\right|_{X_1=0}+\hat{\delta}(|X_1|). \tag{9.122}$$

Using relations (9.51)–(9.52), (9.55), and with the help of Lemma 9.9, we have

$$|Z_1(t)|=|X_1(t)|, \tag{9.123}$$

$$\sup_{\phi_1(t)\leq\theta\leq t}|Z_2(\theta)|\leq \rho_3(\Omega(t)), \tag{9.124}$$

$$\sup_{\phi_2(t)\leq\theta\leq t}|W(\theta)|\leq \rho_4(\Omega(t)), \tag{9.125}$$

where class $\mathcal{K}_\infty$ functions ρ_3 and ρ_4 are

$$\rho_3(s)=s+\rho_1\circ\hat{\alpha}_1(s)+c_1\hat{\alpha}_1(s), \tag{9.126}$$

$$\begin{aligned}\rho_4(s)=\;&\rho_2\circ\hat{\alpha}_1(s)+c_2((1+c_1)\hat{\alpha}_1(s)+\rho_1\circ\hat{\alpha}_1(s))+\left(\hat{\delta}\circ\hat{\alpha}_1(s)+c_1\right)\\&\times(\rho_1\circ\hat{\alpha}_1(s)+\hat{\alpha}_1(s))\frac{1}{\inf_{\theta\geq\phi^{-1}(0)}\phi_1'(\theta)}.\end{aligned} \tag{9.127}$$

Hence,

$$\Xi(t)\leq \rho_5(\Omega(t)), \tag{9.128}$$

with

$$\rho_5(s)=s+\rho_3(s)+\rho_4(s). \tag{9.129}$$

Similarly, using relations (9.84)–(9.86), and with the help of Lemma 9.10, we have for some class $\mathcal{K}_\infty$ function ρ_6 that

$$\Omega(t)\leq \rho_6(\Xi(t)). \tag{9.130}$$

Using (9.128)–(9.130), and with the help of Lemma 9.8, the theorem is proved with

$$\Omega(t)\leq \rho_6\left(G\rho_5(\Omega(0))e^{-gt}\right). \tag{9.131}$$

9.3 ▪ Simulations

In this example we consider the following system:

$$\dot{X}_1(t)=\sin(X_1(t))+X_2(\phi(t)), \tag{9.132}$$

$$\dot{X}_2(t)=U(t), \tag{9.133}$$

where the function $\phi(t)$ is given by

$$\phi(t) = t - \frac{1+t}{1+2t}, \tag{9.134}$$

and hence

$$\phi'(t) = 1 + \frac{1}{(1+2t)^2}, \tag{9.135}$$

$$\phi^{-1}(t) = t + \frac{t+1}{\sqrt{(t+1)^2+1}+t}. \tag{9.136}$$

We choose the initial conditions of the plant as $X_1(0) = 1$ and $X_2(s) = 0$ for all $s \in [\phi(0), 0]$. The controller for this system is

$$\begin{aligned} U(t) = & -c_2(X_2(t) + c_1 P_1(t) + \sin(P_1(t))) \\ & - (c_1 + \cos(P_1(t)))(\sin(P_1(t)) + X_2(t))\frac{d\phi^{-1}(t)}{dt}, \end{aligned} \tag{9.137}$$

where we choose $c_1 = c_2 = 2$ and

$$P_1(t) = X_1(t) + \int_{\phi(t)}^{t} (\sin(P_1(\theta)) + X_2(\theta))\frac{d\theta}{\phi'(\phi^{-1}(\theta))}. \tag{9.138}$$

In Figure 9.1 we show the response of the system in comparison with the uncompensated controller, i.e., the backstepping controller (9.137), which assumes $\phi(t) = t$.

9.4 ▪ Notes and References

An observer design for a class of nonlinear systems with time-varying delay in the measurement state is proposed by Cacace and coauthors in [28].

It is worth noting that we consider the second order case for simplicity of presentation. The results in the present chapter can be extended recursively to the general nth order strict-feedback class with delays in the integrator chain, as we did in Section 2.2, as well as with delays on other states in the system.

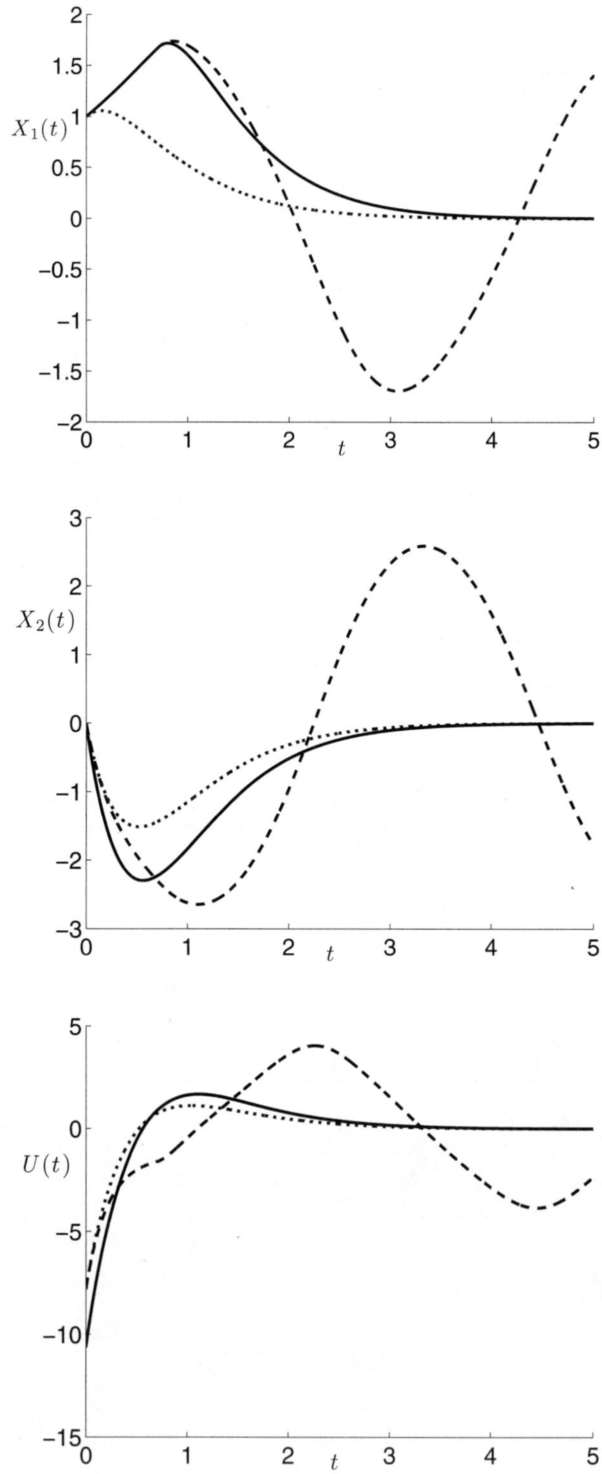

Figure 9.1. *System's response for the simulation example. Dotted lines: system with $\phi(t)=t$ and the uncompensated controller. Dashed lines: system with $\phi(t)$ as in (8.88) and the uncompensated controller. Solid lines: system with $\phi(t)$ as in (8.88) and the delay-compensating controller.*

Part III

State-Dependent Delays

Chapter 10
Predictor Feedback Design When the Delay Is a Function of the State

In this chapter we design the predictor feedback law for nonlinear systems with state-dependent delays. As was made evident throughout Part II of the book, which deals with time-varying delays, the prediction horizon over which the predictor is defined depends on the future values of the delay. Therefore, the state dependence of the delay makes the prediction horizon dependent on the future value of the state, which means that it is impossible to know a priori how far in the future the prediction is needed. In this chapter we resolve this key design challenge and provide the formula for the predictor state. We also highlight that due to a fundamental restriction on the allowable magnitude of the delay function's gradient (the control signal never reaches the plant if the delay rate is larger than one), global stabilization is not possible in general.

10.1 ▪ Nonlinear Predictor Feedback Design for State-Dependent Delay

We consider the system

$$\dot{X}(t) = f\,(X(t), U\,(t - D\,(X(t)))), \tag{10.1}$$

where $X \in \mathbb{R}^n$, $U : [t_0 - D\,(X(t_0)), \infty) \to \mathbb{R}$, $t \geq t_0 \geq 0$, $D \in C^1\left(\mathbb{R}^n; \mathbb{R}_+\right)$, and $f : \mathbb{R}^n \times \mathbb{R} \to \mathbb{R}^n$ is locally Lipschitz with $f(0,0) = 0$.

Let $P(t)$ denote the value of the state X at the time when the control U applied at time t reaches the plant. We refer to P as the "predictor" state. For systems with constant delays, $D = \text{const}$, it simply holds that $P(t) = X(t + D)$. For systems with state-dependent delays the situation is more complex. The time when U reaches the system depends on the value of the state at that time, namely, the following implicit relationship holds: $P(t) = X(t + D(P(t)))$ (and $X(t) = P(t - D(X(t)))$). Nevertheless, even for state-dependent delays, $P(t)$ is simply the predictor of the future state, at the time when the current control will have an effect on the state. Since $P(t)$ is related to $X(t)$ through an implicit relationship, it is evident that a predictor-based compensation of a state-dependent input delay faces a challenge, which is absent in the case where the delay is merely time-varying. We resolve this challenge by performing our design and analysis using transformations of the time variable, $t \mapsto t + D(P(t))$ and $t \mapsto t - D(X(t))$. The difficulty with these transformations, besides not being available explicitly, is that the "prediction horizon" $D(P(t))$ is in general different from the delay $D(X(t))$.

We design a predictor-based controller for the plant (10.1) as

$$U(t) = \kappa(\sigma(t), P(t)), \tag{10.2}$$

where

$$P(\theta) = X(t) + \int_{t-D(X(t))}^{\theta} \frac{f(P(s), U(s))ds}{1 - \nabla D(P(s)) f(P(s), U(s))},$$
$$t - D(X(t)) \le \theta \le t, \tag{10.3}$$
$$\sigma(\theta) = \theta + D(P(\theta)), \quad t - D(X(t)) \le \theta \le t, \tag{10.4}$$

for $t \ge t_0$. The initial predictor $P(\theta)$, $\theta \in [t_0 - D(X(t_0)), t_0]$, is given by (11.3) for $t = t_0$.

It takes some effort to comprehend the mathematical meaning of the relationship (10.3), where $P(\theta)$ appears on both sides of the equation. It is helpful to start from the linear case, $\dot{X}(t) = AX(t) + BU(t-D)$ with a constant delay D. In that case the predictor is given explicitly using the variation of constants formula, with the initial condition $P(t-D) = X(t)$, as $P(t) = \exp(AD)X(t) + \int_{t-D}^{t} \exp(A(t-\theta))BU(\theta)d\theta$. For systems that are nonlinear, and even for linear systems with a state-dependent delay, $P(t)$ cannot be written explicitly, for the same reason for which a nonlinear ODE cannot be solved explicitly. So we represent $P(t)$ implicitly using the nonlinear integral equation (10.3).

The computation of $P(t)$ from (10.3) is straightforward with a discretized implementation in which $P(\theta)$ is assigned values based on the right-hand side of (10.3), which involves earlier values of P and the values of the input U. At each time step the integral in (10.3) can be computed using a method of numerical integration (e.g., the trapezoidal rule) with a total number of discrete points N, given by $N(t) = \left\lfloor \frac{D(X(t))}{h} \right\rfloor$, where h is the time-discretization step and $\lfloor a \rfloor$ denotes the integer part of a. The implementation of the predictor in (10.3) might be unsafe when the discretization method of the predictor in (10.3) is not appropriately chosen [122], whereas a careful discretization yields a safe implementation [119].

To see that $P(t)$ given in (10.3) is the $\sigma(t) - t = D(P(t))$-time-units-ahead predictor of $X(t)$, differentiate (10.3) with respect to θ, set $\theta = t$, and perform a change of variables $\tau = \sigma(t)$ in the ODE for $X(\tau)$ given in (10.1) (where t is replaced by τ) to observe that $P(t)$ satisfies the same ODE in t as $X(\sigma(t))$. Since from (10.3) for $t = t_0$ and $\theta = t_0 - D(X(t_0))$ it follows that $P(t_0 - D(X(t_0))) = X(t_0)$, by defining

$$\phi(t) = t - D(X(t)), \quad t \ge t_0, \tag{10.5}$$
$$\sigma(\theta) = \phi^{-1}(\theta), \quad t - D(X(t)) \le \theta \le t, \tag{10.6}$$

we have that $P(t_0) = X(\sigma(t_0))$. Hence, indeed $P(t) = X(\sigma(t))$ for all $t \ge t_0$.

Noting from (10.5) and (10.6) that $D(X(\sigma(t))) = \sigma(t) - t$, differentiating this equation, we get that

$$\frac{d\sigma(\theta)}{d\theta} = \frac{1}{1 - \nabla D(P(\theta)) f(P(\theta), U(\theta))}, \quad t - D(X(t)) \le \theta \le t. \tag{10.7}$$

The possible division by zero in (10.7) indicates that $\phi(t)$ is not a priori guaranteed to be invertible. Another way to see this challenge is to note that the prediction horizon is governed by $\frac{dD(P(t))}{dt} = \frac{\nabla D(P(t)) f(P(t),U(t))}{1-\nabla D(P(t)) f(P(t),U(t))}$, which, in case of division by zero, can result in the reversal in the direction of the control signal reaching the plant at time $t + D(P(t))$, as displayed in Figure 10.1.

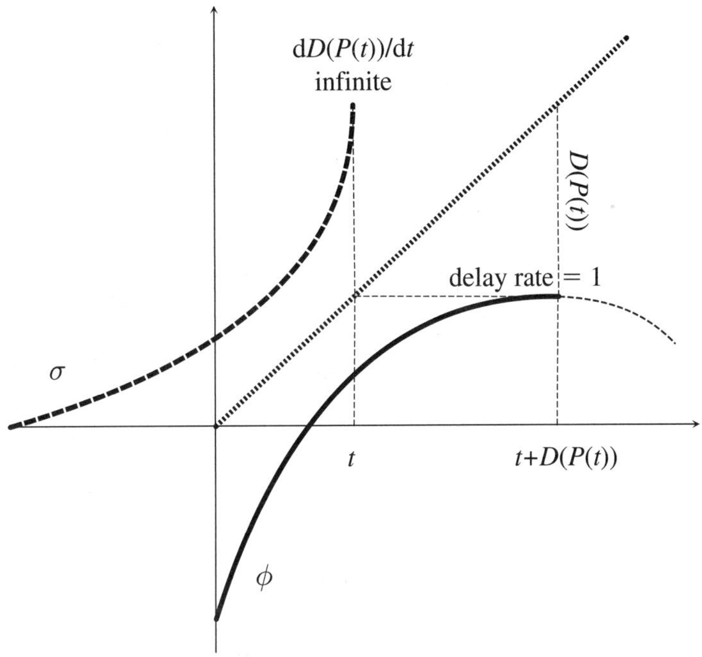

Figure 10.1. *The functions $\sigma(t)$ (dashed line) and $\phi(t)$ (solid line) in the undesirable case where the infinite slope of $D(P(t))$ causes their noninvertibility.*

Motivated by the need to keep the denominator in (10.3) and (10.7) positive, in the next two chapters we consider the condition on the solutions which is given by

$$\mathscr{F}_c: \qquad \nabla D(P(\theta)) f(P(\theta), U(\theta)) < c \quad \text{for all } \theta \geq t_0 - D(X(t_0)) \tag{10.8}$$

for $c \in (0,1]$. We refer to $\mathscr{F}_1$ as the *feasibility condition* of the controller (10.2)–(10.4). As we shall see, this condition is satisfied by restricting the initial condition of the system.

The following example illustrates the fact that global stabilization is not possible even for linear systems.

Example 10.1. We consider a scalar unstable system with a Lyapunov-like delay

$$\dot{X}(t) = X(t) + U\left(t - X(t)^2\right). \tag{10.9}$$

The delay-compensating controller is

$$U(t) = -2P(t), \tag{10.10}$$

where

$$P(\theta) = X(t) + \int_{t-X(t)^2}^{\theta} \frac{(P(s)+U(s))\,ds}{1-2P(s)(P(s)+U(s))}, \quad \theta \geq -X(0)^2. \tag{10.11}$$

In Figure 10.2 we show the response of the system and the function $\phi(t) = t - X(t)^2$ for four different initial conditions of the state and with the initial conditions for the input chosen as $U(\theta) = 0$, $-X(0)^2 \leq \theta \leq 0$. We choose $X(0) = 0.15, 0.25, 0.35, X^*$. With X^* we denote the critical value of $X(0)$ for the given initial condition of the input, such that,

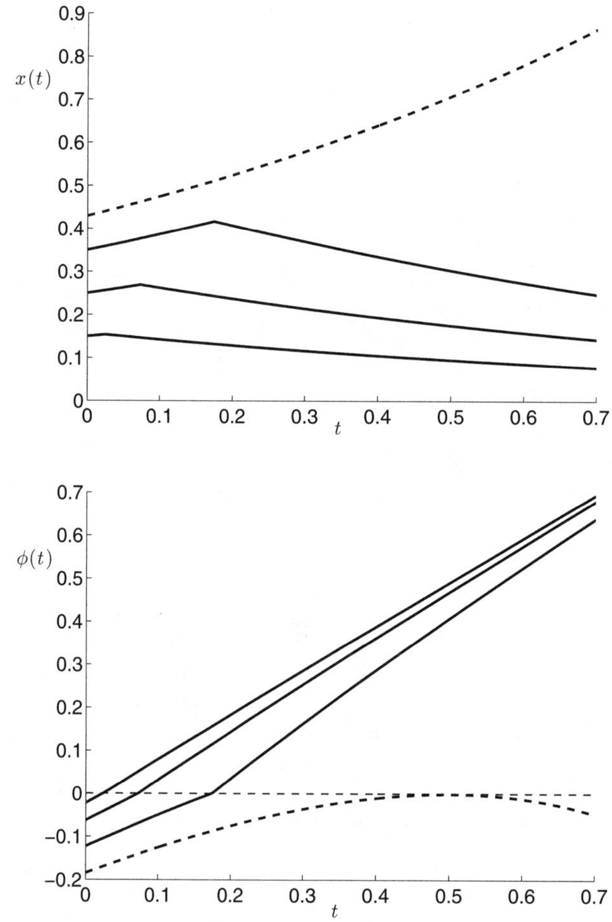

Figure 10.2. *Response of system* (10.9) *with the controller* (10.10)–(10.11) *with initial conditions* $U(\theta) = 0$, $-X(0)^2 \leq \theta \leq 0$ *and four different initial conditions for the state* $X(0) = 0.15, 0.25, 0.35, 0.43$.

for any $X(0) \geq X^*$, the control inputs produced by the feedback law (10.10), (10.11) for positive t never reach the plant. We calculate this time as follows: The function $\phi(t) = t - X(0)^2 e^{2t}$ has a maximum at t^* if $\log\left(\frac{1}{\sqrt{2X(0)^2}}\right) = t^* > 0$. Since $\phi(t^*) = \log\left(\frac{1}{\sqrt{2X(0)^2}}\right) - \frac{1}{2}$ has to be positive for the control to reach the plant, it follows that $X^* = \frac{1}{\sqrt{2e}} = 0.43$. ∎

10.2 ▪ Notes and References

State-dependent delays are ubiquitous, for example, in processes that involve mass transport in a domain that changes with time, such as extrusion processes [35]. In control over networks, it makes sense to send control signals less frequently when the state is small and more frequently when the state is large [53]. Network congestion control typically ignores the dependency of the round-trip time from the queue length, which can lead to the instability of the underlying network [96]. In control of mobile robots the magnitude of the delay depends on the distance of the robot from the operator interface [129]. A priori known functions of time are employed to model state-dependent delays in transmission channels of communication networks, which are used for the remote stabilization of unstable systems [173]. In supply networks, state-dependent delays appear due to transportation of materials [153], [160]. In milling processes, speed-dependent delays arise due to the deformation of the cutting tool [1]. The reaction time of a driver is often modeled as a pure [134] or distributed [152] delay. However, the delay depends on the intensity of the disturbance, the size of the tracking error to which the driver is reacting, the speed of the vehicle, the physical situation of the driver, etc. [52]. In irrigation channels the dynamics of a reach are accurately represented by a time-varying delayed-integrator model [95]. In population dynamics, the time required for the maturation level of a cell to achieve a certain threshold can be modeled as a state-dependent delay [100]. In engine cooling systems the delay in the distribution of the coolant among the consumers depends on the coolant flow [50]. Finally, models of constant delays can be used to approximate state-dependent delays in chemical process control [83], [125].

Chapter 11

Stability Analysis for Forward-Complete Systems with Input Delay

In this chapter we prove stability of the closed-loop system under the predictor feedback design presented in Chapter 10. Due to a fundamental restriction on the allowable magnitude of the delay function's gradient (the control signal never reaches the plant if the delay rate is larger than one), we obtain only a regional stability result, even for the case of forward-complete systems. We establish an estimate of the region of attraction for our control scheme in the state space of the infinite-dimensional closed-loop nonlinear system based on the construction of a strict, time-varying Lyapunov functional in Section 11.1. We present a global result for forward-complete systems under a restrictive Lyapunov-like condition, which has to be a priori verified, that the delay rate be bounded by unity irrespective of the values of the state and input in Section 11.3. We also deal with linear systems, treating them as a special case of the design for nonlinear systems, for which we prove exponential stability. We present several examples, including stabilization of the nonholonomic unicycle subject to distance-dependent input delay in Section 11.2.

11.1 ▪ Stability Analysis for Forward-Complete Nonlinear Systems

We start this section by recalling the predictor feedback design of Chapter 10. The predictor-based controller for the plant

$$\dot{X}(t)=f(X(t),U(t-D(X(t)))), \tag{11.1}$$

where $X\in\mathbb{R}^n$, $U:[t_0-D(X(t_0)),\infty)\to\mathbb{R}$, $t\geq t_0\geq 0$, and $f:\mathbb{R}^n\times\mathbb{R}\to\mathbb{R}^n$ is locally Lipschitz with $f(0,0)=0$, is

$$U(t)=\kappa(\sigma(t),P(t)), \tag{11.2}$$

where

$$P(\theta)=X(t)+\int_{t-D(X(t))}^{\theta}\frac{f(P(s),U(s))ds}{1-\nabla D(P(s))f(P(s),U(s))},$$
$$t-D(X(t))\leq\theta\leq t, \tag{11.3}$$
$$\sigma(\theta)=\theta+D(P(\theta)),\quad t-D(X(t))\leq\theta\leq t, \tag{11.4}$$

for $t\geq t_0$. The initial predictor $P(\theta)$, $\theta\in[t_0-D(X(t_0)),t_0]$, is given by (11.3) for $t=t_0$. To keep the denominator in (11.3) positive, we consider the condition on the solutions which is given by

$$\mathscr{F}_c: \qquad \nabla D(P(\theta))f(P(\theta),U(\theta))<c \quad \text{for all } \theta \geq t_0-D(X(t_0)) \tag{11.5}$$

for $c \in (0,1]$. We refer to $\mathscr{F}_1$ as the *feasibility condition* of the controller (11.2)–(11.4). As we shall see, this condition is satisfied by restricting the initial condition of the system.

We state next the assumptions for the delay function and the plant.

Assumption 11.1. $D \in C^1(\mathbb{R}^n;\mathbb{R}_+)$ *and* ∇D *is locally Lipschitz.*[1]

Assumption 11.2. *The plant* $\dot{X}=f(X,\omega)$ *is strongly forward complete, that is, there exist a smooth positive definite function* R *and class* $\mathscr{K}_\infty$ *functions* α_2, α_3, *and* α_4 *such that*

$$\alpha_2(|X|) \leq R(X) \leq \alpha_3(|X|), \tag{11.6}$$

$$\frac{\partial R(X)}{\partial X} f(X,\omega) \leq R(X)+\alpha_4(|\omega|) \tag{11.7}$$

for all $X \in \mathbb{R}^n$ *and for all* $\omega \in \mathbb{R}$.

Assumption 11.2 guarantees that system (11.1) does not exhibit finite escape time, that is, for every initial condition and every locally bounded input signal the corresponding solution is defined for all $t \geq t_0$. The difference between Assumption 11.2 and Theorem C.13 is that Theorem C.13 guarantees that $R(\cdot)$ is nonnegative and radially unbounded, whereas we assume in addition that $R(\cdot)$ is positive definite (which can be justified by the fact that we assume $f(0,0)=0$).

Assumption 11.3. *The plant* $\dot{X}(t)=f(X(t),\varkappa(t,X(t))+\omega(t))$ *is input-to-state stable with respect to* ω, *and the function* $\varkappa$ *is locally Lipschitz in both arguments and periodic in* t *with* $\varkappa(t,0)=0$ *for all* $t \geq 0$.

Note that under Assumption 11.3, since $\varkappa(t,X)$ is locally Lipschitz and periodic in t with $k(t,0)=0$ for all $t \geq 0$, there exists a class $\mathscr{K}_\infty$ function $\hat{\alpha}$ such that

$$|\varkappa(t,\xi)| \leq \hat{\alpha}(|\xi|) \quad \text{for all } t \geq 0. \tag{11.8}$$

Theorem 11.4. *Consider the plant* (11.1) *together with the control law* (11.2)–(11.4). *Under Assumptions* 11.1, 11.2, *and* 11.3 *there exist a class* $\mathscr{K}$ *function* ψ_{RoA}, *a class* $\mathscr{K}\mathscr{C}_\infty$ *function* ρ, *and a class* $\mathscr{K}\mathscr{L}$ *function* β *such that for all initial conditions for which* U *is locally Lipschitz on the interval* $[t_0-D(X(t_0)),t_0)$ *and which satisfy*

$$B_0(c): \qquad |X(t_0)|+\sup_{t_0-D(X(t_0))\leq\theta\leq t_0}|U(\theta)|<\psi_{\text{RoA}}(c) \tag{11.9}$$

for some $0<c<1$, *there exists a unique solution to the closed-loop system with* X *Lipschitz on* $[t_0,\infty)$, U *Lipschitz on* (t_0,∞), *and*

$$\Omega(t) \leq \beta(\rho(\Omega(t_0),c),t-t_0), \tag{11.10}$$

$$\Omega(t)=|X(t)|+\sup_{t-D(X(t))\leq\theta\leq t}|U(\theta)| \tag{11.11}$$

[1] To ensure uniqueness of solutions.

for all $t \geq t_0$*. Furthermore, there exists a class* $\mathscr{K}$ *function* δ^* *such that, for all* $t \geq t_0$,

$$D(X(t)) \leq D(0) + \delta^*(c), \tag{11.12}$$

$$\left|\dot{D}(X(t))\right| \leq c. \tag{11.13}$$

We prove Theorem 11.4 using Lemmas 11.5–11.12, which are presented next. Note that the definitions of class $\mathscr{KC}$ and $\mathscr{KC}_\infty$ functions are the ones from Appendix C.1.

Lemma 11.5 (backstepping transform of actuator state). *The infinite-dimensional backstepping transformation of the actuator state defined by*

$$W(\theta) = U(\theta) - \kappa(\sigma(\theta), P(\theta)), \quad t - D(X(t)) \leq \theta \leq t, \tag{11.14}$$

together with the predictor-based control law given in relations (11.2)–(11.4)*, transforms the system* (11.1) *to the target system given by*

$$\dot{X}(t) = f\left(X(t), \kappa(t, X(t)) + W(t - D(X(t)))\right), \tag{11.15}$$

$$W(t) = 0 \quad \textit{for all } t \geq t_0. \tag{11.16}$$

Proof. Using (11.2) and the facts that $P(t - D(X(t))) = X(t)$ and $\sigma(t - D(X(t))) = t$, which are immediate consequences of (11.3)–(11.4), we get the statement of the lemma. □

The role of $P(t)$ in our analysis is as an output of a mapping of the state $X(t)$ and input history $U(\theta), \theta \in [t - D, t]$. For instance, in the linear case with a constant delay, this mapping is given explicitly as $P(t) = \exp(AD)X(t) + \int_{t-D}^{t} \exp(A(t-\theta))BU(\theta)d\theta$. In the nonlinear case, the mapping $(X, U) \mapsto P$ is given implicitly by (11.3). The mapping $(X, U) \mapsto P$ is an intermediate step in transforming the original system (X, U) into the target system (X, W), as displayed in Figure 11.1. The transformation $(X, U) \mapsto (X, W)$ is important because the stability analysis can be conducted in the variables (X, W), but not in the original variables (X, U).

Lemma 11.6 (inverse backstepping transform). *The inverse of the infinite-dimensional backstepping transformation defined in* (11.14) *is given by*

$$U(\theta) = W(\theta) + \kappa(\sigma(\theta), \Pi(\theta)), \quad t - D(X(t)) \leq \theta \leq t, \tag{11.17}$$

where

$$\Pi(\theta) = X(t) + \int_{t-D(X(t))}^{\theta} \frac{f(\Pi(s), \kappa(\sigma(s), \Pi(s)) + W(s))}{1 - \nabla D(\Pi(s)) f(\Pi(s), \kappa(\sigma(s), \Pi(s)) + W(s))} ds,$$
$$t - D(X(t)) \leq \theta \leq t. \tag{11.18}$$

Proof. By direct verification, noting also that $\Pi(\theta) = P(\theta)$ for all $t - D(X(t)) \leq \theta \leq t$, where $\Pi(\theta)$ is driven by the transformed input $W(\theta)$, whereas $P(\theta)$ is driven by the input $U(\theta)$. See Figure 11.1. □

It may be slightly puzzling why we present two versions of the predictor, P and Π, which are in fact the same. The reason we use two distinct symbols for the same quantity

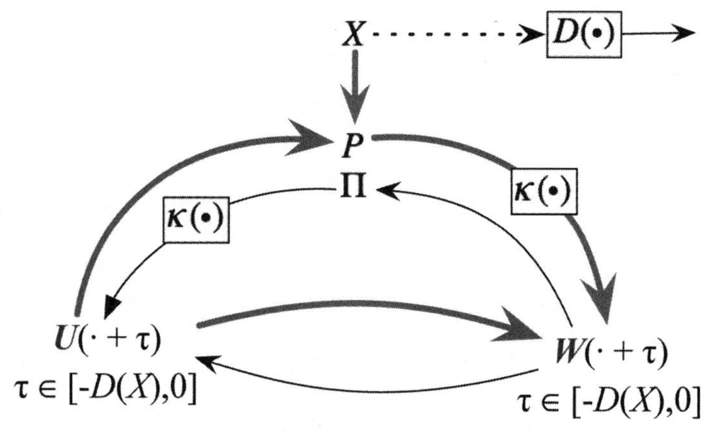

Figure 11.1. *Interconnections between the predictor states P and Π with the transformations W and U in* (11.14) *and* (11.17). *The direct backstepping transformation is defined as* $(X(t), U(\theta)) \mapsto (X(t), W(\theta))$ *and is given in* (11.14), *where $P(\theta)$ is given as a function of $X(t)$ and $U(\theta)$ through relations* (11.3)–(11.4). *Analogously, the inverse transformation is defined as* $(X(t), W(\theta)) \mapsto (X(t), U(\theta))$ *and is given in* (11.17), *where $\Pi(\theta)$ is given as a function of $X(t)$ and $W(\theta)$ through relation* (11.18).

is that, in one case, P is expressed in terms of X and U for the direct backstepping transformation, while, in the other case, Π is expressed in terms of X and W for the inverse backstepping transformation. Since the actual system operates in the (X,U) variables and the analysis is conducted in the (X,W) variables, both the direct and backstepping transformations are important.

Lemma 11.7 (stability estimate for target system). *For any positive constant g, there exist a class $\mathcal{K}_\infty$ function δ_1 and a class $\mathcal{KL}$ function β_2 such that for all solutions of the system satisfying* (11.5) *for $0 < c < 1$, the following hold:*

$$\Xi(t) \leq \beta_2(\rho_*(\Xi(t_0), c), t - t_0), \tag{11.19}$$

$$\Xi(t) = |X(t)| + \sup_{t-D(X(t))\leq\theta\leq t} |W(\theta)| \tag{11.20}$$

for all $t \geq t_0$, where

$$\rho_*(s,c) = \frac{e^{\frac{g}{1-c}}}{1-c} e^{\frac{g}{1-c}(D(0)+\delta_1(s))} s. \tag{11.21}$$

Proof. Based on Assumption 11.3 and Theorem C.16, there exist a C^1 function $S(t, X(t)) : \mathbb{R}_+ \times \mathbb{R}^n \to \mathbb{R}_+$ and class $\mathcal{K}_\infty$ functions α_5, α_6, α_7, and α_8 such that

$$\alpha_5(|X(t)|) \leq S(t, X(t)) \leq \alpha_6(|X(t)|), \tag{11.22}$$

$$\dot{S}(t, X(t)) \leq -\alpha_7(|X(t)|) + \alpha_8(|W(t - D(X(t)))|), \tag{11.23}$$

$$\dot{S}(t, X(t)) = \frac{\partial S(t, X(t))}{\partial t} + \frac{\partial S(t, X(t))}{\partial X(t)} f(X(t), \kappa(t, X(t)) + W(t - D(X(t)))). \tag{11.24}$$

Consider now the following Lyapunov functional for the target system given in (11.15)–(11.16):

$$V(t)=S(t,X(t))+k\int_0^{L(t)}\frac{\alpha_8(r)}{r}dr, \tag{11.25}$$

where

$$\begin{aligned} L(t)&=\sup_{t-D(X(t))\le\theta\le t}\left|e^{g(1+\sigma(\theta)-t)}W(\theta)\right| \\ &=\lim_{n\to\infty}\left(\int_{t-D(X(t))}^{t}e^{2ng(1+\sigma(\theta)-t)}W(\theta)^{2n}d\theta\right)^{\frac{1}{2n}}, \end{aligned} \tag{11.26}$$

with $g>0$. We now upper- and lower-bound $L(t)$ in terms of $\sup_{t-D(X(t))\le\theta\le t}|W(\theta)|$. From (11.5) for $0<c<1$ we get that $\dot\sigma(\theta)\le\frac{1}{1-c}$. Integrating the relation $\dot\sigma(\theta)\le\frac{1}{1-c}$ from $t-D(X(t))$ to θ and, since $\sigma(t-D(X(t)))=t$, we have

$$1+\sigma(\theta)-t\le\frac{1-c+D(X(t))}{1-c},\quad t-D(X(t))\le\theta\le t. \tag{11.27}$$

Since $D\in C^1(\mathbb{R}^n;\mathbb{R}_+)$ there exists a function $\delta_1\in\mathcal{K}_\infty\cap C^1$ such that

$$D(X)\le D(0)+\delta_1(|X|). \tag{11.28}$$

Therefore,

$$L(t)\le\frac{e^{\frac{g}{1-c}(1+D(0)+\delta_1(|X(t)|))}}{1-c}\sup_{t-D(X(t))\le\theta\le t}|W(\theta)|. \tag{11.29}$$

Similarly, using the fact that $\sigma(t-D(X(t)))-t=0$ and since $\sigma(\theta)$ is increasing we get

$$1\le 1+\sigma(\theta)-t,\quad t-D(X(t))\le\theta\le t. \tag{11.30}$$

Therefore, with the help of (11.30) we have that

$$L(t)\ge e^{g}\sup_{t-D(X(t))\le\theta\le t}|W(\theta)|. \tag{11.31}$$

Taking the time derivative of $L(t)$, with (11.16) we get

$$\begin{aligned} \dot L(t)=\lim_{n\to\infty}\frac{1}{2n}&\left(\int_{t-D(X(t))}^{t}e^{2ng(1+\sigma(\theta)-t)}W(\theta)^{2n}d\theta\right)^{\frac{1}{2n}-1} \\ &\times\left(-\left(1-\frac{dD(X(t))}{dt}\right)e^{2ng}W(t-D(X(t)))^{2n}\right. \\ &\left.-2ng\int_{t-D(X(t))}^{t}e^{2ng(1+\sigma(\theta)-t)}W(\theta)^{2n}d\theta\right). \end{aligned} \tag{11.32}$$

Using (11.5) we have $\frac{dD(X(t))}{dt}<1$ and hence $\dot L(t)\le-gL(t)$. With this inequality and (11.23), taking the derivative of (11.25) we get

$$\dot V(t)\le-\alpha_7(|X(t)|)+\alpha_8(|W(t-D(X(t)))|)-kg\alpha_8(L(t)). \tag{11.33}$$

With the help of (11.31) and choosing $k = \frac{2}{g}$ we get $\dot{V}(t) \le -\alpha_7(|X(t)|) - \alpha_8(L(t))$. Using (11.22), the definition of $L(t)$ in (11.26) and (11.25) we conclude that there exists a class $\mathcal{K}$ function γ_1 such that

$$\dot{V}(t) \le -\gamma_1(V(t)). \tag{11.34}$$

Using the comparison principle (Lemma B.7 in Appendix B) and Lemma C.6, there exists a class $\mathcal{KL}$ function β_1 such that $V(t) \le \beta_1(V(t_0), t - t_0)$. Using (11.22), the definition of $V(t)$ in (11.25), and the properties of class $\mathcal{K}$ functions we arrive at

$$|X(t)| + L(t) \le \beta_2(|X(t_0)| + L(t_0), t - t_0) \tag{11.35}$$

for some class $\mathcal{KL}$ function β_2. Using relations (11.29) and (11.31), the lemma is proved. □

There is conservativeness involved in passing between (X, U) and (X, W) in both directions, just like there is a loss to a currency exchange customer both when buying and selling. With Lemma 11.8 we quantify a bound when going from (X, U) to (X, W) via P, and with Lemma 11.9 we quantify a bound when going back from (X, W) to (X, U) via Π.

Lemma 11.8 (bound on predictor in terms of actuator state). *There exists a class $\mathcal{KC}_\infty$ function ρ_1 such that for all solutions of the system satisfying* (11.5) *for* $0 < c < 1$, *the following holds:*

$$|P(\theta)| \le \rho_1\left(|X(t)| + \sup_{t-D(X(t)) \le s \le t} |U(s)|, c\right), \quad t - D(X(t)) \le \theta \le t. \tag{11.36}$$

Proof. Consider the following ODE in θ, which follows by differentiating (11.3):

$$\frac{dP(\theta)}{d\theta} = \frac{f(P(\theta), U(\theta))}{1 - \nabla D(P(\theta)) f(P(\theta), U(\theta))}, \quad t - D(X(t)) \le \theta \le t. \tag{11.37}$$

With the change of variables

$$y = \sigma(\theta), \tag{11.38}$$

we rewrite (11.37) as

$$\frac{dP(\phi(y))}{dy} = f(P(\phi(y)), U(y - D(P(\phi(y))))), \quad t \le y \le \sigma(t). \tag{11.39}$$

Using (11.7) we get

$$\frac{dR(P(\phi(y)))}{d\theta}\frac{d\theta}{dy} \le R(P(\phi(y))) + \alpha_4(|U(y - D(P(\phi(y))))|). \tag{11.40}$$

With (11.5) we have

$$\frac{dR(P(\theta))}{d\theta} \le \frac{1}{1-c}(R(P(\theta)) + \alpha_4(|U(\theta)|)), \quad t - D(X(t)) \le \theta \le t. \tag{11.41}$$

Using the comparison principle (Lemma B.7 in Appendix B) and (11.28) one gets

$$R(P(\theta)) \le e^{\frac{D(0)+\delta_1(|X(t)|)}{1-c}} \left(R(X(t)) + \sup_{t-D(X(t)) \le s \le t} \alpha_4(|U(s)|) \right) \tag{11.42}$$

for $t - D(X(t)) \le \theta \le t$. With standard properties of class $\mathscr{K}_\infty$ functions we get the statement of the lemma where the class $\mathscr{K}\mathscr{C}_\infty$ function ρ_1 is given as

$$\rho_1(s,c) = \alpha_2^{-1}\left((\alpha_3(s) + \alpha_4(s)) e^{\frac{D(0)+\delta_1(s)}{1-c}} \right). \qquad \square \tag{11.43}$$

Lemma 11.9 (bound on predictor in terms of transformed actuator state). *There exists a class $\mathscr{K}$ function γ_4 such that for all solutions of the system satisfying* (11.5) *for* $0 < c < 1$, *the following holds:*

$$|\Pi(\theta)| \le \gamma_4\left(|X(t)| + \sup_{t-D(X(t)) \le s \le t} |W(s)| \right), \quad t - D(X(t)) \le \theta \le t. \tag{11.44}$$

Proof. Under Assumption 11.3 (see Appendix C.14), there exist class $\mathscr{K}\mathscr{L}$ function β_3 and a class $\mathscr{K}$ function γ_2 such that

$$|Y(\tau)| \le \beta_3(|Y(t_0)|, \tau - t_0) + \gamma_2\left(\sup_{s \ge t_0} |\omega(s)| \right), \quad \tau \ge t_0, \tag{11.45}$$

where $Y(\tau)$ is the solution of $\dot{Y}(\tau) = f(Y(\tau), \kappa(\tau, Y(\tau)) + \omega(\tau))$. Using the change of variable (11.38) and definitions (11.18), (11.39), we have that

$$\frac{d\Pi(\phi(y))}{dy} = f(\Pi(\phi(y)), \kappa(y, \Pi(\phi(y))) + W(\phi(y))), \quad t \le y \le \sigma(t). \tag{11.46}$$

Using (11.45) we have for all $t - D(X(t)) \le \theta \le t$

$$|\Pi(\theta)| \le \gamma_3(|X(t)|) + \gamma_2\left(\sup_{t-D(X(t)) \le s \le t} |W(s)| \right), \tag{11.47}$$

where the class $\mathscr{K}$ function γ_3 is defined as $\gamma_3(s) = \beta_3(s, 0)$. With the properties of class $\mathscr{K}$ functions we get the statement of the lemma where $\gamma_4(s) = \gamma_2(s) + \gamma_3(s)$ is of class $\mathscr{K}$. $\square$

Lemma 11.10 (equivalence of norms for original and target system). *There exist a function ρ_2 of class $\mathscr{K}\mathscr{C}_\infty$ and a class $\mathscr{K}_\infty$ function α_9 such that for all solutions of the system satisfying* (11.5) *for* $0 < c < 1$, *the following hold:*

$$|X(t)| + \sup_{t-D(X(t)) \le \theta \le t} |U(\theta)| \le \alpha_9^{-1}\left(|X(t)| + \sup_{t-D(X(t)) \le \theta \le t} |W(\theta)| \right), \tag{11.48}$$

$$|X(t)| + \sup_{t-D(X(t)) \le \theta \le t} |W(\theta)| \le \rho_2\left(|X(t)| + \sup_{t-D(X(t)) \le \theta \le t} |U(\theta)|, c \right) \tag{11.49}$$

for all $t \ge t_0$.

Proof. With the inverse backstepping transformation (11.17) and the bound (11.44) we get the bound (11.48) with $\alpha_9^{-1}(s) = s + \hat{\alpha} \circ \gamma_4(s)$. Using the direct backstepping transformation (11.14) and the bound (11.36) we get the bound (11.49) with $\rho_2(s,c) = s + \hat{\alpha}(\rho_1(s,c))$. □

Lemma 11.11 (ball around the origin within the feasibility region). *There exists a function $\bar{\rho}_c$ of class $\mathcal{KC}_\infty$ such that all the solutions that satisfy*

$$\bar{B}(c): \qquad |X(t)| + \sup_{t-D(X(t))\leq\theta\leq t} |U(\theta)| < \bar{\rho}_c(c,c), \quad t \geq t_0, \tag{11.50}$$

for $0 < c < 1$ *also satisfy* (11.5).

Proof. Since $D \in C^1(\mathbb{R}^n;\mathbb{R}_+)$ and $f : \mathbb{R}^n \times \mathbb{R} \to \mathbb{R}^n$ is locally Lipschitz with $f(0,0) = 0$ there exist class $\mathcal{K}_\infty$ functions δ_2 and α_1 such that[2]

$$|\nabla D(X)| \leq |\nabla D(0)| + \delta_2(|X|), \tag{11.51}$$

$$|f(X,\omega)| \leq \alpha_1(|X| + |\omega|), \tag{11.52}$$

and hence

$$|f(X(t), U(t - D(X(t))))| \leq \alpha_1\left(|X(t)| + \sup_{t-D(X(t))\leq\theta\leq t} |U(\theta)|\right). \tag{11.53}$$

If a solution satisfies for all $t - D(X(t)) \leq \theta \leq t$

$$(|\nabla D(0)| + \delta_2(|P(\theta)|))\alpha_1\left(|P(\theta)| + \sup_{t-D(X(t))\leq s\leq t} |U(s)|\right) < c \tag{11.54}$$

for $0 < c < 1$, then it also satisfies (11.5). Using Lemma 11.8 we conclude that (11.54) is satisfied for $0 < c < 1$ as long as (11.50) holds, where the class $\mathcal{KC}_\infty$ function ρ_c is defined as

$$\rho_c(s,c) = (|\nabla D(0)| + \delta_2(\rho_1(s,c)))\alpha_1((\rho_1(s,c) + s)), \tag{11.55}$$

and with $\bar{\rho}_c$ we denote the inverse function of ρ_c with respect to ρ_c's first argument. □

Lemma 11.12 (estimate of the region of attraction). *There exists a class $\mathcal{K}$ function ψ_{RoA} such that for all initial conditions of the closed-loop system* (11.1)–(11.4) *that satisfy relation* (11.9) *the solutions of the system satisfy* (11.50) *for* $0 < c < 1$ *and hence satisfy* (11.5).

---

[2]Estimate (11.51) is derived based on the nonrestrictive assumption that the delay is a continuously differentiable function of X. Using bound (11.51), one can restrict the gradient of the delay D (which is needed for (11.5)) by restricting the size of the state X. This enables one to estimate the region of attraction of the proposed control law (see Lemma 11.12).

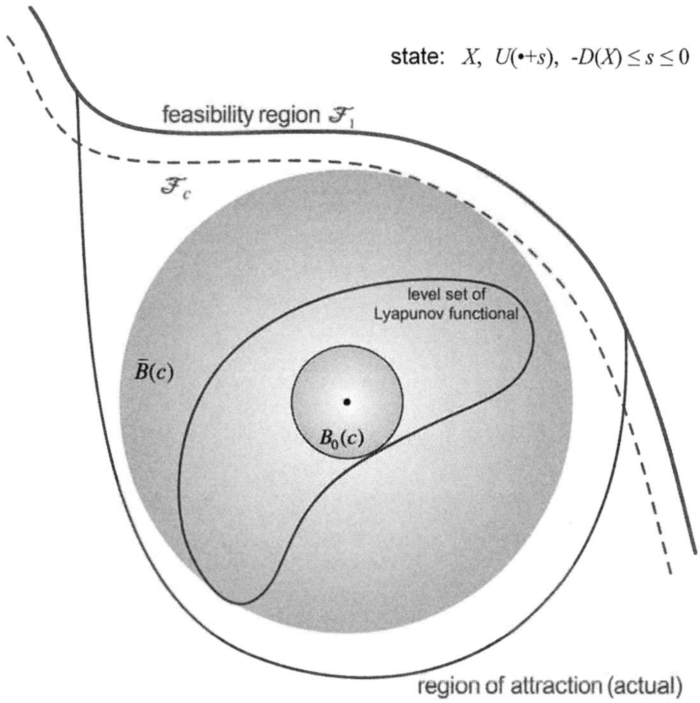

Figure 11.2. *Sets arising in the proof of Theorem* 11.4 *in the infinite-dimensional state space* $\mathbb{R}^n \times C[t - D(X(t)), t)$. $B_0(c)$: *the ball of initial conditions allowed in the proof of the theorem.* $\bar{B}(c)$: *the ball inside which the ensuing solutions are trapped.*

Proof. Using Lemma 11.10, with the help of (11.19), we have that

$$\Omega(t) \leq \alpha_9^{-1}(\beta_2(\rho_*(\rho_2(\Omega(t_0), c), c), t - t_0)), \tag{11.56}$$

where $\Omega(t)$ is defined in (11.11). By defining the class $\mathcal{K}_\infty$ function α_{10} as $\alpha_{10}(s) = \alpha_9^{-1}(\beta_2(s, 0))$, we get

$$\Omega(t) \leq \alpha_{10}(\rho_*(\rho_2(\Omega(t_0)), c), c). \tag{11.57}$$

Hence, for all initial conditions that satisfy the bound (11.9) with any class $\mathcal{K}$ choice $\psi_{\mathrm{RoA}}(c) \leq \bar{\psi}^*{}_{\mathrm{RoA}}(\bar{\rho}_c(c, c), c)$, where $\bar{\psi}^*{}_{\mathrm{RoA}}(s, c)$ is the inverse of the class $\mathcal{KC}_\infty$ function $\psi^*{}_{\mathrm{RoA}}(s, c) = \alpha_{10}(\rho^*(\rho_2(s, c), c))$ with respect to ψ^*_{RoA}'s first argument, the solutions satisfy (11.50). Furthermore, for all those initial conditions, the solutions verify (11.5) for all $\theta \geq t_0 - D(X(t_0))$. Figure 11.2 illustrates the set relationships in the proof. □

***Proof of Theorem* 11.4.** Using (11.56) we get (11.10) with $\beta(s, t) = \alpha_9^{-1}(\beta_2(s, t))$ and $\rho(s, c) = \rho_*(\rho_2(s, c), c)$. From (11.1), the Lipschitzness of U on $[t_0 - D(X(t_0)), t_0)$ guarantees the existence and uniqueness of $X \in C^1[t_0, \sigma^*)$, where $\sigma^* = t_0 + D(X(\sigma^*))$, the system (11.15), (11.16) guarantees the existence and uniqueness of $X \in C^1(\sigma^*, \infty)$, and the boundedness of W and (11.15) guarantee that X is continuous at $t = \sigma^*$. By integrating (11.15) between any two time instants it is shown that X is Lipschitz on $[t_0, \infty)$ with a Lipschitz constant given by a uniform bound on the right-hand side of (11.15). Since

$U(t)=x(\sigma(t),\Pi(t))$, where $\sigma(t)=t+D(\Pi(t))$ and

$$\dot{\Pi}(t)=\frac{f(\Pi(t),x(\sigma(t),\Pi(t)))}{1-\nabla D(\Pi(t))f(\Pi(t),x(\sigma(t),\Pi(t)))} \tag{11.58}$$

for $t\geq t_0$, Assumption 11.1 (Lipschitzness of ∇D), Assumption 11.3 (Lipschitzness of x in both arguments), and (11.5) ensure that the right-hand side of the Π-ODE is Lipschitz, which guarantees that $\Pi\in C^1(t_0,\infty)$. Since $x(t+D(\Pi(t)),\Pi(t))$ is Lipschitz in t on (t_0,∞), so is U. Using Lemma 11.12, (11.28), (11.51), we get (11.12), (11.13) with any class $\mathcal{K}$ function $\delta^*(c)\geq\delta_1(\bar{\rho}_c(c,c))$. □

We note here that in the special case of linear plants, i.e., when system (11.1) is

$$\dot{X}(t)=AX(t)+BU(t-D(X(t))), \tag{11.59}$$

Assumption 11.3 is satisfied when the pair (A,B) is stabilizable and Assumption 11.2 is satisfied for any A by means of the relation

$$\frac{d\,|Y(\tau)|^2}{d\tau}\leq(2|A|+1)\,|Y(\tau)|^2+|B|^2\omega^2(\tau). \tag{11.60}$$

The controller for the linear case is

$$U(t)=KP(t), \tag{11.61}$$

with the predictor $P(t)$ given by

$$P(\theta)=X(t)+\int_{t-D(X(t))}^{\theta}\frac{(AP(s)+BU(s))\,ds}{1-\nabla D(P(s))(AP(s)+BU(s))} \tag{11.62}$$

for $t-D(X(t))\leq\theta\leq t$. The predictor $P(\theta)$ is not given explicitly even for the linear case. We establish next the following result with explicit estimates that highlight the nonlinear role of the delay function $D(X)$ and exponential decay in time.

Theorem 11.13. *Consider the plant* (11.59) *together with the control law* (11.61), (11.62) *and K chosen such that $A+BK$ is Hurwitz, namely, $(A+BK)^T P+P(A+BK)=-Q$ for some $P=P^T>0$ and $Q=Q^T>0$. Under Assumption* 11.1, *for all initial conditions of the plant that satisfy*

$$\Omega(t_0)<\zeta_{\mathrm{RoA}}(c), \tag{11.63}$$

$$\Omega(t)=|X(t)|+\sup_{t-D(X(t))\leq\theta\leq t}|U(\theta)| \tag{11.64}$$

for some $0<c<1$, the following holds:

$$\Omega(t)\leq\zeta_4(\Omega(t_0),c)\,e^{-\lambda(t-t_0)} \tag{11.65}$$

for all $t \geq t_0$, *where* $\lambda > 0$, $\zeta_4 \in \mathcal{KC}_\infty$, *and* $\zeta_{\text{RoA}} \in \mathcal{K}$ *are given by*

$$\zeta_{\text{RoA}}(c) \leq \bar{\zeta}_4\left(\bar{\zeta}_c(c,c),c\right), \tag{11.66}$$

$$\zeta_4(s,c) = (1+|K|r_1)\sqrt{\frac{2}{\min\{\lambda_{\min}(P),k\}}}\sqrt{\zeta_2(s+|K|\zeta_1(s,c),c)}, \tag{11.67}$$

$$\zeta_c(c,s) = (|A|+|B|)(|\nabla D(0)|+\delta_2(\zeta_1(c,s)))(\zeta_1(c,s)+s), \tag{11.68}$$

$$\zeta_1(s,c) = \left(1+\frac{|B|^2}{2|A|+1}\right)^{\frac{1}{2}} e^{\frac{2|A|+1}{2(1-c)}(D(0)+\delta_1(s))}s, \tag{11.69}$$

$$\zeta_2(s,c) = \left(\lambda_{\max}(P)+\frac{k}{(1-c)^2}e^{\frac{2g}{1-c}(1+D(0)+\delta_1(s))}\right)s^2, \tag{11.70}$$

$$k = \frac{1}{g}\left(\frac{2|PB|}{\lambda_{\min}(Q)}+\frac{\lambda_{\min}(Q)}{4}\right), \tag{11.71}$$

$$r_1 = \frac{\lambda_{\max}(P)}{\lambda_{\min}(P)}\left(1+4\frac{\lambda_{\max}(P)|PB|^2}{\lambda_{\min}(Q)^2}\right), \tag{11.72}$$

$$\lambda = \frac{\lambda_{\min}(Q)}{4\max\{k,\lambda_{\max}(P)\}}, \tag{11.73}$$

and g *is an arbitrary positive constant. Furthermore, inequalities* (11.12) *and* (11.13) *hold with* $\delta^*(c) \geq \delta_1(\bar{\zeta}_c(c,c))$.

Proof. For the special case of linear systems, Lemma 11.7 is applied using the Lyapunov function

$$V(t) = X(t)^T P X(t) + kL(t)^2. \tag{11.74}$$

Using the fact that $\dot{L}(t) \leq -gL(t)$ together with equations (11.31) and (11.29) we get

$$\dot{V}(t) \leq -r_2 V(t), \tag{11.75}$$

where $r_2 = \frac{\lambda_{\min}(Q)}{2\max\{k,\lambda_{\max}(P)\}}$. With relations (11.29), (11.31), (11.74), (11.75), and (11.5) for $0 < c < 1$ we establish that

$$|X(t)| + \sup_{t-D(X(t))\leq\theta\leq t}|W(\theta)| \leq \sqrt{\zeta_2\left(|X(t_0)| + \sup_{t_0-D(X(t_0))\leq\theta\leq t_0}|W(\theta)|,c\right)} \times\sqrt{\frac{2}{r_3}e^{-\frac{r_2}{2}(t-t_0)}}. \tag{11.76}$$

Using (11.43) together with (11.60), relation (11.36) holds with $\rho_1(s,c)$ replaced by $\zeta_1(s,c)$. Using (11.44) with $\gamma_4(s) = r_1 s$, Lemma 11.10 applies with

$$\rho_2(s,c) = s + |K|\zeta_1(s,c), \tag{11.77}$$

$$\alpha_9^{-1}(s) = \frac{1}{1+|K|r_1}s. \tag{11.78}$$

As in the proof of Lemma 11.11, condition $\mathcal{F}_c$ for $0 < c < 1$ in (11.5) is satisfied when

$$|X(t)| + \sup_{t-D(X(t))\leq\theta\leq t}|U(\theta)| < \bar{\zeta}_c(c,c), \tag{11.79}$$

where with $\bar{\zeta}_c$ we denote the inverse function of the class $\mathcal{KC}_\infty$ function $\zeta_c(c,s)$ with respect to its first argument. Using Lemma 11.10 together with (11.77), (11.78) and Lemma 11.7 together with (11.76) we get (11.65). Using (11.63), (11.65) with $t = t_0$ we get (11.79). Bounds (11.12)–(11.13) follow from (11.79), (11.28). □

11.2 ▪ Example: Nonholonomic Unicycle Subject to Distance-Dependent Input Delay

We consider the problem of stabilizing a mobile robot modeled as

$$\dot{x}(t) = v(t - D(x(t), y(t)))\cos(\theta(t)), \tag{11.80}$$
$$\dot{y}(t) = v(t - D(x(t), y(t)))\sin(\theta(t)), \tag{11.81}$$
$$\dot{\theta}(t) = \omega(t - D(x(t), y(t))), \tag{11.82}$$

subject to an input delay that grows with the distance relative to the reference position as

$$D(x(t), y(t)) = x(t)^2 + y(t)^2, \tag{11.83}$$

where $(x(t), y(t))$ is the position of the robot, $\theta(t)$ is the heading, $v(t)$ is the speed, and $\omega(t)$ is the turning rate. When $D = 0$ a time-varying stabilizing controller for this system is proposed in [142] as

$$\omega(t) = -5P(t)^2\cos(3\sigma(t)) - P(t)Q(t)\left(1 + 25\cos(3\sigma(t))^2\right) - \Theta(t), \tag{11.84}$$
$$v(t) = -P(t) + 5Q(t)(\sin(3\sigma(t)) - \cos(3\sigma(t))) + Q(t)\omega(t), \tag{11.85}$$
$$P(t) = X(t)\cos(\Theta(t)) + Y(t)\sin(\Theta(t)), \tag{11.86}$$
$$Q(t) = X(t)\sin(\Theta(t)) - Y(t)\cos(\Theta(t)), \tag{11.87}$$

with

$$X = x, \quad Y = y, \quad \Theta = \theta, \quad \sigma(t) = t. \tag{11.88}$$

The proposed method replaces (11.88) with

$$X(t) = x(t) + \int_{t-D(x(t),y(t))}^{t} \dot{\sigma}(s)v(s)\cos(\Theta(s))\,ds, \tag{11.89}$$
$$Y(t) = y(t) + \int_{t-D(x(t),y(t))}^{t} \dot{\sigma}(s)v(s)\sin(\Theta(s))\,ds, \tag{11.90}$$
$$\Theta(t) = \theta(t) + \int_{t-D(x(t),y(t))}^{t} \dot{\sigma}(s)\omega(s)ds, \tag{11.91}$$
$$\sigma(t) = t + D(X(t), Y(t)), \tag{11.92}$$
$$\dot{\sigma}(s) = \frac{1}{1 - 2(X(s)v(s)\cos(\Theta(s)) + Y(s)v(s)\sin(\Theta(s)))}. \tag{11.93}$$

The initial conditions are chosen as $x(0) = y(0) = \theta(0) = 1$ and $\omega(s) = v(s) = 0$ for all $-x(0)^2 - y(0)^2 \leq s \leq 0$. From the given initial conditions we get the initial conditions for the predictors (11.89)–(11.91) as $X(s) = Y(s) = \Theta(s) = 1$ for all $-2 \leq s \leq 0$. From the

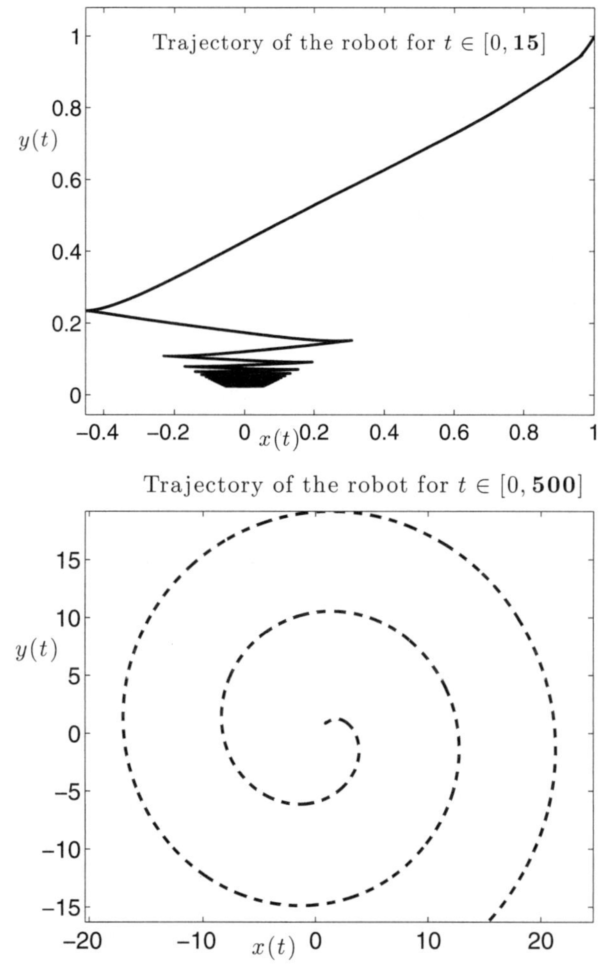

Figure 11.3. *The trajectory of the robot, with the compensated controller* (11.84)–(11.87), (11.89)–(11.93) *(solid line) and the uncompensated controller* (11.84)–(11.87), (11.88) *(dashed line) with initial conditions* $x(0) = y(0) = \theta(0) = 1$ *and* $\omega(s) = v(s) = 0$ *for all* $-x(0)^2 - y(0)^2 \leq s \leq 0$.

above initial conditions for the predictors one can verify that the system initially lies inside the feasibility region. The controller "kicks in" at the time instant t_0 at which $t_0 = x(t_0)^2 + y(t_0)^2$. Since $v(s) = \omega(s) = 0$ for $s < 0$ we conclude that $x(t) = y(t) = \theta(t) = 1$ for all $0 \leq t \leq t_0$, and hence $t_0 = 2$. In Figure 11.3 we show the trajectory of the robot in the xy plane, whereas in Figure 11.4 we show the resulting state-dependent delay and the controls $v(t)$ and $\omega(t)$. In the case of the uncompensated controller (11.84)–(11.87), (11.88), the system is unstable, the delay grows approximately linearly in time, and the vehicle's trajectory is a divergent Archimedean spiral. The compensated controller (11.84)–(11.87), (11.89)–(11.93) recovers the delay-free behavior after 2 seconds. From Figure 11.3 one can also conclude that the heading $\theta(t)$ in the case of the compensated controller converges to zero with damped oscillations, whereas in the case of the uncompensated controller it increases towards negative infinity (the robot moves clockwise on a spiral).

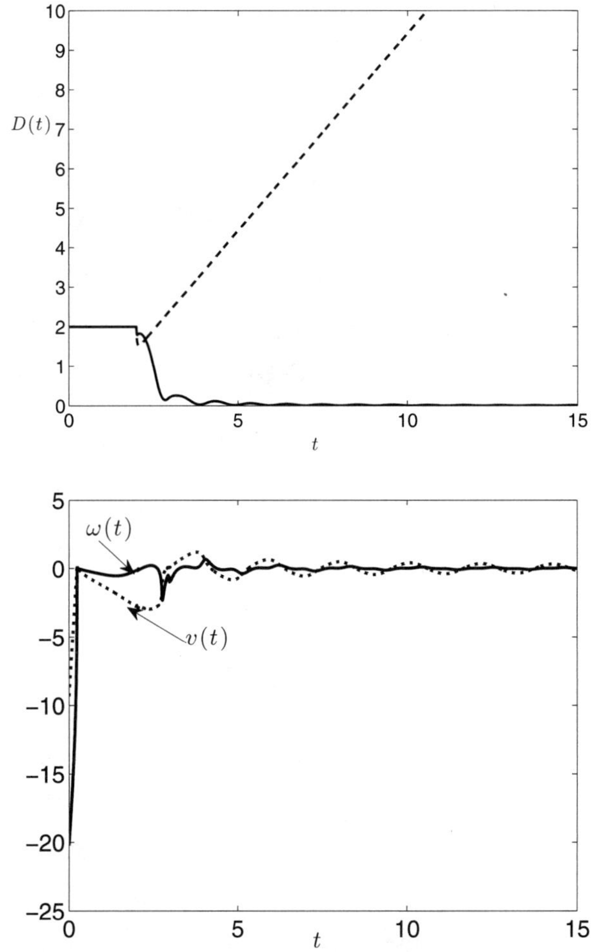

Figure 11.4. *Top: The delay with the controller* (11.84)–(11.87), (11.89)–(11.93) *(solid line) and the controller* (11.84)–(11.87), (11.88) *(dashed line) with initial conditions* $x(0)=y(0)=\theta(0)=1$ *and* $\omega(s)=v(s)=0$ *for all* $-x(0)^2-y(0)^2\le s\le 0$*. Bottom: The control efforts* $v(t)$ *and* $\omega(t)$ *with the controller* (11.84)–(11.87), (11.89)–(11.93) *with initial conditions* $x(0)=y(0)=\theta(0)=1$ *and* $\omega(s)=v(s)=0$ *for all* $-x(0)^2-y(0)^2\le s\le 0$*.*

11.3 ▪ Global Stabilization

The key challenge for the stabilization of systems with state-dependent input delay is to maintain the feasibility condition (11.5), i.e., to keep the delay rate below one. This condition can be satisfied a priori by making the following restrictive but verifiable assumption.

Assumption 11.14. $\nabla D(X)f(X,\omega)<c$ *for some* $0<c<1$ *and all* $(X,\omega)\in\mathbb{R}^{n+1}$.

Corollary 11.15. *Consider the plant* (11.1) *together with the control law* (11.2)–(11.4)*. Under Assumptions* 11.1, 11.2, 11.3, *and* 11.14 *there exist a class* $\mathcal{KL}$ *function* β_g *and a* *class $\mathcal{KC}_\infty$ function ρ_g such that*

$$\Omega(t) \leq \beta_g\left(\rho_g(\Omega(t_0), c), t - t_0\right), \tag{11.94}$$

$$\Omega(t) = |X(t)| + \sup_{t-D(X(t)) \leq \theta \leq t} |U(\theta)| \tag{11.95}$$

for all $t \geq t_0$ and some $0 < c < 1$.

Example 11.16. We consider the scalar system

$$\dot{X}(t) = \frac{X(t) + U(t - D(X(t)))}{U(t - D(X(t)))^2 + 1}, \tag{11.96}$$

with

$$D(X(t)) = \frac{1}{4} \log\left(X(t)^2 + 1\right). \tag{11.97}$$

Taking the time derivative of $D(X(t))$ and using Young's inequality one gets that

$$\frac{dD(X(t))}{dt} \leq \frac{\frac{3}{4}X(t)^2 + \frac{1}{4}U(t - D(X(t)))^2}{\left(X(t)^2 + 1\right)\left(U(t - D(X(t)))^2 + 1\right)} \leq \frac{6}{7}. \tag{11.98}$$

Since system (11.96), (11.97) satisfies Assumption 11.1, 11.2, 11.3, and 11.14, Corollary 11.15 applies. The control law has the form

$$U(t) = -2P(t), \tag{11.99}$$

$$P(t) = \int_{t-\frac{1}{4}\log\left(X(t)^2+1\right)}^{t} \frac{2\left(P(\theta)^2 + 1\right)(P(\theta) + U(\theta))\,d\theta}{\left(2\left(U(\theta)^2 + 1\right)\left(P(\theta)^2 + 1\right) - P(\theta)(P(\theta) + U(\theta))\right)} + X(t). \tag{11.100}$$

In Figure 11.5 we show the response of the system with initial conditions as $X(0) = 1.5$, $U(\theta) = 0$, and $P(\theta) = X(0) + \int_{-\frac{1}{4}\log\left(X(0)^2+1\right)}^{0} \frac{2\left(P(\theta)^2+1\right)P(\theta)d\theta}{\left(P(\theta)^2+2\right)}$ for all θ such that $-\frac{1}{4}\log\left(X(0)^2 + 1\right) \leq \theta \leq 0$. Initially, $X(t)$ runs in open loop and grows exponentially, while $D(t)$ grows roughly linearly, because of (11.97). This goes on until control "kicks in" at $t^* = \frac{1}{4}\log\left(1 + X(0)^2 e^{2t^*}\right) = 0.4835$. For $t > t^*$ the controller starts bringing $X(t)$ back to zero. As $X(t)$ decays according to the target system $\dot{X}(t) = -\frac{X(t)}{1+4X(t)^2}$, the delay $D(t)$ also decays. Starting from $X(t^*)$, $X(t)^2$ first decays roughly linearly in t, making the decay of $D(t)$ logarithmic. When $X(t)$ becomes small, its decay becomes exponential and the decay of $D(t)$ is also exponential. Initially, $P(t)^2$ decays linearly in t and later $P(t)$ decays exponentially. The decay of $U(t) = -2P(t)$ follows the same pattern as $P(t)$. ∎

11.4 ▪ Notes and References

Though the stability results in this chapter are not global, the size of the delay is not limited. By examining the estimates in detail, the reader can observe that when the delay $D(X)|_{X=0}$ is large, namely, when the system is regulated to an equilibrium where

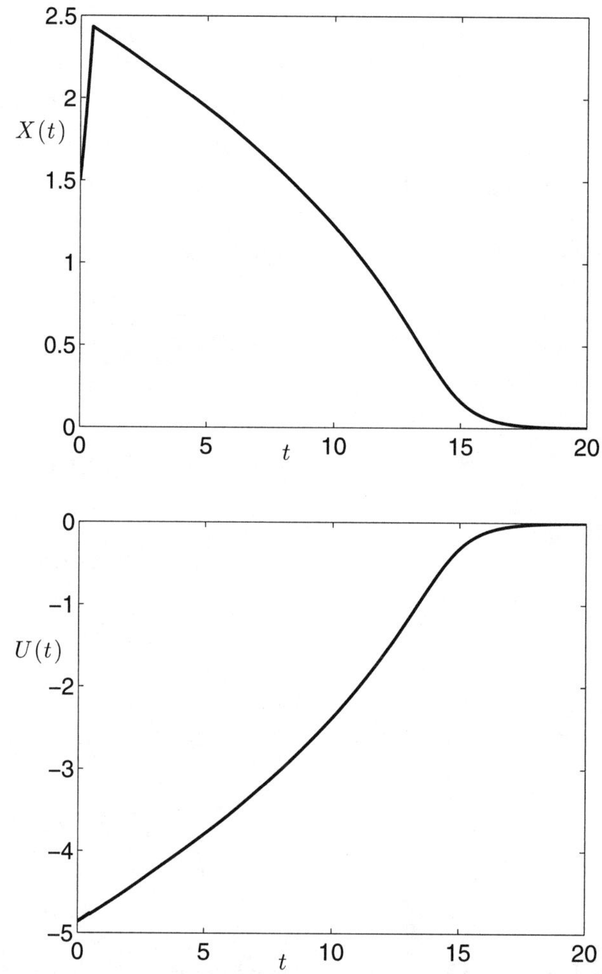

Figure 11.5. *Response of system* (11.96) *with the controller* (11.99)–(11.100) *and initial conditions* $X(0) = 1.5$, $U(\theta) = 0$ *for all* $-\frac{1}{4}\log\left(X(0)^2 + 1\right) \leq \theta \leq 0$.

the delay is necessarily large, the stability estimates dictate that the initial conditions of the state and the input be small. However, no restrictions on $D(X)|_{X=0}$ are imposed. A trade-off exists between the size of the state-dependent delay and the achievable region of attraction in closed loop.

Systems with a state-dependent delay are similar to hyperbolic PDE systems with a state-dependent propagation speed, or with a state-dependent size of the spatial domain. The results on state-dependent input and state delays open an opportunity to tackle such problems. Moreover, one can consider more complicated forms of PDEs, such as diffusion PDEs with state-dependent diffusion coefficients.

Chapter 12

Stability Analysis for Locally Stabilizable Systems with Input Delay

In Section 11.1 we proved a local stability result under assumptions of global stabilizability (Assumption 11.3) and forward-completeness (Assumption 11.2) of the delay-free system. It is reasonable to ask whether a local stability result can be established under a less restrictive assumption of local stabilizability of the delay-free system. In this chapter we provide an affirmative answer to this question. Our proof of this result does not employ a Lyapunov construction and, as such, provides an illustrative alternative to the proof technique in Section 11.1.

In the case of forward-complete systems our results were local due to a fundamental limitation on the allowable magnitude of the delay function's gradient (the control signal never reaches the plant if the delay rate is larger than one) which, since it depends on the state, could be restricted by appropriately restricting the allowable initial conditions. In the case of locally stabilizable systems, the stability results that we obtain are local for an additional reason. One has to further restrict the initial conditions such that, when the control signal "kicks in," the state is within the region of attraction of the nominal design. In achieving that, however, one has to estimate the time for the control signal to reach the plant. We provide an estimate of this time by constructing a bound on the norm of the open-loop solutions (i.e., before the control signal reaches the plant), which allows us to give an estimate of the region of attraction for our control scheme and prove asymptotic stability of the closed-loop system in Section 12.1. We also present an example of a system that can have a finite escape time and whose linearization is not stabilizable.

12.1 ▪ Stability Analysis for Locally Stabilizable Nonlinear Systems

We start this section by recalling the predictor feedback design of Chapter 10. The predictor-based controller for the plant

$$\dot{X}(t) = f(X(t), U(t - D(X(t)))), \tag{12.1}$$

where $f : \mathbb{R}^n \times \mathbb{R} \to \mathbb{R}^n$ is locally Lipschitz with $f(0,0) = 0$, is

$$U(t) = \varkappa(\sigma(t), P(t)), \tag{12.2}$$

where

$$P(\theta)=X(t)+\int_{t-D(X(t))}^{\theta}\frac{f(P(s),U(s))ds}{1-\nabla D(P(s))f(P(s),U(s))},$$
$$t-D(X(t))\leq\theta\leq t, \tag{12.3}$$
$$\sigma(\theta)=\theta+D(P(\theta)),\quad t-D(X(t))\leq\theta\leq t, \tag{12.4}$$

for $t\geq t_0$. The *feasibility condition* of the controller (12.2)–(12.4) is

$$\mathscr{F}_c:\qquad \nabla D(P(\theta))f(P(\theta),U(\theta))<c\quad\text{for all } \theta\geq t_0-D(X(t_0)) \tag{12.5}$$

for $c\in(0,1]$. We recall next the assumption on the delay function D.

Assumption 12.1. *$D\in C^1(\mathbb{R}^n;\mathbb{R}_+)$ and ∇D is locally Lipschitz.*[1]

We note that Assumption 12.1 implies that there exist a function $\delta_1\in\mathscr{K}_\infty\cap C^1$ and a function $\delta_2\in\mathscr{K}_\infty$ such that

$$D(X)\leq D(0)+\delta_1(|X|), \tag{12.6}$$
$$|\nabla D(X)|\leq|\nabla D(0)|+\delta_2(|X|). \tag{12.7}$$

Our new assumption on the plant (12.1) is the following.

Assumption 12.2. *There exists a locally Lipschitz feedback controller $\varkappa:\mathbb{R}_+\times\mathbb{R}^n\mapsto\mathbb{R}$ that satisfies*

$$|\varkappa(t,\xi)|\leq\hat{\alpha}(|\xi|)\quad \textit{for all } t\geq 0 \tag{12.8}$$

for some class $\mathscr{K}_\infty$ function $\hat{\alpha}$, a positive constant R, and a class $\mathscr{KL}$ function β^* such that, for the system $\dot{Y}(t)=f(Y(t),\varkappa(t,Y(t)))$, the following holds for all $|Y(t_0)|\leq R$:

$$|Y(t)|\leq\beta^*(|Y(t_0)|,t-t_0),\quad t\geq t_0. \tag{12.9}$$

Theorem 12.3. *Consider the plant (12.1) together with the predictor controller (12.2)–(12.4) and denote*

$$\Omega(t)=|X(t)|+\sup_{t-D(X(t))\leq\theta\leq t}|U(\theta)|. \tag{12.10}$$

Under Assumptions 12.1 and 12.2 there exist a function ψ_{RoA} of class $\mathscr{K}$ and a class $\mathscr{KL}$ function $\hat{\beta}$ such that for all initial conditions for which U is locally Lipschitz on the interval $[t_0-D(X(t_0)),t_0)$ and which satisfy

$$\Omega(t_0)<\psi_{\mathrm{RoA}}(c) \tag{12.11}$$

for some $0<c<1$, there exists a unique solution to the closed-loop system with X Lipschitz on $[t_0,\infty)$, U Lipschitz on (t_0,∞), and

$$\Omega(t)\leq\hat{\beta}(\Omega(t_0),t-t_0)\qquad \textit{for all } t\geq t_0. \tag{12.12}$$

[1] To ensure uniqueness of solutions.

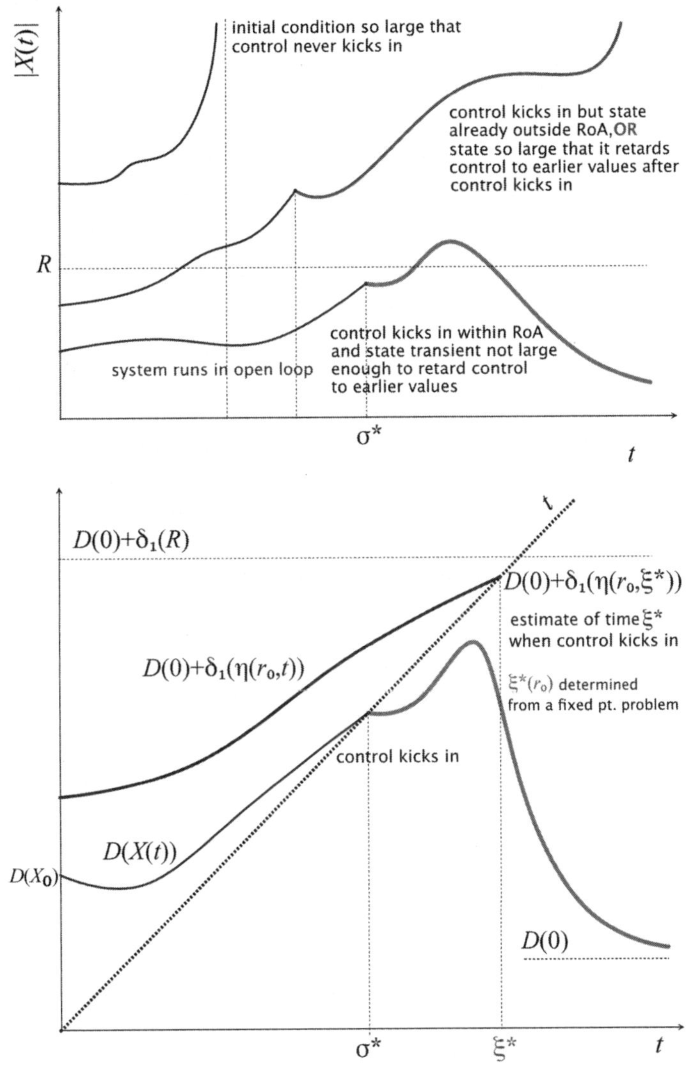

Figure 12.1. *Top: Possible transients of the state $X(t)$. Bottom: An example of a favorable transient of the delay $D(X(t))$ together with an estimate of its upper bound as defined in* (11.28) *and* (12.16).

Furthermore,

$$D(X(t)) \le D(0) + \delta^*(c), \tag{12.13}$$

$$\left|\dot{D}(X(t))\right| \le c \tag{12.14}$$

for all $t \ge t_0$, where $\delta^*(c) = \delta_1\left(\hat{\beta}(\psi_{\mathrm{RoA}}(c), 0)\right)$ and δ_1 is defined in (12.6).

The idea of the proof of Theorem 12.3 is captured by the two plots in Figure 12.1. The top plot in Figure 12.1 depicts four possibilities (not an exhaustive list) that may arise with closed-loop solutions: (a) control never reaches the plant; (b) control reaches the plant but the state has already exited the region of attraction of the delay-free closed-loop system; (c) the state is within the region of attraction of the delay-free closed-loop system when the control reaches the plant, but the subsequent transient leads the state outside of the controller's feasibility region and thus to the reversal of the direction of the control signal; and (d) control reaches the plant while the state is within the region of attraction of the delay-free closed-loop system and the solution remains within the controller's feasibility region, so that the control signal is never retarded to earlier values, and the delay remains compensated for all subsequent times. The proof of Theorem 12.3 estimates a set of initial conditions from which all the solutions belong to category (d).

The bottom plot in Figure 12.1 depicts the strategy of the proof of Theorem 12.3. The exact time σ^* when the control reaches the plant is not known analytically. We find an upper bound $\xi^* \geq \sigma^*$ by using an upper bound $D(X) \leq D(0) + \delta_1(|X|)$ on the delay and by estimating an upper bound on the open-loop solution $|X(t)| \leq \eta(r_0, t - t_0)$, where $r_0 := \Omega(t_0)$. The upper bound ξ^* is then determined from the fixed-point problem $\xi^* - t_0 = D(0) + \delta_1(\eta(r_0, \xi^* - t_0))$. The solution $\xi^*(r_0)$ to the fixed-point problem is a function of the size of the initial condition r_0. By reducing r_0 sufficiently, we can ensure that the control signal reaches the plant before $|X(t)|$ has exceeded R, namely, before $D(X(t))$ has exceeded the known bound $D(0) + \delta_1(R)$.

The actual proof of Theorem 12.3 consists of Lemmas 12.4–12.8, which are presented next.

Lemma 12.4 (a bound on open-loop solutions). *For the plant* (12.1) *there exists a function* $\eta(r,s) : \mathbb{B} \mapsto [0, \infty)$, *where*

$$\mathbb{B} = \left\{ (r,s) \in \left(\mathbb{R}_+\right)^2 : 0 \leq s < \lim_{z \to \infty} \int_r^z \frac{d\mu}{\alpha_1(\mu)} \right\}, \tag{12.15}$$

with the properties

(a) $\eta(r,s)$ *is increasing in both of its arguments* r *and* s;

(b) $\eta(r,s)$ *is continuous in its domain of definition and, moreover,* $\lim_{r \to 0} \eta(r,s) = 0$ *uniformly in* s,

such that for all $t \leq \hat{t} = \min\left\{\sigma^*, t_0 + \lim_{z\to\infty} \int_{r_0}^z \frac{d\mu}{\alpha_1(\mu)}\right\}$, *where* $\sigma^* = t_0 + D(X(\sigma^*))$, *it holds that*

$$|X(t)| \leq \eta(r_0, t - t_0), \tag{12.16}$$

where $r_0 = \Omega(t_0) = |X(t_0)| + \sup_{t_0 - D(X(t_0)) \leq \theta \leq t_0} |U(\theta)|$.

Proof. The existence of the function η is proved by construction. A representative form of η is shown graphically in Figure 12.2. Since for all $t \leq \hat{t}$ the system runs in open-loop, it holds that

$$\frac{d\,|X(t)|}{dt} \leq \alpha_1\left(|X(t)| + \sup_{t_0 - D(X(t_0)) \leq \theta \leq t_0} |U(\theta)| \right), \quad t_0 \leq t \leq \hat{t}. \tag{12.17}$$

Setting $y(t) = |X(t)| + \sup_{t_0 - D(X(t_0)) \leq \theta \leq t_0} |U(\theta)|$, we get that

$$\dot{y}(t) \leq \alpha_1(y(t)), \quad t_0 \leq t \leq \hat{t}. \tag{12.18}$$

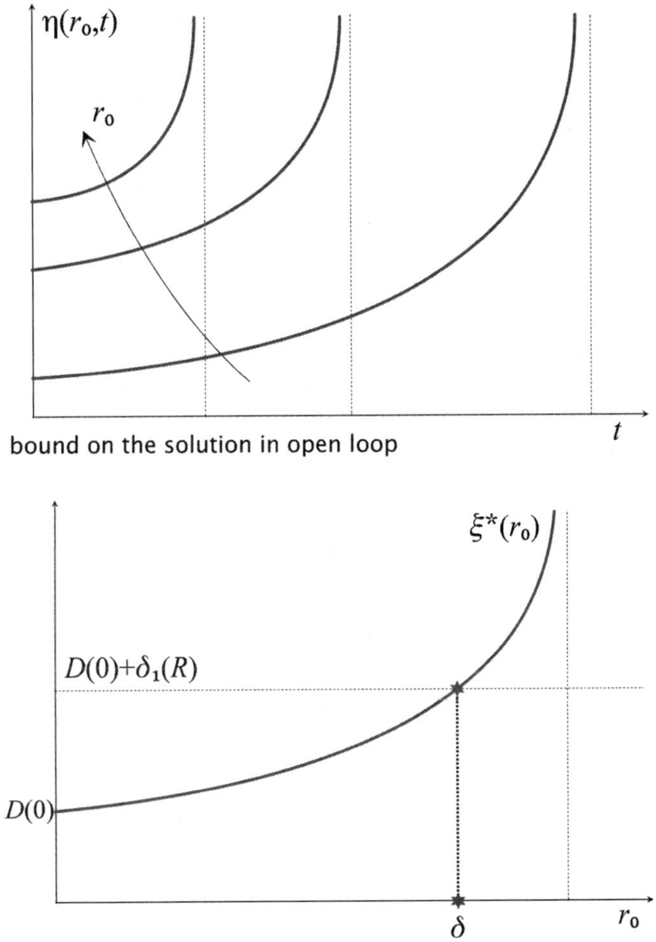

Figure 12.2. *Top: A representative form of the function $\eta(r_0,t)$ in Lemma 12.4. Bottom: Restricting the initial conditions in Lemma 12.6 to $r_0 \in [0,\delta]$ so that when the control kicks in, which is no later than $\xi^*(r_0)$, the delay is no greater than $D(0)+\delta_1(R)$ and, consequently, the plant state is no greater than R.*

Consider now the ODE

$$\dot{z}(t) = \alpha_1(z(t)), \quad z(t_0) = y(t_0) \geq 0. \tag{12.19}$$

Define for any $a > 0$ the function $g(z)$ for all $z \in (0,\infty)$ as

$$g(z) = \int_a^z \frac{ds}{\alpha_1(s)}. \tag{12.20}$$

The function g is continuous and strictly increasing for all $z \in (0,\infty)$ since $g'(z) = \frac{1}{\alpha_1(z)}$. The range of the function g is (b,d), where $b = \lim_{z\to 0^+} \int_a^z \frac{ds}{\alpha_1(s)}$ and $d = \lim_{z\to\infty} \int_a^z \frac{ds}{\alpha_1(s)}$. Note that b can be $-\infty$ and d can be $+\infty$. From (12.19) we obtain $\int_{z(t_0)}^{z(t)} \frac{ds}{\alpha_1(s)} = t - t_0$. By defining $z(t_0) = z_0$ (since from now on we treat $z(t_0)$ as a parameter) we get

$g(z(t)) = g(z_0) + t - t_0$ for all $t \geq t_0$ and $z_0 > 0$. Thus, $z(t) = g^{-1}(g(z_0) + t - t_0)$ for all $t \geq t_0$ and $z_0 > 0$. Define for all $s \geq 0$

$$\eta(r,s) = \begin{cases} g^{-1}(g(r)+s), & r > 0, \\ 0, & r = 0, \end{cases} \tag{12.21}$$

From (12.21) we observe that $\eta : \{(r,s) \in (\mathbb{R}_+)^2 : 0 \leq g(r) + s < d\} = \mathbb{B} \mapsto [0,\infty)$ and, moreover, η is increasing in both of its arguments since g is increasing. From its definition, η is continuous in its domain. To see this, first note that η is continuous in its domain with an exception maybe at the points $(0,s)$. However, since from (12.19) it holds that $z_0 = 0 \Rightarrow z(t) = 0$ for all $t > t_0$, by continuity of $z(t)$ with respect to its initial conditions, $z_0 \to 0 \Rightarrow z(t) \to 0$ for all $t \in (t_0, l), l \to \infty$. Therefore, $\lim_{r\to 0} \eta(r,s) = 0 = \eta(0,s)$ uniformly in s. Using the comparison principle (Lemma B.7 in Appendix B) and (12.18) we get the statement of the lemma. □

Lemma 12.5 (for small initial conditions, the upper bound on the delay as a function of the time when the control kicks in is a contraction). *There exists a sufficiently small $\bar{r}_0$ such that for all $r_0 \in [0,\bar{r}_0]$ the mapping $T_{r_0}(s^*) = D(0) + \delta_1(\eta(r_0, s^*))$ is a contraction in $[D(0), D(0) + \delta_1(R)]$.*

Proof. We start by setting $s^* = \xi^* - t_0$. Based on Lemma 12.4, since $L(r_0) = \lim_{z\to\infty} \int_{r_0}^{z} \frac{d\mu}{\alpha_1(\mu)}$ satisfies $\lim_{r_0\to 0} L(r_0) \to \infty$ there exists a sufficiently small $\hat{r}_0$ such that for all initial conditions $r_0 \in [0,\hat{r}_0]$ it holds that

$$\xi^* < t_0 + \lim_{z\to\infty} \int_{r_0}^{z} \frac{d\mu}{\alpha_1(\mu)}, \tag{12.22}$$

and hence, $\eta(r_0, \xi^* - t_0)$ is continuous in $(r_0, \xi^*) \in [0,\hat{r}_0] \times [t_0 + D(0), t_0 + D(0) + \delta_1(R)]$. Since $D \in C^1(\mathbb{R}^n; \mathbb{R}_+)$ (which allows us to choose $\delta_1 \in \mathcal{K}_\infty \cap C^1$) and by continuity of η and the fact that $\lim_{r\to 0} \eta(r,s) = 0$ uniformly in s, there exists a sufficiently small $r_0^* > 0$ such that for all $r_0 \in [0, r_0^*]$,

$$T_{r_0}[D(0), D(0) + \delta_1(R)] \subseteq [D(0), D(0) + \delta_1(R)] \tag{12.23}$$

(e.g., $\delta_1(\eta(r_0^*, D(0) + \delta_1(R))) = \delta_1(R)$), i.e., the mapping T_{r_0} maps the set $[D(0), D(0) + \delta_1(R)]$ into itself. Differentiating (12.21) with respect to s and using the fact that $g'(z) = \frac{1}{\alpha_1(z)}$ we get

$$\begin{aligned} \frac{\partial \eta(r,s)}{\partial s} &= \frac{1}{g'(g^{-1}(g(r)+s))} \\ &= \alpha_1(\eta(r_0, s)). \end{aligned} \tag{12.24}$$

Hence,

$$\frac{dT_{r_0}(s^*)}{ds^*} = \delta_1'(\eta(r_0, s^*))\alpha_1(\eta(r_0, s^*)), \tag{12.25}$$

where $s^* = \xi^* - t_0$. Using the facts that $\delta_1 \in C^1$, $\alpha_1 \in \mathcal{K}_\infty$, and η is continuous with $\lim_{r\to 0} \eta(r,s) = 0$ uniformly in s, there exists a sufficiently small r_0^{**} such that for all

$r_0 \in [0, r_0^{**}]$ and $\xi^* \in [t_0 + D(0), t_0 + D(0) + \delta_1(R)]$

$$\left|\frac{dT_{r_0}(s^*)}{ds^*}\right| = \left|\delta_1'(\eta(r_0, s^*))\alpha_1(\eta(r_0, s^*))\right| < 1, \tag{12.26}$$

where $s^* = \xi^* - t_0$. By Lemma 3.1 in [75], $|T_{r_0}(s_1^*) - T_{r_0}(s_2^*)| \leq L|s_1^* - s_2^*|$ for all $s_1^*, s_2^* \in [D(0), D(0)+\delta_1(R)]$, where $s_1^* = \xi_1^* - t_0$, $s_2^* = \xi_2^* - t_0$, and L satisfies $L < 1$. Theorem B.1 in [75] guarantees that the mapping $T_{r_0}(\xi^* - t_0)$ is a contraction in $[D(0), D(0) + \delta_1(R)]$ for all $r_0 \in [0, \bar{r}_0], \bar{r}_0 = \min\{\hat{r}_0, r_0^*, r_0^{**}\}$. □

Lemma 12.6 (for small initial conditions, plant state is within the delay-free region of attraction when control kicks in). *There exists a sufficiently small $\delta > 0$ such that for all solutions of the closed-loop system satisfying* (12.5) *for $0 < c < 1$ and for all initial conditions of the plant* (12.1) *satisfying*

$$r_0 \leq \delta, \tag{12.27}$$

there exists a $\sigma^* \geq t_0$ such that $\sigma^* = t_0 + D(X(\sigma^*))$ and, moreover,

$$|X(\sigma^*)| \leq R, \tag{12.28}$$

that is, r_0 is inside the region of attraction of the controller.

Proof. As long as the solutions of the closed-loop system satisfy (12.5) for $0 < c < 1$, the function $\phi(t)$ is increasing, with $\phi(t_0) = t_0 - D(X(t_0))$. Hence, there exists a time σ^* at which $\sigma^* = t_0 + D(X(\sigma^*))$, and moreover, based on (12.6), σ^* is finite when $|X(\sigma^*)|$ is finite. Consider now the fixed-point problem

$$T_{r_0}(\xi^* - t_0) = D(0) + \delta_1(\eta(r_0, \xi^* - t_0)) = \xi^* - t_0. \tag{12.29}$$

Since by Lemma 12.5 $T_{r_0}(\xi^* - t_0)$ is a contraction in $[D(0), D(0)+\delta_1(R)]$, the fixed-point problem (12.29) has a unique solution $\xi^*(r_0) \in [t_0 + D(0), t_0 + D(0) + \delta_1(R)]$. Differentiating (12.29) with respect to r_0 we get that

$$\begin{aligned} \xi^{*'}(r_0) = {} & \delta_1'(\eta(r_0, \xi^* - t_0))\alpha_1(\eta(r_0, \xi^* - t_0))\xi^{*'}(r_0) \\ & + \delta_1'(\eta(r_0, \xi^* - t_0))\frac{\alpha_1(\eta(r_0, \xi^* - t_0))}{\alpha_1(r_0)}. \end{aligned} \tag{12.30}$$

Using (12.26), the fact that $\delta_1 \in \mathcal{K}_\infty \cap C^1$, and the fact that

$$\delta_1'(\eta(r_0, \xi^* - t_0))\frac{\alpha_1(\eta(r_0, \xi^* - t_0))}{\alpha_1(r_0)} > 0 \tag{12.31}$$

for all $r_0 > 0$, we have that $\xi^{*'}(r_0) > 0$, and hence $\xi^*(r_0)$ is increasing. Since $\xi^{*'}(r_0)$ is continuous for $r_0 > 0$ we conclude that $\xi^*(r_0)$ is also continuous for $r_0 > 0$. Since $\lim_{r \to 0} \eta(r, s) = 0$ uniformly in s and $\delta_1 \in \mathcal{K}_\infty \cap C^1$, we have from (12.29) that given any $\epsilon > 0$ there exists a sufficiently small r_0 such that $|\xi^*(r_0) - t_0 - D(0)| < \epsilon$, and hence ξ^* is continuous also at zero. Since the function ξ^* is monotonically increasing, it is invertible. Denote $\delta = \min\{\bar{r}_0, \xi^{*-1}(t_0 + D(0) + \delta_1(R))\}$, as depicted in Figure 12.2 (bottom). Then, thanks to (12.29),

$$\eta(r_0, \xi^*(r_0) - t_0) = \delta_1^{-1}(\xi^*(r_0) - D(0) - t_0) \leq R \tag{12.32}$$

for all $r_0 \in [0, \delta]$. The proof is completed if $\xi^* \geq \sigma^*$, since then bound (12.28) holds. Assume that $\xi^* < \sigma^*$, and since $\phi(t)$ is increasing we have that $\phi(\xi^*) < \phi(\sigma^*)$. Since $\xi^* < \sigma^*$, using Lemma 12.4 we have that $|X(t)| \leq \eta(r_0, t - t_0)$ for all $t \leq \xi^*$. Therefore,

$$t - D(X(t)) \geq t - D(0) - \delta_1(\eta(r_0, t - t_0)), \quad t \leq \xi^*. \tag{12.33}$$

Consequently,

$$\phi(\xi^*) = \xi^* - D(X(\xi^*)) \geq t_0 = \phi(\sigma^*), \tag{12.34}$$

which contradicts the assumption. Thus $\xi^* \geq \sigma^*$. Using (12.32) and the fact that the function $\eta(r, s)$ is increasing in s we get (12.28). □

Lemma 12.7 (stability estimate). *There exists a class $\mathcal{K}_\infty$ function $\hat{\alpha}$ such that for all solutions of the closed-loop system* (12.1), (12.2)–(12.4) *that satisfy* (12.5) *for $0 < c < 1$ it holds that*

$$\begin{aligned} |X(t)| + \sup_{t - D(X(t)) \leq \theta \leq t} |U(\theta)| \leq {} & 2\beta^*(\eta(r_0, \sigma^* - t_0), \max\{0, t - \sigma^*\}) \\ & + \hat{\alpha}(\beta^*(\eta(r_0, \sigma^* - t_0), \max\{0, t - \sigma^*\})) \end{aligned} \tag{12.35}$$

for all $t \geq t_0$.

Proof. Under Assumption 12.2 and using Lemma 12.6 we get

$$|X(t)| \leq \beta^*(|X(\sigma^*)|, t - \sigma^*), \quad t \geq \sigma^*. \tag{12.36}$$

Using Lemma 12.4 and the fact that, without loss of generality, $\beta^*(s, 0) \geq s$, with (12.36) we get

$$|X(t)| \leq \beta^*(\eta(r_0, \sigma^* - t_0), \max\{0, t - \sigma^*\}), \quad t \geq t_0. \tag{12.37}$$

Since $U(\theta) = \varkappa(\sigma(\theta), X(\sigma(\theta)))$ for all $\theta \geq t_0$, with (12.36) and (12.8) we get

$$\sup_{t - D(X(t)) \leq \theta \leq t} |U(\theta)| \leq \hat{\alpha}(\beta^*(|X(\sigma^*)|, t - \sigma^*)), \quad t \geq \sigma^*. \tag{12.38}$$

Moreover, for $t \leq \sigma^*$ we have

$$\sup_{t - D(X(t)) \leq \theta \leq t} |U(\theta)| \leq \sup_{t - D(X(t)) \leq \theta \leq t_0} |U(\theta)| + \sup_{t_0 \leq \theta \leq t} |U(\theta)|. \tag{12.39}$$

Hence, for all $t \leq \sigma^*$

$$\sup_{t - D(X(t)) \leq \theta \leq t} |U(\theta)| \leq \sup_{t_0 - D(X(t_0)) \leq \theta \leq t_0} |U(\theta)| + \hat{\alpha}(\beta^*(|X(\sigma^*)|, 0)). \tag{12.40}$$

Combining bounds (12.37), (12.38), and (12.40) we get the statement of the lemma. □

Lemma 12.8 (ball of initial conditions that guarantees controller feasibility). *There exists a class $\mathcal{K}$ function $\mu : [0,1) \mapsto [0,\infty)$ such that for all initial conditions of the plant that satisfy* (12.11) *with $\psi_{\mathrm{RoA}}(c) \le \min\{\delta, \mu(c)\}$, the solutions of the closed-loop system* (12.1), (12.2)–(12.4) *satisfy* (12.5) *for $0 < c < 1$.*

Proof. With the change of variables $y = \sigma(\theta)$ we have that $t_0 \le y \le \sigma^*$ for all $t_0 - D(X(t_0)) \le \theta \le t_0$. Comparing the ODE for P, which is written in the y variable as

$$\frac{dP(\phi(y))}{dy} = f\left(P(\phi(y)), U\left(y - D\left(P(\phi(y))\right)\right)\right), \quad t \le y \le \sigma(t), \tag{12.41}$$

with (12.1) and using Lemma 12.4 we get

$$|P(\theta)| \le \eta(r_0, \sigma^* - t_0) \quad \text{for all } t_0 - D(X(t_0)) \le \theta \le t_0. \tag{12.42}$$

By noting that $P(\theta) = X(\sigma(\theta))$ for all $\theta \ge t_0$, with the help of bound (12.37) we have that

$$|P(\theta)| \le \beta^*\left(\eta(r_0, \sigma^* - t_0), 0\right), \quad \theta \ge t_0. \tag{12.43}$$

Therefore, using the fact that $\beta^*(s,0) \ge s$, from (12.42) and (12.43) we get that

$$|P(\theta)| \le \beta^*\left(\eta(r_0, \sigma^* - t_0), 0\right) \quad \text{for all } \theta \ge t_0 - D(X(t_0)). \tag{12.44}$$

Using (12.38) and (12.40) we get that

$$\sup_{t - D(X(t)) \le \theta \le t} |U(\theta)| \le r_0 + \hat{\alpha}\left(\beta^*\left(\eta(r_0, \sigma^* - t_0), 0\right)\right), \quad t \ge t_0. \tag{12.45}$$

Using the fact that $f : \mathbb{R}^n \times \mathbb{R} \to \mathbb{R}^n$ is locally Lipschitz with $f(0,0) = 0$, there exists a class $\mathcal{K}_\infty$ function α_1 such that

$$|f(X, \omega)| \le \alpha_1(|X| + |\omega|). \tag{12.46}$$

Combining (12.7), (12.46) we conclude that the solutions of the system lie inside the feasibility region as long as

$$\left(|\nabla D(0)| + \delta_2(|P(\theta)|)\right) \alpha_1\left(|P(\theta)| + \sup_{t - D(X(t)) \le s \le t} |U(s)|\right) < c \tag{12.47}$$

holds for all $t - D(X(t)) \le \theta \le t$ and for all $t \ge t_0$. With the help of the fact that $\sigma^* \le \xi^*(r_0)$ and using bounds (12.42), (12.44), and (12.45), (12.47) holds if

$$\begin{aligned} F(r_0) :=& \left(|\nabla D(0)| + \delta_2\left(\beta^*\left(\eta(r_0, \xi^*(r_0) - t_0), 0\right)\right)\right) \\ & \times \alpha_1\left(\beta^*\left(\eta(r_0, \xi^*(r_0) - t_0), 0\right) + r_0 + \hat{\alpha}\left(\beta^*\left(\eta(r_0, \xi^*(r_0) - t_0), 0\right)\right)\right) \\ \le& \; c. \end{aligned} \tag{12.48}$$

Condition (12.48) is satisfied if (12.11) holds with any class $\mathcal{K}$ choice $\psi_{\mathrm{RoA}}(c) \le \min\{\delta, \mu(c)\}$, where $\mu(c) = \sup_{\mu^* \in \mathbb{A}} \mu^*$ and $\mathbb{A} = \left\{\mu^* \in \mathbb{R}_+ : F(\mu^*) \le c\right\}$. Using the fact that $\xi^*(\cdot)$ is a monotonically increasing function, $F(\cdot)$ is also increasing with $F(0) = 0$. Hence, $\mu(c) = \sup_{\mu^* \in \mathbb{A}} \mu^*$ is continuous and increasing, with $\mu(0) = 0$. □

***Proof of Theorem* 12.3.** Bound (12.12) follows from Lemma 12.7 and the facts that $\xi^*(r)$ is increasing, $\sigma^* \leq \xi^*$, and $D(0)+\delta_1(\eta(r_0,\xi^*(r)-t_0))=\xi^*(r)-t_0$, with

$$\begin{aligned}\hat{\beta}(r,t-t_0)=&\,2\beta^*(\eta(r,\xi^*(r)-t_0),\max\{0,t-\xi^*(r)\})\\&+\hat{\alpha}(\beta^*(\eta(r,\xi^*(r)-t_0),\max\{0,t-\xi^*(r)\})).\end{aligned} \tag{12.49}$$

The rest of the results follow as in the proof of Theorem 11.4. □

Example 12.9. We consider the scalar system

$$\dot{X}(t)=X(t)^4+2X(t)^5+\left(X(t)^2+X(t)^3\right)U\left(t-X(t)^2\right). \tag{12.50}$$

The origin of (12.50) is neither locally exponentially stabilizable nor globally asymptotically stabilizable (because it is not reachable for $X(0)<-1$). In the delay-free case, the controller $U(t)=-X(t)$ yields a closed-loop system $\dot{X}(t)=-X(t)^3+2X(t)^5$, which is locally asymptotically stable, with $R=\frac{1}{\sqrt{2}}$. We assume for simplicity of the analysis that $U(\theta)=0$ for all $\theta\leq 0$. Denoting $y_0=8\log\left(|X(0)+\frac{1}{2}|\right)-8\log(|X(0)|)-\frac{4X(0)^2-X(0)+\frac{1}{3}}{X(0)^3}$, the controller "kicks in" at the time $\sigma^*=(X^*)^2$, where X^* satisfies

$$8\log\left(\left|X^*+\frac{1}{2}\right|\right)-8\log(|X^*|)-\frac{4(X^*)^2-X^*+\frac{1}{3}}{(X^*)^3}-y_0=(X^*)^2. \tag{12.51}$$

Let $X(0)=0.543$. Solving (12.51) we get $\sigma^*=0.46$ and $X^*=\sqrt{\sigma^*}=0.678$, which is almost at $R=\frac{1}{\sqrt{2}}$. The predictor controller is $U(t)=-P(t)$,

$$\begin{aligned}P(t)=&\int_{t-X(t)^2}^{t}\frac{\left(P(\theta)^4+2P(\theta)^5+\left(P(\theta)^2+P(\theta)^3\right)U(\theta)\right)d\theta}{1-2P(\theta)\left(P(\theta)^4+2P(\theta)^5+\left(P(\theta)^2+P(\theta)^3\right)U(\theta)\right)}\\&+X(t).\end{aligned} \tag{12.52}$$

In Figure 12.3 we show that the predictor feedback achieves local asymptotic stabilization. ■

12.2 ▪ Notes and References

The class of nonlinear systems considered in the present chapter, i.e., systems that are only locally stabilizable in the absence of the delay, is the least restrictive class of systems for which one can pursue stabilization. Systems that are not forward complete but are globally stabilizable in the absence of the delay belong to this class, such as, for example, systems in the strict-feedback form [92]. For this class of systems global stabilization is not possible in the presence of an input delay since they might exhibit a finite escape time before the control signal reaches the plant. The analysis of the closed-loop stability under predictor feedback presented in this chapter applies to this class of systems. The stability analysis for a specific scalar example of a nonlinear system in this class, but with a constant input delay, was presented in [86].

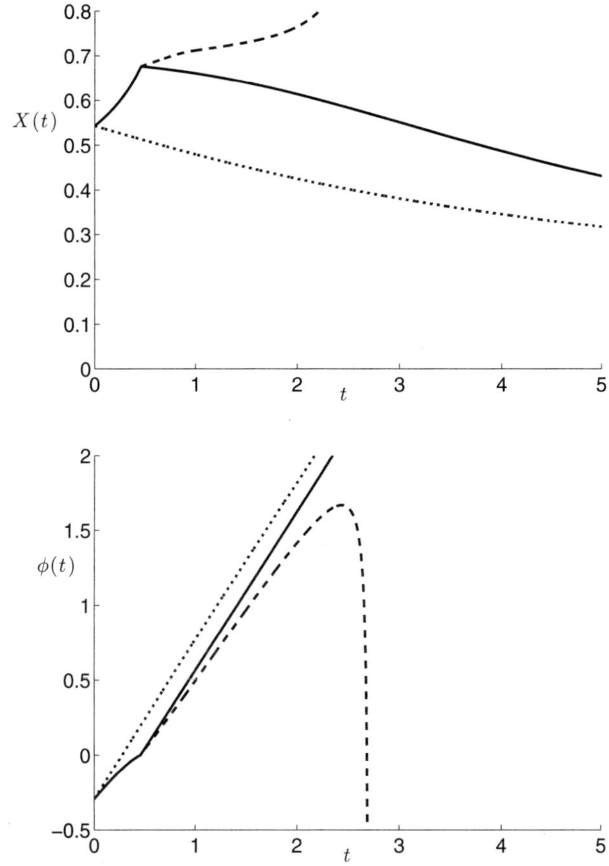

Figure 12.3. *The state $X(t)$, the control effort $U(t)$, and the function $\phi(t) = t - X(t)^2$ of system (12.50) with the delay-compensating controller (solid line), the uncompensated controller (dashed line), and the nominal controller for a system without delay (dotted line) for $X(0) = 0.543$ and $U(s) = 0$ for all $s \leq 0$.*

Chapter 13

Nonlinear Systems with State Delay

In this chapter we extend the technique for compensating state-dependent delays from systems with delayed inputs to systems with delayed states. We focus on predictor feedback design for nonlinear systems in the strict-feedback form having a state-dependent state delay on the virtual input. The two key challenges are the definition of the predictor state and the fact that the predictor design does not follow immediately from the delay-free design. We resolve these challenges and design a predictor-based control law in Section 13.1. As in the case of input delays, due to the fundamental limitation of the allowable magnitude of the delay function's gradient, we obtain only regional results. For forward-complete nonlinear systems we establish asymptotic stability of the resulting infinite-dimensional nonlinear system for general nonnegative-valued delay functions of the state and give an estimate the region of attraction of the proposed controller in Section 13.2. We also provide two examples in Section 13.3, including one dealing with the control of cooling systems.

13.1 ▪ Problem Formulation and Controller Design

We consider the system

$$\dot{X}_1(t)=f_1(t,X_1(t),X_2(t-D(X_1(t)))), \tag{13.1}$$

$$\dot{X}_2(t)=f_2(t,X_1(t),X_2(t))+U(t), \tag{13.2}$$

where $X_1\in\mathbb{R}^n$, X_2, $U\in\mathbb{R}$, and $t\geq t_0\geq 0$. We assume that $f_1:[t_0,\infty)\times\mathbb{R}^{n+1}\mapsto\mathbb{R}^n$ is locally Lipschitz with $f_1(t,0,0)=0$ for all $t\geq t_0$ and that there exists a class $\mathcal{K}_\infty$ function $\hat{\alpha}$ such that

$$|f_1(t,X_1,X_2)|\leq\hat{\alpha}(|X_1|+|X_2|)\quad\text{for all } t\geq t_0. \tag{13.3}$$

We further assume that $f_2:[t_0,\infty)\times\mathbb{R}^{n+1}\mapsto\mathbb{R}$ is locally Lipschitz with respect to $(X_1,X_2)\in\mathbb{R}^{n+1}$ with $f_2(t,0,0)=0$ for all $t\geq t_0$. The goal of the paper is to show that for (13.1), (13.2) there exist functions $P(\theta)$ and $\sigma(\theta)$, where $t-D(X_1(t))\leq\theta\leq t$, such that the controller

$$\begin{aligned}U(t)=&-f_2(t,X_1(t),X_2(t))-c_2(X_2(t)-x(\sigma(t),P_1(t)))\\&+\frac{\frac{\partial x(\sigma,P_1)}{\partial\sigma}+\frac{\partial x(\sigma,P_1)}{\partial P_1}f_1(\sigma(t),P_1(t),X_2(t))}{1-\nabla D(P_1(t))f_1(\sigma(t),P_1(t),X_2(t))},\end{aligned} \tag{13.4}$$

where c_2 is an arbitrary positive constant and

$$P_1(\theta)=X_1(t)+\int_{t-D(X_1(t))}^{\theta}\frac{f_1(\sigma(s),P_1(s),X_2(s))\,ds}{1-\nabla D(P_1(s))f_1(\sigma(s),P_1(s),X_2(s))},$$
$$t-D(X_1(t))\le\theta\le t, \tag{13.5}$$
$$\sigma(\theta)=\theta+D(P_1(\theta)),\quad t-D(X_1(t))\le\theta\le t, \tag{13.6}$$

for $t\ge t_0$, compensates for the state-dependent state delay and achieves asymptotic stability of the resulting closed-loop system. We refer to the quantity $P_1(\theta)$ given in (13.5) as "predictor" since $P_1(t)$ is the $D(P_1(t))$-time-units-ahead predictor of $X_1(t)$, i.e., $P_1(t)=X_1(t+D(P_1(t)))$. This fact can be seen as follows. Differentiating relation (13.5) with respect to θ and setting $\theta=t$ we get

$$\frac{dP_1(t)}{dt}=\frac{f_1(\sigma(t),P_1(t),X_2(t))}{1-\nabla D(P_1(t))f_1(\sigma(t),P_1(t),X_2(t))}. \tag{13.7}$$

Performing a change of variables $\tau=\sigma(t)$ in the ODE for $X_1(\tau)$ given by $\frac{dX_1(\tau)}{d\tau}=f_1(\tau,X_1(\tau),X_2(\tau-D(X_1(\tau))))$, we have that

$$\frac{dX_1(\sigma(t))}{dt}=\frac{d\sigma(t)}{dt}f_1(\sigma(t),X_1(\sigma(t)),X_2(t)). \tag{13.8}$$

From (13.8) one observes that $P_1(t)$ satisfies the same ODE in t as $X_1(\sigma(t))$ because from (13.6)–(13.8) it follows that for all $t-D(X_1(t))\le\theta\le t$

$$\frac{d\sigma(\theta)}{d\theta}=\frac{1}{1-\nabla D(X_1(\sigma(\theta)))f_1(\sigma(\theta),X_1(\sigma(\theta)),X_2(\theta))}, \tag{13.9}$$

provided that $P_1(t)=X_1(\sigma(t))$. Since from (13.5) for $t=t_0$ and $\theta=t_0-D(X_1(t_0))$ it follows that $P_1(t_0-D(X_1(t_0)))=X_1(t_0)$, by defining

$$\phi(t)=t-D(X_1(t)),\quad t\ge t_0, \tag{13.10}$$
$$\sigma(\theta)=\phi^{-1}(\theta),\quad t-D(X_1(t))\le\theta\le t, \tag{13.11}$$

we have that $P_1(t_0)=X_1(\sigma(t_0))$. Noting from relations (13.10) and (13.11) that $D(X_1(\sigma(t)))=\sigma(t)-t$, differentiating this relation, we get (13.9). Comparing (13.7) with (13.8) we conclude with the help of (13.9) that indeed $P_1(t)=X_1(\sigma(t))$ for all $t\ge t_0$.

Motivated by the need to keep the denominator in (13.5) and in (13.9) positive, throughout the paper we consider the condition on the solutions, which is given by

$$\mathscr{G}_c:\quad \nabla D(P_1(\theta))f_1(\sigma(\theta),P_1(\theta),X_2(\theta))<c \quad \text{for all } \theta\ge t_0-D(X_1(t_0)) \tag{13.12}$$

for $c\in(0,1]$. We refer to $\mathscr{G}_1$ as the *feasibility condition* of the controller (13.4)–(13.5).

13.2 ▪ Stability Analysis for Forward-Complete Systems

Throughout the section we make the following assumptions concerning the plant (13.1)–(13.2).

Assumption 13.1. *$D \in C^1(\mathbb{R}^n;\mathbb{R}_+)$ and ∇D is locally Lipschitz.*[1]

Assumption 13.2. *There exist a smooth positive definite function R and class $\mathcal{K}_\infty$ functions α_1, α_2, and α_3 such that for the plant $\dot{X} = f_1(t,X,\omega)$, the following hold:*

$$\alpha_1(|X|) \le R(t,X) \le \alpha_2(|X|), \tag{13.13}$$
$$\frac{\partial R(t,X)}{\partial t} + \frac{\partial R(t,X)}{\partial X} f_1(t,X,\omega) \le R(t,X) + \alpha_3(|\omega|) \tag{13.14}$$

for all $X,\omega \in \mathbb{R}^{n+1}$ and $t \ge t_0$.

Assumption 13.2 guarantees that the plant $\dot{X} = f_1(t,X,\omega)$ with ω as input is forward complete.

Assumption 13.3. *There exist functions $\varkappa \in C^1(\mathbb{R}_+ \times \mathbb{R}^n;\mathbb{R})$ and $\hat{\rho} \in \mathcal{K}_\infty$, such that the plant $\dot{X}(t) = f_1(t,X(t),\varkappa(t,X(t)) + \omega(t))$ is input-to-state stable with respect to ω, and $\varkappa$ is uniformly bounded with respect to its first argument, that is,*

$$|\varkappa(t,X)| \le \hat{\rho}(|X|) \quad \textit{for all } t \ge 0. \tag{13.15}$$

Theorem 13.4. *Consider the plant (13.1)–(13.2) together with the control law (13.4)–(13.6). Under Assumptions 13.1, 13.2, and 13.3 there exist a class $\mathcal{K}$ function ξ_{RoA}, a class $\mathcal{KL}$ function β, and a class $\mathcal{KC}_\infty$ function ξ_1 such that for all initial conditions for which X_2 is locally Lipschitz on the interval $[t_0 - D(X_1(t_0)), t_0]$ and which satisfy*

$$|X_1(t_0)| + \sup_{t_0 - D(X_1(t_0)) \le \theta \le t_0} |X_2(\theta)| < \xi_{\mathrm{RoA}}(c) \tag{13.16}$$

for some $0 < c < 1$, there exists a unique solution to the closed-loop system with $X_1 \in C^1[t_0,\infty)$, $X_2 \in C^1(t_0,\infty)$ and

$$\Omega(t) \le \beta(\xi_1(\Omega(t_0),c), t - t_0), \tag{13.17}$$
$$\Omega(t) = |X_1(t)| + \sup_{t - D(X_1(t)) \le \theta \le t} |X_2(\theta)| \tag{13.18}$$

for all $t \ge t_0$. Furthermore, there exists a class $\mathcal{K}$ function δ^*, such that for all $t \ge t_0$ the following hold:

$$D(X_1(t)) \le D(0) + \delta^*(c), \tag{13.19}$$
$$\left|\dot{D}(X_1(t))\right| \le c. \tag{13.20}$$

The proof of Theorem 13.4 is based on Lemmas 13.5–13.12, which are presented next. Note that the definitions of class $\mathcal{KC}$ and $\mathcal{KC}_\infty$ functions are the ones from Appendix C.1.

Lemma 13.5 (backstepping transform of the delayed state). *The infinite-dimensional backstepping transformation of the state X_2 defined by*

$$Z_2(\theta) = X_2(\theta) - \varkappa(\sigma(\theta), P_1(\theta)), \quad t - D(X_1(t)) \le \theta \le t, \tag{13.21}$$

[1] To ensure uniqueness of solutions.

together with the predictor-based control law given in relations (13.4)–(13.5) *transforms the system* (13.1)–(13.2) *to the target system given by*

$$\dot{X}_1(t) = f_1(t, X_1(t), \kappa(t, X_1(t)) + Z_2(t - D(X_1(t)))), \tag{13.22}$$

$$\dot{Z}_2(t) = -c_2 Z_2(t). \tag{13.23}$$

Proof. Using (13.1) and the facts that $P_1(t - D(X_1(t))) = X_1(t)$ and that $\sigma\big(t - D(X_1(t))\big) = t$, which are immediate consequences of (13.5) and (13.6), we get (13.22). Setting $\theta = t$ in (13.21) and taking the derivative with respect to t of the resulting equation we get (13.23) using (13.4), (13.7), and (13.9). □

Lemma 13.6 (inverse backstepping transform). *The inverse of the infinite-dimensional backstepping transformation defined in* (13.21) *is given by*

$$X_2(\theta) = Z_2(\theta) + \kappa(\sigma(\theta), \Pi_1(\theta)), \quad t - D(X_1(t)) \le \theta \le t, \tag{13.24}$$

where

$$\Pi_1(\theta) = \int_{t-D(X_1(t))}^{\theta} \frac{f_1(\sigma(s), \Pi_1(s), \kappa(\sigma(s), \Pi_1(s)) + Z_2(s))\, ds}{1 - \nabla D(\Pi_1(s)) f_1(\sigma(s), \Pi_1(s), \kappa(\sigma(s), \Pi_1(s)) + Z_2(s))} + X_1(t), \quad t - D(X_1(t)) \le \theta \le t. \tag{13.25}$$

Proof. The proof is by direct verification, noting also that $\Pi_1(\theta) = P_1(\theta)$ for all $t - D(X_1(t)) \le \theta \le t$, where $\Pi_1(\theta)$ is driven by the transformed state $Z_2(\theta)$, whereas $P_1(\theta)$ is driven by the state $X_2(\theta)$ for $t - D(X_1(t)) \le \theta \le t$. See Figure 13.1. □

Lemma 13.7 (stability estimate for target system). *There exists a class $\mathcal{KL}$ function β^* such that for all solutions of the system* (13.1), (13.2) *satisfying* (13.12) *for* $0 < c < 1$, *the*

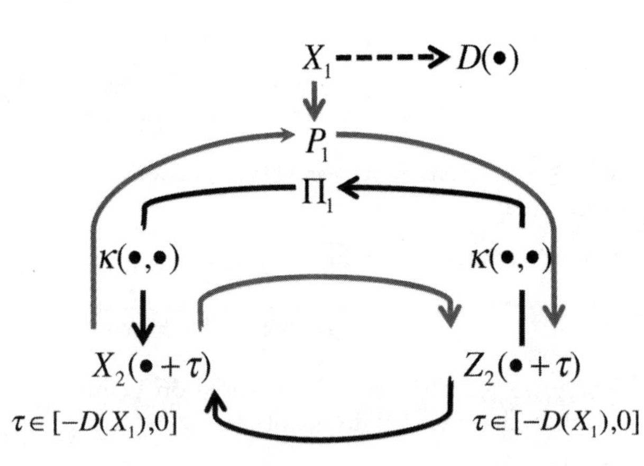

Figure 13.1. *Interconnections between the predictor states P_1 and Π_1 with the transformations Z_2 and X_2 in* (13.21) *and* (13.24). *The direct backstepping transformation is defined as $(X_1(t), X_2(\theta)) \mapsto (X_1(t), Z_2(\theta))$ and is given in* (13.21), *where $P_1(\theta)$ is given as a function of $X_1(t)$ and $X_2(\theta)$ through relation* (13.5). *Analogously, the inverse transformation is defined as $(X_1(t), Z_2(\theta)) \mapsto (X_1(t), X_2(\theta))$ and is given in* (13.24), *where $\Pi_1(\theta)$ is given as a function of $X_1(t)$ and $Z_2(\theta)$ through relation* (13.24).

following holds for all $t \geq t_0$:

$$\Xi(t) \leq \beta^*(\Xi(t_0), t - t_0), \tag{13.26}$$

$$\Xi(t) = |X_1(t)| + \sup_{t-D(X_1(t)) \leq \theta \leq t} |Z_2(\theta)|. \tag{13.27}$$

Proof. Solving (13.23), and using the fact that $\sigma(t_0) = t_0 + D(X_1(\sigma(t_0)))$ and the fact that $\phi(t)$ is increasing for all $t \geq t_0$, we get for all $t \geq \sigma(t_0)$,

$$\sup_{t-D(X_1(t)) \leq \theta \leq t} |Z_2(\theta)| \leq \sup_{t_0-D(X_1(t_0)) \leq \theta \leq t_0} |Z_2(\theta)| e^{-c_2(t-D(X_1(t))-t_0)}, \tag{13.28}$$

where we also used the trivial inequality $|Z_2(t_0)| \leq \sup_{t_0-D(X_1(t_0)) \leq \theta \leq t_0} |Z_2(\theta)|$. Similarly, for all $t_0 \leq t \leq \sigma(t_0)$ we get

$$\sup_{t-D(X_1(t)) \leq \theta \leq t} |Z_2(\theta)| \leq \sup_{t_0-D(X_1(t_0)) \leq \theta \leq t_0} |Z_2(\theta)| + \sup_{t_0 \leq \theta \leq t \leq \sigma(t_0)} |Z_2(\theta)| \tag{13.29}$$

and hence, combining (13.29) with (13.23), we get for all $t_0 \leq t \leq \sigma(t_0)$

$$\sup_{t-D(X_1(t)) \leq \theta \leq t} |Z_2(\theta)| \leq 2 \sup_{t_0-D(X_1(t_0)) \leq \theta \leq t_0} |Z_2(\theta)|. \tag{13.30}$$

Therefore, using (13.28) and (13.30) we get for all $t \geq t_0$

$$\sup_{t-D(X_1(t)) \leq \theta \leq t} |Z_2(\theta)| \leq 2 \sup_{t_0-D(X_1(t_0)) \leq \theta \leq t_0} |Z_2(\theta)| e^{c_2(D(0)+\delta_1(|X_1(t)|))} \times e^{-c_2(t-t_0)}. \tag{13.31}$$

Under Assumption 13.3 (see Definition C.14) there exist class $\mathcal{KL}$ function $\hat{\beta}$ and class $\mathcal{K}$ function $\hat{\gamma}$ such that for all $t \geq s \geq t_0$

$$|X_1(t)| \leq \hat{\beta}(|X_1(s)|, t-s) + \hat{\gamma}\left(\sup_{s \leq \tau \leq t} |Z_2(\tau - D(X_1(\tau)))|\right). \tag{13.32}$$

Setting $s = t_0$ we have for all $t \geq t_0$ that

$$|X_1(t)| \leq \hat{\beta}(|X_1(t_0)|, 0) + \hat{\gamma}\left(\sup_{\phi(t_0) \leq \theta \leq t_0} |Z_2(\theta)| + \sup_{t_0 \leq \theta \leq \phi(t)} |Z_2(\theta)|\right), \tag{13.33}$$

and hence

$$|X_1(t)| \leq \hat{\beta}(|X_1(t_0)|, 0) + \hat{\gamma}\left(2 \sup_{\phi(t_0) \leq \theta \leq t_0} |Z_2(\theta)|\right) \quad \text{for all } t \geq t_0. \tag{13.34}$$

Setting $s = \frac{t+t_0}{2}$ in (13.32) we have that

$$|X_1(t)| \leq \hat{\beta}\left(\left|X_1\left(\frac{t+t_0}{2}\right)\right|, \frac{t-t_0}{2}\right) + \hat{\gamma}\left(\sup_{\phi\left(\frac{t+t_0}{2}\right) \leq \theta \leq \phi(t)} |Z_2(\theta)|\right). \tag{13.35}$$

We estimate now $\sup_{\phi(\frac{t+t_0}{2})\leq\theta\leq\phi(t)}|Z_2(\theta)|$. Solving (13.23) we get for all $t\geq 2\sigma(t_0)-t_0$

$$\sup_{\phi(\frac{t+t_0}{2})\leq\theta\leq\phi(t)}|Z_2(\theta)|\leq 2\sup_{t_0-D(X_1(t_0))\leq\theta\leq t_0}|Z_2(\theta)|e^{-c_2(\phi(\frac{t+t_0}{2})-t_0)}. \tag{13.36}$$

With the help of relations (13.23) and (13.36) we get for all $t_0\leq t\leq 2\sigma(t_0)-t_0$

$$\begin{aligned}\sup_{\phi(\frac{t+t_0}{2})\leq\theta\leq\phi(t)}|Z_2(\theta)|&\leq\sup_{\phi(\frac{t+t_0}{2})\leq\theta\leq t_0}|Z_2(\theta)|+\sup_{t_0\leq\theta\leq\phi(t)}|Z_2(\theta)|\\&\leq 2\sup_{t_0-D(X_1(t_0))\leq\theta\leq t_0}|Z_2(\theta)|.\end{aligned} \tag{13.37}$$

Hence, using the fact that $\phi(\frac{t+t_0}{2})=\frac{t+t_0}{2}-D(X_1(\frac{t+t_0}{2}))$ we get for all $t\geq t_0$

$$\begin{aligned}\sup_{\phi(\frac{t+t_0}{2})\leq\theta\leq\phi(t)}|Z_2(\theta)|\leq 2&\sup_{t_0-D(X_1(t_0))\leq\theta\leq t_0}|Z_2(\theta)|e^{c_2(D(0)+\delta_1(|X_1(\frac{t+t_0}{2})|))}\\&\times e^{-\frac{c_2}{2}(t-t_0)}.\end{aligned} \tag{13.38}$$

Setting $s=t_0$ and replacing t by $\frac{t+t_0}{2}$ we get from (13.32) that

$$\left|X_1\left(\frac{t+t_0}{2}\right)\right|\leq\hat{\beta}\left(|X_1(t_0)|,\frac{t-t_0}{2}\right)+\hat{\gamma}\left(\sup_{\phi(t_0)\leq\theta\leq\phi(\frac{t+t_0}{2})}|Z_2(\theta)|\right). \tag{13.39}$$

Since

$$\begin{aligned}\sup_{\phi(t_0)\leq\theta\leq\phi(\frac{t+t_0}{2})}|Z_2(\theta)|&\leq\sup_{\phi(t_0)\leq\theta\leq t_0}|Z_2(\theta)|+\sup_{t_0\leq\theta\leq\phi(t)}|Z_2(\theta)|\\&\leq 2\sup_{t_0-D(X_1(t_0))\leq\theta\leq t_0}|Z_2(\theta)|,\end{aligned} \tag{13.40}$$

we arrive at

$$\left|X_1\left(\frac{t+t_0}{2}\right)\right|\leq\hat{\beta}(|X_1(t_0)|,0)+\hat{\gamma}\left(2\sup_{t_0-D(X_1(t_0))\leq\theta\leq t_0}|Z_2(\theta)|\right) \tag{13.41}$$

for all $t\geq t_0$. Combining (13.31), (13.34), (13.35), (13.38), (13.41), we get the statement of the lemma with

$$\begin{aligned}\beta^*(s,t-t_0)=\hat{\beta}\left(\hat{\beta}(s,0)+\hat{\gamma}(2s),\frac{t-t_0}{2}\right)+2se^{c_2\left(D(0)+\delta_1\left(\hat{\beta}(s,0)+\hat{\gamma}(2s)\right)\right)}e^{-c_2(t-t_0)}\\+\hat{\gamma}\left(2se^{c_2\left(D(0)+\delta_1\left(\hat{\beta}(s,0)+\hat{\gamma}(2s)\right)\right)}e^{-\frac{c_2}{2}(t-t_0)}\right).\quad\square\end{aligned} \tag{13.42}$$

Lemma 13.8 (bound on predictor in terms of system's states). *There exists a class $\mathcal{KC}_\infty$ function ξ such that for all solutions of the system* (13.1), (13.2) *satisfying* (13.12) *for* $0<c<1$, *the following holds for all* $t-D(X_1(t))\leq\theta\leq t$*:*

$$|P_1(\theta)|\leq\xi\left(|X_1(t)|+\sup_{t-D(X_1(t))\leq\tau\leq t}|X_2(\tau)|,c\right). \tag{13.43}$$

Proof. Consider the following ODE in θ, which follows by differentiating (13.5) together with the initial condition $P_1(t-D(X_1(t)))=X_1(t)$:

$$\frac{dP_1(\theta)}{d\theta}=\frac{f_1(\sigma(\theta),P_1(\theta),X_2(\theta))}{1-\nabla D(P_1(\theta))f_1(\sigma(\theta),P_1(\theta),X_2(\theta))},\\ t-D(X_1(t))\leq\theta\leq t. \tag{13.44}$$

With the change of variables

$$y=\sigma(\theta), \tag{13.45}$$

we rewrite (13.44) as

$$\frac{dP_1(\phi(y))}{dy}=f_1(y,P_1(\phi(y)),X_2(y-D(P_1(\phi(y))))),\quad t\leq y\leq\sigma(t). \tag{13.46}$$

Using (13.14) we get

$$\frac{dR(y,P_1(\phi(y)))}{d\theta}\frac{d\theta}{dy}\leq R(y,P_1(\phi(y)))+\alpha_3(|X_2(y-D(P_1(\phi(y))))|). \tag{13.47}$$

With (13.12) we have for all $t-D(X_1(t))\leq\theta\leq t$ that

$$\frac{dR(\sigma(\theta),P_1(\theta))}{d\theta}\leq\frac{1}{1-c}(R(\sigma(\theta),P_1(\theta))+\alpha_3(|X_2(\theta)|)). \tag{13.48}$$

Under Assumption 13.1 there exists a function $\delta_1\in\mathcal{K}_\infty\cap C^1$ such that

$$D(X_1)\leq D(0)+\delta_1(|X_1|), \tag{13.49}$$

and hence, using the comparison principle (Lemma B.7 in Appendix B), we have from (13.48) for all $t-D(X_1(t))\leq\theta\leq t$ that

$$R(\sigma(\theta),P_1(\theta))\leq\left(R(t,X_1(t))+\sup_{t-D(X_1(t))\leq\tau\leq t}\alpha_3(|X_2(\tau)|)\right)\\ \times e^{\frac{D(0)+\delta_1(|X_1(t)|)}{1-c}}. \tag{13.50}$$

With standard properties of class $\mathcal{K}_\infty$ functions we get the statement of the lemma with $\xi\in\mathcal{KC}_\infty$ given by

$$\xi(s,c)=\alpha_1^{-1}\left((\alpha_2(s)+\alpha_3(s))e^{\frac{D(0)+\delta_1(s)}{1-c}}\right).\quad\square \tag{13.51}$$

Lemma 13.9 (bound on predictor in terms of transformed system's states). *There exists a class $\mathcal{K}$ function γ such that for all solutions of the system* (13.1), (13.2) *satisfying* (13.12) *for $0<c<1$, the following holds:*

$$|\Pi_1(\theta)|\leq\gamma\left(|X_1(t)|+\sup_{t-D(X_1(t))\leq\tau\leq t}|Z_2(\tau)|\right),\\ t-D(X_1(t))\leq\theta\leq t. \tag{13.52}$$

Proof. Let $Y(s)$ be the solution of $\frac{dY(s)}{ds} = f_1(s,Y(s),\varkappa(s,Y(s))+\omega(s))$ for $s \geq s_0 \geq 0$. Under Assumption 13.3 (see Definition C.14), there exist a class $\mathcal{KL}$ function $\hat{\beta}$ and a class $\mathcal{K}$ function $\hat{\gamma}$ such that

$$|Y(s)| \leq \hat{\beta}(|Y(s_0)|, s-s_0) + \hat{\gamma}\left(\sup_{s_0 \leq r \leq s} |\omega(r)|\right) \quad \text{for all } s \geq s_0. \tag{13.53}$$

Using the change of variable (13.45) and definition (13.25), we have that

$$\frac{d\Pi_1(\phi(y))}{dy} = f_1(y, \Pi_1(\phi(y)), \varkappa(y, \Pi_1(\phi(y))) + Z_2(\phi(y))),$$
$$t \leq y \leq \sigma(t). \tag{13.54}$$

Since $\Pi_1(\phi(y))$ satisfies the same ODE in y as the ODE for $Y(s)$ in s we have that

$$|\Pi_1(\phi(y))| \leq \hat{\beta}(|X_1(t)|, y-t) + \hat{\gamma}\left(\sup_{t \leq y \leq \sigma(t)} |Z_2(\phi(y))|\right)$$
$$\text{for all } t \leq y \leq \sigma(t). \tag{13.55}$$

Using (13.45) (which can be also written as $\theta = \phi(y)$) with the fact that $\hat{\beta}(s,r) \leq \hat{\beta}(s,0)$ for all $r \geq 0$, we get from (13.55)

$$|\Pi_1(\theta)| \leq \hat{\beta}(|X_1(t)|, 0) + \hat{\gamma}\left(\sup_{t-D(X_1(t)) \leq \tau \leq t} |Z_2(\tau)|\right),$$
$$t - D(X_1(t)) \leq \theta \leq t. \tag{13.56}$$

With the properties of class $\mathcal{K}$ functions we get the statement of the lemma where $\gamma(s) = \hat{\beta}(s,0) + \hat{\gamma}(s)$. ☐

Lemma 13.10 (equivalence of norms for original and target system). *There exist a function ξ_1 of class $\mathcal{KC}_\infty$ and a class $\mathcal{K}_\infty$ function α_4 such that for all solutions of the system* (13.1), (13.2) *satisfying* (13.12) *for* $0 < c < 1$, *the following hold:*

$$\Xi(t) \leq \xi_1(\Omega(t), c), \tag{13.57}$$
$$\Omega(t) \leq \alpha_4(\Xi(t)) \tag{13.58}$$

for all $t \geq t_0$, where $\Xi(t)$ and $\Omega(t)$ are defined in (13.27) *and in* (13.18), *respectively.*

Proof. Using the direct backstepping transformation (13.21) and the bounds (13.15), (13.43) we get the bound (13.57) with $\xi_1(s,c) = s + \hat{\rho}(\xi(s,c))$. Using the inverse backstepping transformation (13.24) and the bounds (13.15), (13.52) we get the bound (13.58) with $\alpha_4(s) = s + \hat{\rho}(\gamma(s))$. ☐

Lemma 13.11 (ball around the origin within the feasibility region). *There exists a function $\bar{\xi}_c$ of class $\mathcal{KC}_\infty$ such that all solutions of the system* (13.1), (13.2) *that satisfy*

$$|X_1(t)| + \sup_{t-D(X_1(t)) \leq \theta \leq t} |X_2(\theta)| < \bar{\xi}_c(c,c) \tag{13.59}$$

for $0 < c < 1$ *also satisfy* (13.12).

Proof. Since $D \in C^1\left(\mathbb{R}^n;\mathbb{R}_+\right)$ there exists a class $\mathcal{K}_\infty$ function δ_2 such that

$$|\nabla D(X_1)| \leq |\nabla D(0)| + \delta_2(|X_1|). \tag{13.60}$$

If a solution satisfies for all $t - D(X_1(t)) \leq \theta \leq t$

$$(|\nabla D(0)| + \delta_2(|P_1(\theta)|))\hat{\alpha}\left(|P_1(\theta)| + \sup_{t-D(X_1(t))\leq\theta\leq t} |X_2(\theta)|\right) \leq c \tag{13.61}$$

for $0 < c < 1$, then it also satisfies (13.12). Using Lemma 13.8, (13.61) is satisfied for $0 < c < 1$ as long as the bound (13.59) holds, where the class $\mathcal{KC}_\infty$ function ξ_c is given by

$$\xi_c(s,c) = (|\nabla D_1(0)| + \delta_2(\xi(s,c)))\hat{\alpha}(\xi(s,c)+s), \tag{13.62}$$

and we denote by $\bar{\xi}_c$ the inverse function of ξ_c with respect to ξ_c's first argument. □

Lemma 13.12 (estimate of the region of attraction). *There exists a class $\mathcal{K}$ function ξ_{RoA} such that for all initial conditions of the closed-loop system* (13.1), (13.2), (13.4), (13.5) *that satisfy* (13.16), *the solutions of the system* (13.1), (13.2) *satisfy* (13.59) *for $0 < c < 1$ and hence satisfy* (13.12).

Proof. Using Lemma 13.10, with the help of (13.26), we have that

$$\Omega(t) \leq \alpha_4(\beta^*(\xi_1(\Omega(t_0),c), t - t_0)), \tag{13.63}$$

where

$$\Omega(t) = |X_1(t)| + \sup_{t-D(X_1(t))\leq\theta\leq t} |X_2(\theta)|. \tag{13.64}$$

Hence, for all initial conditions that satisfy the bound (13.16) with any class $\mathcal{K}$ choice $\xi_{\mathrm{RoA}}(c) \leq \bar{\xi}^*{}_{\mathrm{RoA}}\left(\bar{\xi}_c(c,c),c\right)$, where $\bar{\xi}^*{}_{\mathrm{RoA}}(s,c)$ is the inverse of the function $\xi^*{}_{\mathrm{RoA}}(s,c) = \alpha_4(\beta^*(\xi_1(s,c),0)) \in \mathcal{KC}_\infty$ with respect to ξ^*_{RoA}'s first argument, the solutions satisfy (13.59). Moreover, for all of those initial conditions, the solutions verify (13.12) for all $\theta \geq t_0 - D(X_1(t_0))$. □

***Proof of Theorem* 13.4.** Using (13.63) we get (13.17) with $\beta(s,t) = \alpha_4(\beta^*(s,t))$. From (13.1), Assumption 13.1, and the Lipschitzness of X_2 in $[t_0 - D(X_1(t_0)), t_0]$ guarantee the existence and uniqueness of $X_1 \in C^1[t_0, \sigma(t_0))$, where $\sigma(t_0) = t_0 + D(X_1(\sigma(t_0)))$. The target system (13.22) and (13.23) guarantees the existence and uniqueness of $X_1 \in C^1(\sigma(t_0), \infty)$. The continuity of X_2 at t_0 and (13.1) guarantee that $X_1 \in C^1[t_0, \infty)$. Using the fact that Π_1 satisfies the ODE

$$\dot{\Pi}_1 = \frac{f_1(\sigma(t), \Pi_1(t), \kappa(\sigma(t), \Pi_1(t)) + Z_2(t))}{1 - \nabla D(\Pi_1(t)) f_1(\sigma(t), \Pi_1(t), \kappa(\sigma(t), \Pi_1(t)) + Z_2(t))} \tag{13.65}$$

for $t \geq t_0$, where $\sigma(t) = t + D(\Pi_1(t))$, the local Lipschitzness of ∇D, relation (13.12), and Assumption 13.3 ($\kappa \in C^1([t_0,\infty) \times \mathbb{R}^n; \mathbb{R})$) guarantee that $\Pi_1 \in C^1(t_0, \infty)$. Using (13.24), with the help of (13.23) and the fact that $\kappa \in C^1([t_0,\infty) \times \mathbb{R}^n; \mathbb{R})$, we get that

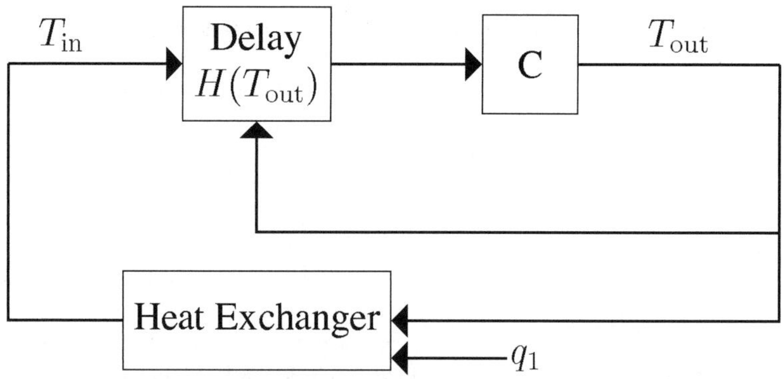

Figure 13.2. *A typical marine cooling system with one consumer.*

$X_2 \in C^1(t_0,\infty)$. Using Lemma 13.12 together with (13.49) and (13.60) we get (13.19), (13.20) with any class $\mathcal{K}$ choice $\delta^*(c) \geq \bar{\delta}_1(\xi_c(c,c))$. □

13.3 ▪ Examples

Example 13.13 (application to cooling systems). In marine transportation of materials the design of control laws for the ship's cooling system is of paramount importance due to the significant potential for energy optimization [50]. In Figure 13.2 we show a typical marine cooling circuit with one consumer, denoted by C, and a heat exchanger. We denote by T_{in} the input temperature towards the consumer, i.e., the output temperature of the heat exchanger. Due to the transportation time of the coolant (typically water) from the heat exchanger to the consumer C, the actual input temperature T_{in} in the consumer is delayed by H, namely, $T_{\text{in}}(t-H)$. The delay time H depends on the flow rate q_2 which can be controlled through a pump. In order to design a feedback law q_2 we take into account that the flow rate q_2 has to be proportional to the temperature at the other end of the consumer, which we denote by T_{out}. This is because it makes sense to increase the flow rate if the outer temperature of the consumer C is increasing. A simple choice is $q_2 = k_1 T_{\text{in}} + k_2$. The control objective is to regulate the temperature T_{out} to a constant set-point, say $T_{\text{eq}} > 0$. This is achieved by controlling the flow rate q_1 at the input of the heat exchanger through a pump.

Denoting by $T_{\text{out}} = X_1$, $T_{\text{in}} = X_2$, $q_2 = k_1 X_1 + k_2$, $H = \frac{b}{q_2} = D$, and $q_1 = U$ and neglecting the effect of the hydraulics in the system (since the hydraulic dynamics are assumed to be much faster than the heat dynamics [50]), the equations that describe the thermodynamics of the cooling circuit of Figure 13.2 are

$$\dot{X}_1(t) = a\left(X_1(t) - X_2\left(t - \frac{b}{k_1 X_1(t) + k_2}\right)\right)(k_1 X_1(t) + k_2), \tag{13.66}$$

$$\dot{X}_2(t) = (k_1 X_1(t) + k_2)(X_1(t) - X_2(t)) - U(t), \tag{13.67}$$

where $a < 0$ and $b, k_1, k_2 > 0$. Since the coolant flows only in one direction, $q_1 > 0$, and hence $U > 0$. Since the coolant is typically water, both T_{out} and T_{in} cannot fall below zero, and hence $X_1, X_2 > 0$. In addition since the consumer always adds heat (due to its functioning), $T_{\text{out}} \geq T_{\text{in}}(t-H)$ and $T_{\text{out}} \geq T_{\text{in}}$, and hence $X_1(t) - X_2\left(t - \frac{b}{k_1 X_1 + k_2}\right) \geq 0$ and $X_1(t) - X_2(t) \geq 0$ for all $t \geq 0$. Assumption 13.2 is satisfied (in the domain of interest) since X_1 remains bounded because $q_2 = k_1 X_1(t) + k_2 \geq 0$, which follows from the fact that if $q_2 = 0$, then $\dot{X}_1(t) = 0$ (so q_2 cannot cross from being positive to being negative). Assumption 13.1 is satisfied for all $X_1 > 0$ and Assumption 13.3 is satisfied since for $q_2 > 0$, (13.66) is input-to-state stable from the "disturbance" $w = X_2\left(t - \frac{b}{k_1 X_1(t) + k_2}\right)$. We choose the control law $\varkappa(X_1) = X_1 + \frac{c_1}{a} \frac{X_1}{k_1 X_1 + k_2}$, and hence the predictor-based control law for this system becomes

$$U(t) = (k_1 X_1(t) + k_2)(X_1(t) - X_2(t)) + c_2 \left(X_2(t) - P_1(t) - \frac{c_1}{a} \frac{P_1(t) - T_{\text{eq}}}{k_1 P_1(t) + k_2} \right)$$
$$- \left(1 + \frac{c_1}{a k_1} \frac{T_{\text{eq}} + \frac{k_2}{k_1}}{\left(P_1(t) + \frac{k_2}{k_1}\right)^2} \right)$$
$$\times \frac{a(P_1(t) - X_2(t))(k_1 P_1(t) + k_2)}{1 + \frac{b k_1}{(k_1 P_1(t) + k_2)^2} a(P_1(t) - X_2(t))(k_1 P_1(t) + k_2)}, \tag{13.68}$$

where

$$P_1(t) = \int_{t - \frac{b}{k_1 X_1(t) + k_2}}^{t} \frac{a(P_1(\theta) - X_2(\theta))(k_1 P_1(\theta) + k_2)}{1 + \frac{b k_1}{(k_1 P_1(\theta) + k_2)^2} a(P_1(\theta) - X_2(\theta))(k_1 P_1(\theta) + k_2)} d\theta$$
$$+ X_1(t). \tag{13.69}$$

We choose the parameters of the plant and of the controller as $a = -1$ and $c_1 = c_2 = b = k_1 = k_2 = 1$ and the initial conditions as $X_1(0) = 1$ and $X_2(\theta) = 0.2$ for all $-\frac{b}{k_1 X_1(0) + k_2} \leq \theta \leq 0$. We show in Figure 13.3 the temperatures T_{out}, T_{in} together with the input flow q_1. We compare the response of the system with the predictor-based controller (13.68) and with no control. In the case of the open-loop response we observe that the temperatures T_{out} ,T_{in} converge to the same value but not at the desired set-point. This is what one expects since the system has an equilibrium at $X_1 = X_2$. In contrast, the predictor-based controller regulates the temperatures T_{out}, T_{in} at the desired set-point T_{eq}. ■

Example 13.14. We consider the system

$$\dot{s}(t) = v(t - r_1 \sin^2(\omega s(t))), \tag{13.70}$$
$$\dot{v}(t) = a(t), \tag{13.71}$$

where the state variables are denoted by s and v and the control variable is denoted by $a(t)$. This system has some resemblance to the model considered in [169] for the "soft" automatic landing. In view of condition (13.12) one should expect that global stabilization is not achievable. This fact can be viewed in terms of the system dynamics (13.70)–(13.71) and in particular from relation $v(t - r_1 \sin^2(\omega s(t)))$. Subsystem (13.70) is forward complete from v since it is linear, and the delay $r_1 \sin^2(\omega s(t))$ satisfies Assumption 13.1. Hence, Theorem 13.4 applies. We choose the parameters of the plant as $r_1 = 0.3$ and $\omega = 15$, the initial conditions of the plant as $s(0) = 1$ and $v(\theta) = 0.1$ for all $-r_1 \sin^2(\omega s(0)) \leq \theta \leq 0$, and the parameters of the nominal controller as $c_1 = c_2 = 0.5$. The predictor-based controller is given by

$$a(t) = -c_2(v(t) + c_1 P_1(t)) - c_1 \frac{v(t)}{1 - r_1 \omega \sin(\omega P_1(t)) \cos(\omega P_1(t)) v(t)}, \tag{13.72}$$

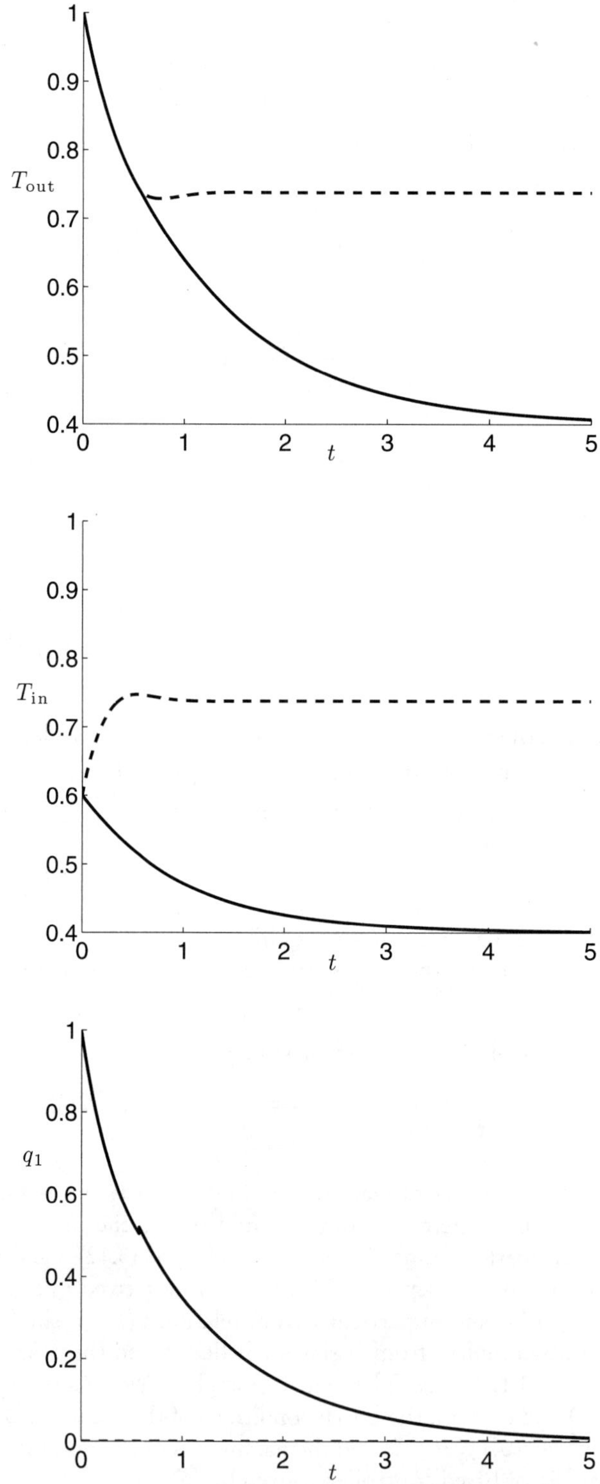

Figure 13.3. *Response of the cooling model* (13.66)–(13.67) *with the predictor-based controller* (13.68) *(solid line) and in open-loop (dashed line).*

where for all $t - r_1 \sin^2(\omega s(t)) \leq \theta \leq t$

$$P_1(\theta) = s(t) + \int_{t-r_1 \sin^2(\omega s(t))}^{\theta} \frac{v(s)ds}{1 - r_1 \omega \sin(\omega P_1(s)) \cos(\omega P_1(s)) v(s)}. \tag{13.73}$$

The control signal reaches the state s at the time instant t^*, where t^* is such that $0.3 \sin^2(15(0.1t^* + 1)) = t^* = 0.0887$. In Figure 13.4 we show the response of the system. As Theorem 13.4 predicts $s(t), v(t)$ converge to zero, whereas $\phi(t)$ and $\sigma(t)$ remain increasing for all times. From Figure 13.4 we also observe that at the time instants where $0.3 \sin^2(15s(t_1)) = 0$ we have that $\phi(t) = t = \sigma(t)$. Moreover, the peaks of the control signal $a(t)$ occur at the time instants where $\sigma(t)$ increases rapidly. ∎

13.4 ▪ Notes and References

The stability of systems with state-dependent delays on the state is studied by Cooke and Huang [33], Verriest [168], and Walther [169].

Although we consider plants with only state-dependent state delay, the results of this paper can be extended to the case of simultaneous state-dependent input and state delays. The tools that one has to use are the ones from Chapter 9. However, the stability analysis will be much more involved: one has to satisfy not only one, but two (one for each delay) feasibility conditions. Since we obtain only regional results one may wonder if the results of this paper can be applied to locally stabilizable plants. The answer to this question is positive by applying techniques similar to those from Chapter 12.

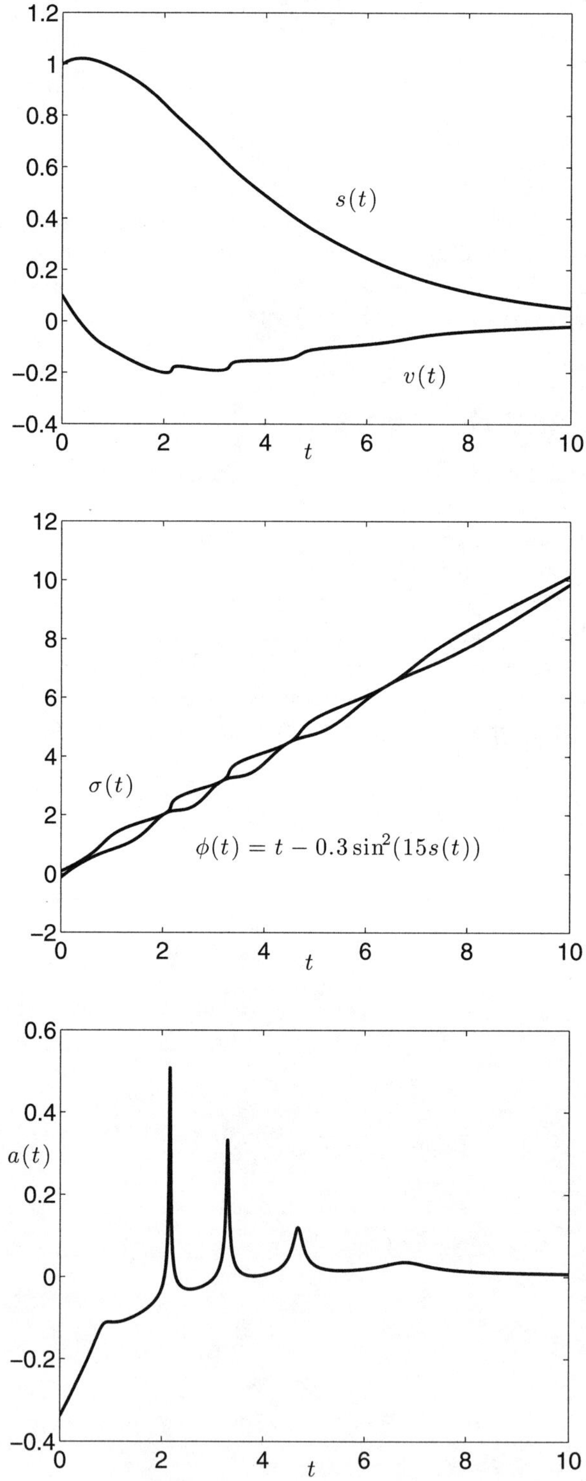

Figure 13.4. *Response of system* (13.70)–(13.71) *with the controller* (13.72)–(13.73) *and initial conditions as* $s(0) = 1$ *and* $v(\theta) = 0.1$ *for all* $-r_1 \sin^2(\omega s(0)) \le \theta \le 0$.

Chapter 14

Robustness of Nonlinear Constant-Delay Predictors to Time- and State-Dependent Delay Perturbations

We consider forward-complete nonlinear systems that are locally exponentially stabilizable in the absence of delay (by a possibly time-varying control law), for which we employ the predictor-based design. The predictor controller is designed assuming constant input delay and using only an estimation of the unmeasured (since the delay is not known) infinite-dimensional actuator state. We prove robustness in the H_2 norm of the actuator state, of the constant-delay predictor-based feedback, under simultaneous time-varying and state-dependent perturbations on the delay. Specifically, using the nonlinear infinite-dimensional backstepping transformation, we construct a Lyapunov functional for the closed-loop system that is composed of the plant, the predictor feedback, and the observer for the actuator state. With the constructed Lyapunov functional, we prove in Section 14.1 that the closed-loop system remains locally asymptotically stable when the perturbation and its rate are small. We illustrate the robustness properties of the predictor feedback with an example of a DC motor which is controlled through a network in Section 14.2. The network induces a delay which is composed of a known constant part and an unknown time-varying perturbation on this nominal value. In addition, the delay is subject to a state-dependent perturbation that depends on the armature current.

14.1 ▪ Robustness to Time- and State-Dependent Delay Perturbations for Nonlinear Systems

We consider nonlinear plants of the form

$$\dot{X}(t)=f\left(X(t),U\left(t-\hat{D}-\delta\left(t,X(t)\right)\right)\right), \tag{14.1}$$

where $f : C^2(\mathbb{R}^n \times \mathbb{R};\mathbb{R}^n)$ satisfies $f(0,0)=0$, $\hat{D}>0$, under the nominal, constant-delay predictor feedback

$$U(t)=\kappa\left(t+\hat{D},\hat{P}(t)\right), \tag{14.2}$$

where for all $t-\hat{D}\leq\theta\leq t$

$$\hat{P}(\theta)=X(t)+\int_{t-\hat{D}}^{\theta} f\left(\hat{P}(s),U(s)\right)ds \tag{14.3}$$

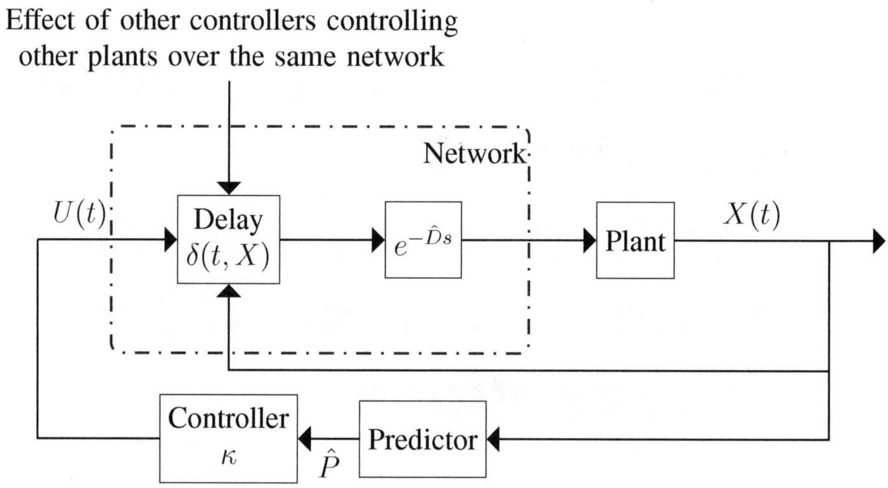

Figure 14.1. *Control over a network, with delay that varies with time (as a result of other users' activities) and may be state-dependent. The designer only knows a nominal, constant delay value $\hat{D}$. The delay fluctuation $\delta(t,X)$ is unknown.*

is the estimate of the actual predictor $P^*(\theta)$, defined as

$$P^*(\theta) = X(t) + \int_{t-\hat{D}-\delta(t,X(t))}^{\theta} \frac{f(P^*(s),U(s))}{1-G(s)} ds, \tag{14.4}$$

$$G(s) = \delta_t(\sigma(s),P^*(s)) + \nabla\delta(\sigma(s),P^*(s)) f(P^*(s),U(s)) \tag{14.5}$$

for all $t-\hat{D}-\delta(t,X(t)) \leq \theta \leq t$, where δ_t corresponds to the partial derivative of δ with respect to its first argument and σ, the actual prediction time (which should be compared with the estimated prediction time $t+\hat{D}$), is

$$\sigma(\theta) = t + \int_{t-\hat{D}-\delta(t,X(t))}^{\theta} \frac{ds}{1-G(s)}. \tag{14.6}$$

An example of such a model together with the predictor-based controller is shown in Figure 14.1.

Let us now make clear why P^* is the actual predictor state of X. Define the actual delayed time as

$$\phi(t) = t - \hat{D} - \delta(t,X(t)) \tag{14.7}$$

and the actual prediction time as

$$\begin{aligned} \sigma(t) &= \phi^{-1}(t), \\ &= t + \hat{D} + \delta(\sigma(t),X(\sigma(t))). \end{aligned} \tag{14.8}$$

Then we show that the signal in (14.4) satisfies $P^*(t) = X(\sigma(t))$. Differentiating (14.8) we get that

$$\dot{\sigma}(t) = 1 + \delta_t(\sigma(t),X(\sigma(t)))\dot{\sigma}(t) + \nabla\delta(\sigma(t),X(\sigma(t)))X'(\sigma(t))\dot{\sigma}(t), \tag{14.9}$$

where the notation X' corresponds to the derivative of X with respect to its argument, and since $\sigma(t)=\phi^{-1}(t)$, using (14.1) we have

$$\dot{\sigma}(t)=\frac{1}{1-F(t)}, \tag{14.10}$$
$$F(t)=\delta_t(\sigma(t),X(\sigma(t)))+\nabla\delta(\sigma(t),X(\sigma(t)))f(X(\sigma(t)),U(t)). \tag{14.11}$$

Define the change of variables $t=\sigma(\theta)$, where $\phi(t)\leq\theta\leq t$. With the help of (14.10) rewrite (14.1) in terms of θ as

$$\frac{dP^*(\theta)}{d\theta}=\frac{f(P^*(\theta),U(\theta))}{1-G(\theta)} \quad \text{for all } \phi(t)\leq\theta\leq t, \tag{14.12}$$

where we substitute $X(\sigma(\theta))$ with $P^*(\theta)$. Integrating (14.12) from $\phi(t)$ to θ we get (14.4) by using the fact that $P^*(\phi(t))=X(\sigma(\phi(t)))=X(t)$. Noting that $\sigma(\phi(t))=t$ we also get (14.6). A more detailed discussion about definition (14.4) can be found in Chapter 10.

The predictor state (14.3) is the certainty equivalent predictor for system (14.1). This becomes clear by setting $\delta=0$ in (14.4). Note that $\hat{P}(\theta)$ (or $P^*(\theta)$) should be viewed as the output of an operator, parametrized by t, acting on $P(s)$ and $U(s)$, $t-\hat{D}\leq s\leq\theta$ (or $t-\hat{D}-\delta(t,X(t))\leq s\leq\theta$), in the same way that the solution $X(t)$ to an ODE can be viewed as the output of an operator, parametrized by t_0, acting on $X(s)$ and the input $U(s)$, $t_0\leq s\leq t$. However, $\hat{P}$ is given implicitly since the plant is nonlinear (for the same reason that the solution $X(t)$ to a nonlinear ODE is given implicitly). In the case of a linear plant $\dot{X}(t)=AX(t)+BU(t-\hat{D})$, equation (14.3) for the predictor state can be solved explicitly as $\hat{P}(\theta)=e^{A(\theta-t+\hat{D})}X(t)+\int_{t-\hat{D}}^{\theta}e^{A(\theta-s)}BU(s)ds$, and hence $\hat{P}(t)=e^{A\hat{D}}X(t)+\int_{t-\hat{D}}^{t}e^{A(t-\theta)}BU(\theta)d\theta$. This is the standard predictor (used in Chapter 2), which is obtained using the variations of constants formula. An equivalent representation of the signal $\hat{P}(\theta)$ is

$$\hat{p}(x,t)=X(t)+\hat{D}\int_0^x f(\hat{p}(y,t),\hat{u}(y,t))dy \tag{14.13}$$

for all $x\in[0,1]$, where $\hat{u}$ is the estimation of the actuator state $U(\theta)$, $t-\hat{D}-\delta(t,X(t))\leq\theta\leq t$, which satisfies

$$\hat{D}\hat{u}_t(x,t)=\hat{u}_x(x,t), \tag{14.14}$$
$$\hat{u}(1,t)=U(t), \tag{14.15}$$

that is,

$$\hat{u}(x,t)=U(t+\hat{D}(x-1)) \quad \text{for all } x\in[0,1]. \tag{14.16}$$

With this definition, $\hat{p}(x,t)$ is the output of an operator, parametrized by t, that acts on $\hat{p}(y,t)$ and $\hat{u}(y,t)$, $y\in[0,x]$. With this representation $\hat{p}(1,t)=\hat{P}(t)$.

Note also that from relation (14.4) we see that for P^* to be well-defined the denominator in (14.4) must satisfy the following condition for all $\theta\geq t_0-\hat{D}-\delta(t_0,X(t_0))$:

$$c>\delta_t(\sigma(\theta),P^*(\theta))+\nabla\delta(\sigma(\theta),P^*(\theta))f(P^*(\theta),U(\theta)) \tag{14.17}$$

for $c \in (0,1]$, which is a condition on the perturbation δ, the initial conditions, and the solutions of the system. As it turns out later on, this condition is satisfied by appropriately restricting the perturbation δ and the initial conditions of the plant. We proceed with the assumptions on the delay-free plant.

Assumption 14.1. *The plant $\dot{X} = f(X,\omega)$ is strongly forward complete, that is, there exist a smooth positive definite function R and class $\mathcal{K}_\infty$ functions α_1, α_2, and α_3, such that*

$$\alpha_1(|X|) \le R(X) \le \alpha_2(|X|), \tag{14.18}$$

$$\frac{\partial R(X)}{\partial X} f(X,\omega) \le R(X) + \alpha_3(|\omega|) \tag{14.19}$$

for all $X \in \mathbb{R}^n$ and for all $\omega \in \mathbb{R}$.

Assumption 14.1 guarantees that for every initial condition and every measurable locally essentially bounded input signal, the corresponding solution of the system exists for all times. Forward-completeness is a natural requirement for nonlinear plants with input delay. In the absence of this assumption, i.e., when the plant exhibits a finite escape time, the control signal might reach the plant too late. The difference with standard forward-completeness from [2] is that R in Assumption 14.1 is positive definite, in accordance to the fact that $f(0,0) = 0$.

Assumption 14.2. *There exist positive constants μ, r^*, b, λ^*, a function $\hat{\alpha}$ which belongs to class $\mathcal{K}_\infty$, and a function $\kappa : C^3\left(\mathbb{R}_+ \times \mathbb{R}^n; \mathbb{R}\right)$ satisfying for all $t \ge 0$*

$$\left|\frac{\partial^{i+j}\kappa(t,\xi)}{\partial^i t \partial^j \xi}\right| \le \begin{cases} \hat{\alpha}(|\xi|), & 0 \le i \le 3,\ j = 0, \\ \mu + \hat{\alpha}(|\xi|), & 0 \le i \le 3,\ j = 1,\dots,3-i \end{cases}, \tag{14.20}$$

such that for the plant $\dot{X}(t) = f(X(t), \kappa(t,X(t)))$ the following holds for all $X(t_0) \in D_{r^*}$

$$|X(t)| \le b|X(t_0)|e^{-\lambda^*(t-t_0)} \quad \text{for all } t \ge t_0, \tag{14.21}$$

where $D_{r^*} = \{X \in \mathbb{R}^n \,|\, |X| \le r^*\}$.

Assumption 14.2 implies that the delay-free closed-loop system, that is, system $\dot{X}(t) = f(X(t), \kappa(t,X(t)))$, is locally exponentially stable, with a region of attraction given by the set D_{r^*}.

Theorem 14.3. *Consider the closed-loop system consisting of the plant* (14.1) *and the control law* (14.2), (14.3). *Under Assumptions* 14.1, 14.2 *there exist positive constants c_1, c^{**}, class $\mathcal{K}_\infty$ functions $\hat{\mu}$, α^*, a class $\mathcal{KC}_\infty$ function ζ, and a class $\mathcal{KL}$ function β such that if the perturbation δ satisfies*

$$|\delta(t,\xi)| + |\delta_t(t,\xi)| + |\nabla\delta(t,\xi)| \le c_1 + \hat{\mu}(|\xi|) \tag{14.22}$$

for all $(t,\xi) \in [t_0,\infty) \times \mathbb{R}^n$, then for all initial conditions which satisfy

$$\Pi(t_0) < c^{**}, \tag{14.23}$$

where

$$\Pi(t) = |X(t)| + \int_{t-\hat{D}}^{t} \alpha^*(|U(\theta)|)d\theta + \int_{t-\hat{D}}^{t} \ddot{U}(\theta)^2 d\theta + \int_{t-\hat{D}-\max\{0,\delta(t,X(t))\}}^{t} \dot{U}(\theta)^2 d\theta, \tag{14.24}$$

it holds that

$$\Pi(t) \leq \beta(\zeta(\Pi(t_0), R), t - t_0) \quad \text{for all } t \geq t_0, \tag{14.25}$$

for some $0 < R < \min\left\{r^*, c, \hat{D}\right\}$ *and* $0 < c < 1$.

The proof of our main result is based on Lemmas 14.4–14.8, which are presented next. Note that the definition of class $\mathcal{KC}$ and $\mathcal{KC}_\infty$ functions is the one from Appendix C.1. We introduce first the backstepping transformation.

Lemma 14.4. *Define the backstepping transformation*

$$\hat{w}(x,t) = \hat{u}(x,t) - \kappa\left(t + \hat{D}x, \hat{p}(x,t)\right), \tag{14.26}$$

together with its inverse,

$$\hat{u}(x,t) = \hat{w}(x,t) + \kappa\left(t + \hat{D}x, \hat{\rho}(x,t)\right), \tag{14.27}$$

where $\hat{\rho}$ *is given for all* $x \in [0,1]$ *by*[1]

$$\hat{\rho}(x,t) = X(t) + \hat{D}\int_0^x f\left(\hat{\rho}(y,t), \hat{w}(y,t) + \kappa\left(t + \hat{D}y, \hat{\rho}(y,t)\right)\right)dy. \tag{14.28}$$

System (14.1) *together with the control law* (14.2), (14.3) *can be written in the following form:*

$$\dot{X}(t) = f\left(X(t), \kappa(t, X(t)) + \hat{w}(0,t) + \tilde{u}(0,t)\right), \tag{14.29}$$

$$\hat{D}\hat{w}_t(x,t) = \hat{w}_x(x,t) + r_1(x,t)\tilde{f}(t), \tag{14.30}$$

$$\hat{w}(1,t) = 0, \tag{14.31}$$

where

$$\tilde{f}(t) = f\left(\hat{\rho}(0,t), \tilde{u}(0,t) + \hat{u}(0,t)\right) - f\left(\hat{\rho}(0,t), \hat{u}(0,t)\right), \tag{14.32}$$

$$r_1(x,t) = -\hat{D}\frac{\partial \kappa\left(t + \hat{D}x, \hat{\rho}(x,t)\right)}{\partial \hat{\rho}} e^{\hat{D}\int_0^x \frac{\partial f(\hat{\rho}(y,t), \hat{u}(y,t))}{\partial \hat{\rho}} dy}. \tag{14.33}$$

The observer error

$$\tilde{u}(x,t) = u(x,t) - \hat{u}(x,t) \tag{14.34}$$

[1] An important observation at this point is that $\hat{p}$ in (14.13) and $\hat{\rho}$ in (14.28) are identical. To see this, observe that, through the backstepping transformation, both $\hat{p}$ and $\hat{\rho}$ satisfy the same ODEs in the spatial variable x, with the same initial condition X. However, $\hat{p}$ is expressed in terms of the original variables $(X, \hat{u})$ in the direct backstepping transformation, whereas $\hat{\rho}$ is expressed in terms of the transformed variables $(X, \hat{w})$ and is used in the inverse transformation.

satisfies

$$\tilde{u}_t(x,t)=\pi(x,t)\tilde{u}_x(x,t)-\left(1-\hat{D}\pi(x,t)\right)r(x,t), \tag{14.35}$$
$$\tilde{u}(1,t)=0 \tag{14.36}$$

and

$$\tilde{u}_{xt}(x,t)=\pi(x,t)\tilde{u}_{xx}(x,t)+\pi_x(x,t)\tilde{u}_x(x,t)-\left(1-\hat{D}\pi(x,t)\right)r_x(x,t)+\hat{D}\pi_x(x,t)r(x,t), \tag{14.37}$$
$$\tilde{u}_x(1,t)=\left(\frac{1}{\pi(1,t)}-\hat{D}\right)r(1,t), \tag{14.38}$$

where

$$\pi(x,t)=\frac{1+x(\dot{\sigma}(t)-1)}{\sigma(t)-t}, \tag{14.39}$$

the function σ *is defined in* (14.8), *and*

$$r(x,t)=\frac{1}{\hat{D}}\hat{w}_x(x,t)+\frac{\partial x\left(t+\hat{D}x,\hat{p}(x,t)\right)}{\partial t}+\frac{\partial x\left(t+\hat{D}x,\hat{p}(x,t)\right)}{\partial\hat{p}}f\left(\hat{p}(x,t),\hat{u}(x,t)\right). \tag{14.40}$$

Furthermore,

$$\hat{D}\hat{w}_{xt}(x,t)=\hat{w}_{xx}(x,t)+r_2(x,t)\tilde{f}(t), \tag{14.41}$$
$$\hat{w}_x(1,t)=-r_1(1,t)\tilde{f}(t) \tag{14.42}$$

and

$$\hat{D}\hat{w}_{xxt}(x,t)=\hat{w}_{xxx}(x,t)+r_3(x,t)\tilde{f}(t), \tag{14.43}$$
$$\begin{aligned}\hat{w}_{xx}(1,t)=&-r_2(1,t)\tilde{f}(t)+r_4(t)\tilde{f}(t)+\tilde{f}^T(t)r_5(t)\tilde{f}(t)-r_1(1,t)\tilde{f}_{\hat{p}}(t)\\&\times f\left(\hat{p}(0,t),\tilde{u}(0,t)+\hat{u}(0,t)\right)-r_1(1,t)\hat{D}r(0,t)\tilde{f}_{\hat{u}}+\hat{D}r_1(1,t)\\&\times r(0,t)\left(1-\hat{D}\pi(0,t)\right)\frac{\partial f\left(\hat{p}(0,t),\tilde{u}(0,t)+\hat{u}(0,t)\right)}{\partial\hat{u}}\\&-\hat{D}r_1(1,t)\pi(0,t)\tilde{u}_x(0,t)\frac{\partial f\left(\hat{p}(0,t),\tilde{u}(0,t)+\hat{u}(0,t)\right)}{\partial\hat{u}},\end{aligned} \tag{14.44}$$

where

$$\tilde{f}_{\hat{p}}(t)=\frac{\partial f\left(\hat{p}(0,t),\tilde{u}(0,t)+\hat{u}(0,t)\right)}{\partial\hat{p}}-\frac{\partial f\left(\hat{p}(0,t),\hat{u}(0,t)\right)}{\partial\hat{p}}, \tag{14.45}$$
$$\tilde{f}_{\hat{u}}(t)=\frac{\partial f\left(\hat{p}(0,t),\tilde{u}(0,t)+\hat{u}(0,t)\right)}{\partial\hat{u}}-\frac{\partial f\left(\hat{p}(0,t),\hat{u}(0,t)\right)}{\partial\hat{u}} \tag{14.46}$$

and

$$\begin{aligned}
r_2(x,t) = & -\hat{D}^2 \frac{\partial^2 x\left(t+\hat{D}x, \hat{\rho}(x,t)\right)}{\partial \hat{\rho} \partial t} + \hat{D} f^T\left(\hat{\rho}(x,t), \hat{u}(x,t)\right) \\
& \times \frac{\partial^2 x\left(t+\hat{D}x, \hat{\rho}(x,t)\right)}{\partial \hat{\rho}^2} e^{\hat{D}\int_0^x \frac{\partial f(\hat{\rho}(y,t),\hat{u}(y,t))}{\partial \hat{\rho}} dy} - \hat{D}^2 \\
& \times \frac{\partial x\left(t+\hat{D}x, \hat{\rho}(x,t)\right)}{\partial \hat{\rho}} \frac{\partial f\left(\hat{\rho}(x,t), \hat{u}(x,t)\right)}{\partial \hat{\rho}} \\
& \times e^{\hat{D}\int_0^x \frac{\partial f(\hat{\rho}(y,t),\hat{u}(y,t))}{\partial \hat{\rho}} dy}, \qquad (14.47) \\
r_3(x,t) = & \, r_{2_x}(x,t), \qquad (14.48) \\
r_4(t) = & \, \hat{D}^2 \frac{\partial^2 x\left(t+\hat{D}, \hat{\rho}(1,t)\right)}{\partial t \partial \hat{\rho}} e^{\hat{D}\int_0^1 \frac{\partial f(\hat{\rho}(x,t),\hat{u}(x,t))}{\partial \hat{\rho}} dx} + \hat{D}^2 f^T(\hat{\rho}, \hat{u}) \\
& \times \frac{\partial^2 x\left(t+\hat{D}, \hat{\rho}(1,t)\right)}{\partial^2 \hat{\rho}} e^{\hat{D}\int_0^1 \frac{\partial f(\hat{\rho}(x,t),\hat{u}(x,t))}{\partial \hat{\rho}} dx} \\
& + \frac{\partial x\left(t+\hat{D}, \hat{\rho}(1,t)\right)}{\partial \hat{\rho}} e^{\hat{D}\int_0^1 \frac{\partial f(p(x,t),\hat{u}(x,t))}{\partial p} dx} \hat{D}^2 \\
& \times \left(\int_0^1 \frac{\partial^2 f\left(\hat{\rho}(x,t), \hat{u}(x,t)\right)}{\partial^2 \hat{\rho}} f(\hat{\rho}(x,t), \hat{u}) dx \right. \\
& \left. + \hat{D} \int_0^1 \frac{\partial^2 f\left(\hat{\rho}(x,t), \hat{u}(x,t)\right)}{\partial \hat{\rho} \partial \hat{u}} r(x,t) dx \right), \qquad (14.49) \\
r_5(t) = & \, e^{\hat{D}\int_0^1 \frac{\partial f(\hat{\rho}(y,t),\hat{u}(y,t))}{\partial \hat{\rho}} dy}{}^T \frac{\partial^2 x\left(t+\hat{D}, \hat{\rho}(1,t)\right)}{\partial^2 \hat{\rho}} \\
& \times \hat{D}^3 e^{\hat{D}\int_0^1 \frac{\partial f(\hat{\rho}(x,t),\hat{u}(x,t))}{\partial \hat{\rho}} dx}, \qquad (14.50)
\end{aligned}$$

where $\frac{\partial^2 f(\hat{\rho}(x,t),\hat{u}(x,t))}{\partial^2 \hat{\rho}} f(\hat{\rho}(x,t), \hat{u}(x,t))$ *corresponds to* $Q = \left\{q_{i,j}\right\}_{1\leq i,j\leq n}$, *with* $q_{i,j} = \frac{\partial^2 f_i(\hat{\rho}(x,t),\hat{u}(x,t))}{\partial \hat{\rho}_j \partial \hat{\rho}} f(\hat{\rho}(x,t), \hat{u}(x,t))$, *where* $f = (f_1, \ldots, f_n)^T$ *and* $\hat{\rho} = (\hat{\rho}_1, \ldots, \hat{\rho}_n)^T$.

Proof. We rewrite system (14.1) in the form

$$\dot{X}(t) = f(X(t), u(0,t)), \qquad (14.51)$$
$$u_t(x,t) = \pi(x,t) u_x(x,t), \qquad (14.52)$$
$$u(1,t) = U(t), \qquad (14.53)$$

where the actual actuator state $u(x,t)$ (for which an observer is given in (14.14)–(14.15)) satisfies (14.52)–(14.53) and

$$u(x,t) = U(\phi(t + x(\sigma(t) - t))) \quad \text{for all } x \in [0,1], \qquad (14.54)$$

where ϕ is defined in (14.7), or, incorporating δ, as

$$u(x,t) = U\Big(t + x\left(\hat{D} + \delta\left(\sigma(t), X(\sigma(t))\right)\right) - \hat{D} \\ - \delta\left(t + x\left(\hat{D} + \delta\left(\sigma(t), X(\sigma(t))\right)\right)\right)\Big). \tag{14.55}$$

The rest of the lemma is proved with lengthy but straightforward calculations and using Lemma 14.10 from Section 14.3.1, the backstepping transformation (14.26), its inverse (14.27), and (14.13), (14.14)–(14.15), (14.51)–(14.53), (14.28), (14.34). □

For π to be a meaningful propagation speed it should be positive and uniformly bounded from below and above. Using (14.39), one can conclude that δ, the initial conditions, and the solutions of the system should satisfy, in addition to (14.17) which guarantees that $0 < \dot{\sigma}(t)$, also

$$0 < \hat{D} + \delta(\sigma(\theta), P^*(\theta)) \quad \text{for all } \theta \geq t_0 - D(X(t_0)), \tag{14.56}$$

where $D(X(t_0)) = \hat{D} + \delta(t_0, X(t_0))$, which guarantees $0 < \sigma(t) - t$. The two conditions (14.17), (14.56) incorporate the functions σ and P^*, that is, they are not expressed in terms of the perturbation δ and the functional Π. We derive next a sufficient condition for (14.17), (14.56) to be satisfied, in terms of Π.

The proofs of Lemmas 14.5–14.7, which are presented next, are technical and require the use of additional technical lemmas. Since we want to focus on the conceptual ideas and not on the technical details of our robustness results, we present the proofs of Lemmas 14.5–14.7 in Section 14.3.

Lemma 14.5. *There exist positive constants c_1, c^*, such that if the perturbation δ satisfies* (14.22), *then for all solutions of the system satisfying,*

$$\Pi(t) < c^*, \tag{14.57}$$

they also satisfy

$$(c_1 + \hat{\mu}(|P^*(\theta)|))(1 + |f(P^*(\theta), U(\theta))|) < R \tag{14.58}$$

for all $\phi(t) \leq \theta \leq t$, where R satisfies

$$0 < R < \min\left\{r^*, c, \hat{D}\right\} \tag{14.59}$$

for some $0 < c < 1$, and hence conditions (14.17) *for $0 < c < 1$ and* (14.56) *are satisfied.*

Proof. See Section 14.3.2. □

Lemma 14.6. *There exist a continuously differentiable, positive definite function S^*, a class $\mathcal{K}_\infty$ function α^*, and positive constants λ, c_1, c^* such that if the perturbation δ satisfies*

(14.22), *then for all solutions of the system satisfying* (14.57), *the Lyapunov function*

$$
\begin{aligned}
V(t) = S^*(X(t)) &+ g_{11}\int_0^1 e^{g_1 x}|\tilde{u}(x,t)|dx + g_6\int_0^1 e^{g_2 x}\tilde{u}_x(x,t)^2 dx \\
&+ g_9\hat{D}\int_0^1 e^{g_{10}x}\left|\hat{w}_x(x,t)\right|dx + g_{12}\hat{D}\int_0^1 e^{g_3 x}\alpha^*(|\hat{w}(x,t)|)dx \\
&+ g_8\hat{D}\int_0^1 e^{g_5 x}\hat{w}_{xx}(x,t)^2 dx + g_7\hat{D}\int_0^1 e^{g_4 x}\hat{w}_x(x,t)^2 dx,
\end{aligned}
\tag{14.60}
$$

where $g_i > 0$, $i = 1,\dots,12$, *satisfies*

$$
V(t) \le V(t_0)e^{-\lambda(t-t_0)} \quad \textit{for all } t \ge t_0. \tag{14.61}
$$

Proof. See Section 14.3.3. □

The next two lemmas relate the Lyapunov function V with the norm of the system in the original variables, represented with PDEs, and the norm in PDE representation with the norm in standard delay form.

Lemma 14.7. *There exist a positive constant* c^*, *a class* $\mathcal{KC}_\infty$ *function* α_{24}, *and a class* $\mathcal{K}_\infty$ *function* α_{25} *such that for all solutions of the system satisfying* (14.57) *the following holds:*

$$
\alpha_{24}(\Gamma(t),R) \le V(t) \le \alpha_{25}(\Gamma(t)), \tag{14.62}
$$

where

$$
\begin{aligned}
\Gamma(t) = |X(t)| &+ \int_0^1 \alpha^*\left(|\hat{u}(x,t)|\right)dx + \int_0^1 u_x(x,t)^2 dx \\
&+ \int_0^1 \hat{u}_x(x,t)^2 dx + \int_0^1 \hat{u}_{xx}(x,t)^2 dx.
\end{aligned}
\tag{14.63}
$$

Proof. See Section 14.3.4. □

Lemma 14.8. *There exist positive constants* c^*, c_1 *and class* $\mathcal{KC}_\infty$ *functions* ζ_1 *and* ζ_2 *such that if the perturbation* δ *satisfies* (14.22), *then for all solutions of the system satisfying* (14.57) *the following holds:*

$$
\zeta_1(\Pi(t),R) \le \Gamma(t) \le \zeta_2(\Pi(t),R). \tag{14.64}
$$

Proof. Using Lemma 14.5 and (14.8), (14.10) we get that

$$
\frac{1}{\hat{D}+R} \le \frac{1}{\sigma(t)-t} \le \frac{1}{\hat{D}-R}, \tag{14.65}
$$

$$
\frac{1}{1+R} \le \dot{\sigma}(\theta) \le \frac{1}{1-R} \quad \text{for all } \phi(t) \le \theta \le t. \tag{14.66}
$$

With relations (14.16), (14.54), (14.24), (14.63) and applying the appropriate change of variables in the integrals, the proof is immediate using (14.65), (14.66). □

***Proof of Theorem* 14.3.** Using Lemma 14.8 we conclude that (14.57) is satisfied if $\Gamma(t) \leq \zeta_1(c^*, R)$, and hence, with Lemma 14.7, (14.57) is satisfied if

$$V(t) \leq \alpha_{24}(\zeta_1(c^*, R), R) \tag{14.67}$$

is satisfied. Assume for the moment that (14.67) is satisfied. With Lemmas 14.6, 14.7, and 14.8, relation (14.67) is satisfied if c^{**} in (14.23) is such that

$$c^{**} \leq \bar{\zeta}_2\left(\alpha_{25}^{-1}(\alpha_{24}(\zeta_1(c^*, R), R)), R\right), \tag{14.68}$$

where $\bar{\zeta}_2$ denotes the inverse function of ζ_2 with respect to its first argument for each value of its second argument. Using (14.61), with some routine class $\mathcal{K}$ majorizations that involve Lemmas 14.7 and 14.8, we get estimate (14.25). □

14.2 ▪ Example: Control of a DC Motor over a Network

We consider the following model of a field-controlled DC motor [154] with negligible shaft damping:

$$\frac{d\omega(t)}{dt} = \theta i_{\mathrm{f}}(t) i_{\mathrm{a}}(t), \tag{14.69}$$

$$\frac{di_{\mathrm{a}}(t)}{dt} = -b i_{\mathrm{a}}(t) + k - c i_{\mathrm{f}}(t)\omega(t), \tag{14.70}$$

$$\frac{di_{\mathrm{f}}(t)}{dt} = -a i_{\mathrm{f}}(t) + U\left(t - \hat{D} - \rho(t, i_{\mathrm{f}}(t), i_{\mathrm{a}}(t), \omega(t))\right), \tag{14.71}$$

where i_{f}, i_{a} are field and armature currents, respectively, ω is the angular velocity, and a, b, c, θ are positive constants. The equilibria of the unforced system are $(\omega, i_{\mathrm{a}}, i_{\mathrm{f}}) = \left(\omega_0, \frac{k}{b}, 0\right)$ for some constant ω_0. The system is feedback linearizable for $(\omega, i_{\mathrm{a}}, i_{\mathrm{f}}) \in D$, $D = \left\{(\omega, i_{\mathrm{a}}, i_{\mathrm{f}}) \in \mathbb{R}^3 | \omega > 0 \text{ and } i_{\mathrm{a}} > \frac{k}{2b}\right\}$. A delay-free design, based on full-state linearization, is (Chapter 13.3 in [75])

$$U(t) = \frac{1}{\gamma}(-K_1 Z_1(t) - K_2 Z_2(t) - K_3 Z_3(t) - \alpha), \tag{14.72}$$

$$Z_1(t) = \theta i_{\mathrm{a}}(t)^2 + c\omega(t)^2 - \theta\frac{k^2}{b^2} - c\omega_0^2, \tag{14.73}$$

$$Z_2(t) = 2\theta i_{\mathrm{a}}(t)(k - b i_{\mathrm{a}}(t)), \tag{14.74}$$

$$Z_3(t) = 2\theta(k - 2b i_{\mathrm{a}}(t))(-b i_{\mathrm{a}}(t) + k - c i_{\mathrm{f}}(t)\omega(t)), \tag{14.75}$$

$$\gamma = -2c\theta(k - 2b i_{\mathrm{a}}(t))\omega(t), \tag{14.76}$$

$$\begin{aligned} \alpha = {} & 2ca\theta(k - 2b i_{\mathrm{a}}(t)) i_{\mathrm{f}}(t)\omega(t) - 2b\theta(3k - 4b i_{\mathrm{a}}(t) - 2c i_{\mathrm{f}}(t)\omega(t)) \\ & \times(-b i_{\mathrm{a}}(t) + k - c i_{\mathrm{f}}(t)\omega(t)) - 2c\theta(k - 2b i_{\mathrm{a}}(t)) i_{\mathrm{f}}(t)^2\omega(t). \end{aligned} \tag{14.77}$$

Shifting the equilibrium $\left(\omega_0, \frac{k}{b}, 0\right)$ to the origin and setting $X_1 = \omega - \omega_0$, $X_2 = i_{\mathrm{a}} - \frac{k}{b}$, $X_3 = i_{\mathrm{f}}$, $\delta(t, X(t)) = \rho(t, i_{\mathrm{f}}(t), i_{\mathrm{a}}(t), \omega(t))$, $X(t) = (X_1(t), X_2(t), X_3(t))$ we get

$$\dot{X}_1(t) = \theta X_2(t) X_3(t) + \frac{\theta k}{b} X_3(t), \tag{14.78}$$

$$\dot{X}_2(t) = -bX_2(t) - cX_3(t)X_1(t) - c\omega_0 X_3(t), \tag{14.79}$$

$$\dot{X}_3(t) = -aX_3(t) + U\left(t - \hat{D} - \delta(t, X(t))\right). \tag{14.80}$$

The motor is controlled through a network that induces a constant delay $\hat{D}$ (e.g., [154]). The known constant delay is subject to a time-varying perturbation due to the effect of transmission of control signals to other motors through the network. We further assume that the perturbation δ increases when the armature current increases. Define the estimated predictors of X_1, X_2, and X_3 as

$$\hat{P}_1(t) = X_1(t) + \theta \int_{t-\hat{D}}^{t} \left(\hat{P}_2(s)\hat{P}_3(s) + \frac{k}{b}\hat{P}_3(s)\right) ds, \tag{14.81}$$

$$\hat{P}_2(t) = X_2(t) + \int_{t-\hat{D}}^{t} \left(-b\hat{P}_2(s) - c\hat{P}_1(s)\hat{P}_3(s) - c\omega_0 \hat{P}_3(s)\right) ds, \tag{14.82}$$

$$\hat{P}_3(t) = X_3(t) + \int_{t-\hat{D}}^{t} \left(-a\hat{P}_3(s) + U(s)\right) ds, \tag{14.83}$$

respectively. Setting in (14.72)–(14.77) $\omega = X_1 + \omega_0$, $i_a = X_2 + \frac{k}{b}$, $i_f = X_3$ and replacing X_1, X_2, X_3 by the predictors (14.81)–(14.83) we get the nominal predictor feedback.

We choose the set-point for the angular velocity of the motor as $\omega_0 = 1.5$, the nominal delay $\hat{D} = 1$, and the parameters of the plant as $a = b = c = k = \theta = 1$. The delay perturbation is $\delta(t, X_2(t)) = 0.5\left(X_2(t) + \frac{k}{b}\right)^2 + 0.2\sin(t)^2$. The initial conditions for the plant and the actuator state are chosen as $X_1(0) = -1$, $X_2(0) = -0.2$, $X_3(0) = 0.1$, and $U(\theta) = 0$, $-1 - 0.5(X_2(0)+1)^2 \leq \theta \leq 0$, respectively. The parameters of the controller are chosen as $K_1 = -1$, $K_2 = K_3 = -3$, such as the linearizable, delay-free system (i.e., the delay-free plant in the Z coordinates) has three eigenvalues at -1, and the initial estimate of the actuator state as $U(\theta) = 0$, $-1 \leq \theta \leq 0$.

In Figure 14.2 we show the field and armature currents, and in Figure 14.3 the input voltage and the angular velocity of the motor. The nominal predictor feedback achieves local stabilization of the closed-loop system at the desired equilibrium, despite the presence of the perturbation.

14.3 ▪ Proofs

Before we present the proofs of Lemmas 14.5–14.7 we devote the following subsection to the presentation of a series of technical lemmas which are used towards the proofs of Lemmas 14.5–14.7.

14.3.1 ▪ Technical Lemmas

Lemma 14.9. *There exists a class $\mathcal{KC}_\infty$ function $\hat{\zeta}_1$ such that if the perturbation δ and the solutions of the system satisfy* (14.17) *for* $0 < c < 1$ *and* (14.56), *the following holds for all* $\phi(t) \leq \theta \leq t$:

$$|P^*(\theta)| \leq \hat{\zeta}_1\left(|X(t)| + \sup_{\phi(t) \leq s \leq t} |U(s)|, R\right). \tag{14.84}$$

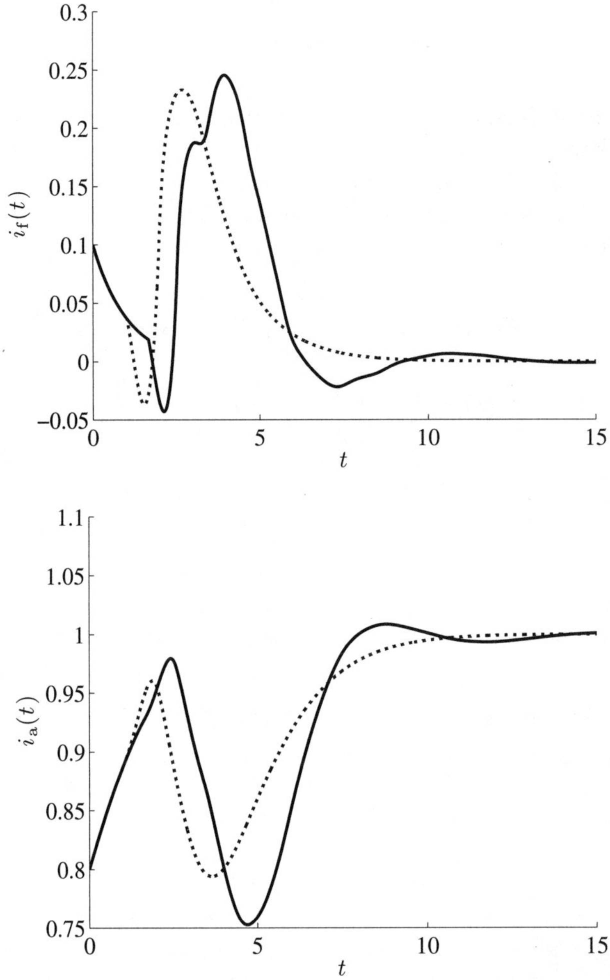

Figure 14.2. *The field (top) and armature (bottom) currents for the network controlled DC motor* (14.69)–(14.71) *under nominal predictor feedback and input delay perturbations* $\delta(t, i_a(t)) = 0.5 i_a(t)^2 + 0.2\sin(t)^2$ *(solid line),* $\delta(t, i_a(t)) = 0$ *(dotted line). The initial conditions are* $i_f(0) = 0.1$, $i_a(0) = 0.8$, $\omega(0) = 1$, *and* $U(\theta) = 0$, $-1 - \delta(0, i_a(0)) \leq \theta \leq 0$.

Proof. The proof was provided in Chapter 11 (Lemma 11.8). ☐

Lemma 14.10. *The predictor* $\hat{p}$ *in* (14.13) *satisfies*

$$\hat{D}\hat{p}_t(x,t) = \hat{p}_x(x,t) + \hat{D} e^{\hat{D}\int_0^x \frac{\partial f(\hat{p}(y,t),\hat{u}(y,t))}{\partial \hat{p}} dy} \tilde{f}(t), \tag{14.85}$$

where $\tilde{f}$ *is defined in* (14.32).

Proof. Differentiating (14.13) with respect to t, x and using (14.14)–(14.15) with the fact that $\hat{p}(0,t) = X(t)$ we get

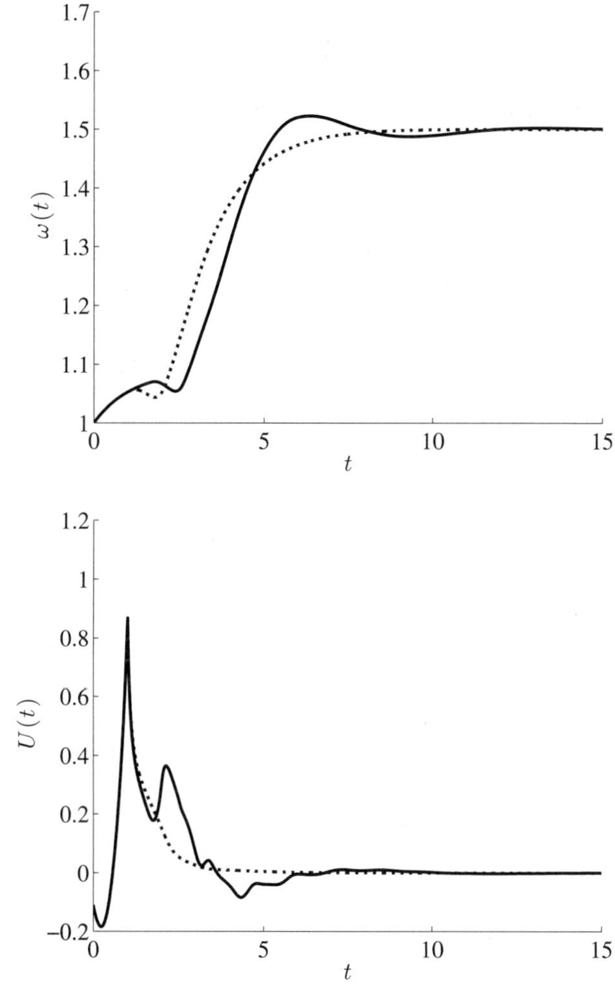

Figure 14.3. *The angular velocity (top) and the field voltage (bottom) for the network controlled DC motor* (14.69)-(14.71) *under nominal predictor feedback and input delay perturbations* $\delta(t, i_a(t)) = 0.5 i_a(t)^2 + 0.2\sin(t)^2$ *(solid line),* $\delta(t, i_a(t)) = 0$ *(dotted line). The initial conditions are* $i_f(0) = 0.1$, $i_a(0) = 0.8$, $\omega(0) = 1$, *and* $U(\theta) = 0$, $-1 - \delta(0, i_a(0)) \leq \theta \leq 0$.

$$\Psi(x,t) = \hat{D} f(\hat{p}(0,t), \tilde{u}(0,t) + \hat{u}(0,t)) + \hat{D} \int_0^x \frac{\partial f\left(\hat{p}(y,t), \hat{u}(y,t)\right)}{\partial \hat{p}} \hat{D} \hat{p}_t(y,t) dy$$
$$+ \hat{D} \int_0^x \frac{\partial f\left(\hat{p}(y,t), \hat{u}(y,t)\right)}{\partial \hat{u}} \hat{u}_y(y,t) dy - \hat{D} f(\hat{p}(x,t), \hat{u}(x,t)), \tag{14.86}$$
$$\Psi(x,t) = \hat{D} \hat{p}_t(x,t) - \hat{p}_x(x,t). \tag{14.87}$$

Since $f(\hat{p}(x,t), \hat{u}(x,t)) = \int_0^x \frac{\partial f\left(\hat{p}(y,t), \hat{u}(y,t)\right)}{\partial \hat{p}} \hat{p}_y(y,t) dy + \int_0^x \frac{\partial f\left(\hat{p}(y,t), \hat{u}(y,t)\right)}{\partial \hat{u}} \hat{u}_y(y,t) dy + f(\hat{p}(0,t), \hat{u}(0,t))$, we get

$$\Psi(x,t)=\hat{D}\int_0^x \frac{\partial f\left(\hat{p}(y,t),\hat{u}(y,t)\right)}{\partial \hat{p}}\Psi(y,t)+\hat{D}\tilde{f}(t). \tag{14.88}$$

Solving (14.88) for Ψ, the lemma is proved. □

Lemma 14.11. *There exists a class $\mathscr{K}_\infty$ function α_6 such that for all $x\in[0,1]$*

$$|\hat{p}(x,t)|\leq \alpha_6\left(|X(t)|+\int_0^1 \alpha^*(|\hat{u}(x,t)|)dx\right). \tag{14.89}$$

Proof. Differentiating (14.13) with respect to x, we get under Assumption 14.1 that

$$\hat{D}\frac{\partial R\left(\hat{p}(x,t)\right)}{\partial \hat{p}}f\left(\hat{p}(x,t),\hat{u}(x,t)\right)\leq \hat{D}R\left(\hat{p}(x,t)\right)+\hat{D}\alpha_3\left(\left|\hat{u}(x,t)\right|\right). \tag{14.90}$$

Since $\hat{p}$ satisfies $\hat{p}_x(x,t)=\hat{D}f\left(\hat{p}(x,t),\hat{u}(x,t)\right)$ we arrive at

$$\frac{dR\left(\hat{p}(x,t)\right)}{dx}\leq \hat{D}R\left(\hat{p}(x,t)\right)+\hat{D}\alpha_3\left(\left|\hat{u}(x,t)\right|\right). \tag{14.91}$$

Using the comparison principle (Lemma B.7 in Appendix B) and the fact that $\hat{p}(0,t)=X(t)$ we get

$$R\left(\hat{p}(x,t)\right)\leq e^{\hat{D}}R(X(t))+\hat{D}e^{\hat{D}}\int_0^1 \alpha_3\left(\left|\hat{u}(x,t)\right|\right)dx. \tag{14.92}$$

With Assumption 14.1 the proof is complete with any $\alpha^*(s)$ such that $e^{\hat{D}}\hat{D}\alpha_3(s)<\alpha^*(s)$. □

Lemma 14.12. *There exists class $\mathscr{K}_\infty$ functions $\alpha_{11}\dots\alpha_{13}$ such that for all $x\in[0,1]$*

$$|\hat{w}(x,t)|\leq \alpha_{11}(\Omega(t)), \tag{14.93}$$
$$|\hat{w}_x(x,t)|\leq |\hat{u}_x(x,t)|+\alpha_{12}(\Omega(t)), \tag{14.94}$$
$$\int_0^1 \hat{w}_{xx}(x,t)^2dx\leq 6\int_0^1 \hat{u}_{xx}(x,t)^2dx+\alpha_{13}(\Omega(t)), \tag{14.95}$$

where

$$\Omega(t)=|X(t)|+\int_0^1 \alpha^*(|\hat{u}(x,t)|)dx+\int_0^1 \hat{u}_x(x,t)^2dx. \tag{14.96}$$

Proof. The proof of the lemma is based on algebraic manipulations and routine class $\mathscr{K}$ majorizations using the direct (14.26) backstepping transformation together with relations (14.13) for the predictor state and Lemma 14.11. For the reader's benefit we prove

(14.93) and (14.94). The rest can be proved similarly. From (14.26) and (14.20) we get that $|\hat{w}(x,t)| \leq |\hat{u}(x,t)| + \hat{\alpha}\left(|\hat{p}(x,t)|\right)$. Using the fact that

$$\sup_{x\in[0,1]} |\hat{u}(x,t)| \leq |\hat{u}(1,t)| + \int_0^1 |\hat{u}_x(x,t)|dx, \tag{14.97}$$

with relation (14.2) and Lemma 14.11 we get (14.93). For proving (14.94) we proceed as follows. Differentiating (14.26) we get

$$\begin{aligned}\hat{w}_x(x,t) = \hat{u}_x(x,t) + \hat{D}\frac{\partial \varkappa\left(t+\hat{D}x, \hat{p}(x,t)\right)}{\partial t} + \hat{D}\\ \times \frac{\partial \varkappa\left(t+\hat{D}x, \hat{p}(x,t)\right)}{\partial \hat{p}} f\left(\hat{p}(x,t), \hat{u}(x,t)\right).\end{aligned} \tag{14.98}$$

Combining (14.20), (14.118) with Lemma 14.11 and (14.97) we arrive at (14.94) with appropriate class $\mathscr{K}$ majorizations. □

Lemma 14.13. *There exist positive constants M^*, c^* such that for all solutions of the system satisfying* (14.57) *the following holds for all $x \in [0,1]$:*

$$|\hat{p}(x,t)| \leq M^*\left(|X(t)| + \int_0^1 \alpha^*(|\hat{w}(x,t)|)dx\right). \tag{14.99}$$

Proof. Under Assumption 14.2 and choosing $c^* < R$, from Theorem C.8 there exist a continuously differentiable function $S : \mathbb{R}_+ \times D_R \to \mathbb{R}^n$, where $D_R = \{X \in \mathbb{R}^n \,||X| < R\}$, and positive constants M_1, M_2, M_3, and M_4 such that for all $X \in D_R$

$$M_1|X|^2 \leq S(t,X) \leq M_2|X|^2, \tag{14.100}$$

$$\begin{aligned}\frac{\partial S(t,X(t))}{\partial t} + \frac{\partial S(t,X(t))}{\partial X} f\left(X(t), \varkappa(t,X(t))\right)\\ \leq -M_3|X(t)|^2,\end{aligned} \tag{14.101}$$

$$\left|\frac{\partial S(t,X)}{\partial X}\right| \leq M_4|X|. \tag{14.102}$$

Since $f \in C^2(\mathbb{R}^n \times \mathbb{R};\mathbb{R})$ for all $X \in D_R$ and every $\omega \in \mathbb{R}$ such that $|\omega| \leq M$ for some positive constant M, there exists an increasing function in both arguments $L \in C(\mathbb{R}_+^2;\mathbb{R}_+)$ such that along the solutions of $\dot{X}(t) = f\left(X(t), \varkappa(t,X(t)) + \omega(t)\right)$ it holds that

$$\begin{aligned}\dot{S} \leq& -M_3|X(t)|^2 + \frac{\partial S(t,X)}{\partial X}\left(f\left(X(t), \varkappa(t,X(t)) + \omega(t)\right)\right.\\ &\left. - f\left(X(t), \varkappa(t,X(t))\right)\right)\\ \leq& -M_3|X(t)|^2 + M_4|X(t)|L(R,M)|\omega(t)|,\end{aligned} \tag{14.103}$$

where we used Lemma 3.1 in [75]. With $S^* = \sqrt{S}$ we get

$$\dot{S^*}(t,X(t)) \leq -\frac{M_3}{2\sqrt{M_1}}|X(t)| + \frac{M_4 L(R,M)}{2\sqrt{M_1}}|\omega(t)|. \tag{14.104}$$

Differentiating (14.28) with respect to x we get that

$$\hat{p}_x(x,t) = \hat{D} f\left(\hat{p}(x,t), \varkappa\left(t+\hat{D}x, \hat{p}(x,t)\right) + \hat{w}(x,t)\right). \tag{14.105}$$

Using the fact that for all $x \in [0,1]$, $|\hat{w}(x,t)| \leq \int_0^1 |\hat{w}_x(x,t)| dx$ (which follows from (14.31)), relation (14.94) together with (14.57) and Lemma 14.8 gives that for all $x \in [0,1]$, $|\hat{w}(x,t)| \leq M$, with $M = \zeta_2(c^*) + \alpha_{12}(\zeta_2(c^*))$. Using a change of variables in (14.105) as $x' = t + \hat{D}x$ and comparing the resulting ODE in x' for $\hat{p}$ with the ODE in t for $\dot{X}(t) = f(X(t), \varkappa(t, X(t)) + \omega(t))$, with (14.100) and after appropriately majorizing $s < \alpha^*(s)$, the proof is complete with $M^*(R,M) = \frac{\sqrt{M_2}}{\sqrt{M_1}} + \hat{D}\frac{M_4 L(R,M)}{2M_1}$, and hence with $c^* = \min\left\{R, c_1^*\right\}$, where c_1^* satisfies

$$M^*(R, M(c_1^*))\left(c_1^* + \alpha_{11}\left(\zeta_2(c_1^*)\right)\right) < R. \qquad \square \tag{14.106}$$

Lemma 14.14. *There exist class $\mathcal{KC}_\infty$ functions $\alpha_{14} \ldots \alpha_{16}$ and a positive constant c^* such that for all solutions of the systems satisfying* (14.57), *the following hold:*

$$|\hat{u}(x,t)| \leq \alpha_{14}(Y(t), R), \tag{14.107}$$

$$|\hat{u}_x(x,t)| \leq |\hat{w}_x(x,t)| + \alpha_{15}(Y(t), R), \tag{14.108}$$

$$\int_0^1 \hat{u}_{xx}(x,t)^2 \leq 6 \int_0^1 \hat{w}_{xx}(x,t)^2 dx + \alpha_{16}(Y(t), R), \tag{14.109}$$

for all $x \in [0,1]$, where

$$Y(t) = |X(t)| + \int_0^1 \alpha^*(|\hat{w}(x,t)|) dx + \int_0^1 \hat{w}_x(x,t)^2 dx. \tag{14.110}$$

Proof. Choose c^* as in Lemma 14.13. Then the proof of the lemma is based on algebraic manipulations and routine class $\mathcal{K}$ majorizations using the inverse transformation (14.27), relation (14.28) for the predictor state, and Lemma 14.13. $\square$

Lemma 14.15. *There exist class $\mathcal{KC}_\infty$ functions $\alpha_{17} \ldots \alpha_{23}$ and positive constants $c^*, \mu_1, \mu_2, \mu_3, \mu_4, \mu_5$ such that for all solutions of the system satisfying* (14.57) *the following hold for all $x \in [0,1]$:*

$$|r(x,t)| \leq \frac{1}{\hat{D}} |\hat{w}_x(x,t)| + \alpha_{17}(Y(t), R), \tag{14.111}$$

$$|r_1(x,t)| \leq \mu_1 + \alpha_{18}(Y(t), R), \tag{14.112}$$

$$|r_2(x,t)| \leq \mu_2 + \alpha_{19}(Y(t), R), \tag{14.113}$$

$$\int_0^1 r_3(x,t)^2 dx \leq \mu_3 + \alpha_{20}(Y(t), R), \tag{14.114}$$

$$|r_4(t)| \leq \mu_4 + \alpha_{21}(Y(t), R), \tag{14.115}$$

$$|r_5(t)| \leq \mu_5 + \alpha_{22}(Y(t), R), \tag{14.116}$$

$$\int_0^1 r_x(x,t)^2 dx \leq \alpha_{23}(Y(t), R) + \frac{6}{\hat{D}^2} \int_0^1 \hat{w}_{xx}(x,t)^2 dx, \tag{14.117}$$

where $Y(t)$ is defined in (14.110).

Proof. Let c^* be as in Lemma 14.13. The proof is based on (14.33)–(14.50) combined with (14.20), the fact that f is twice differentiable, and calculations similar to those in the proof of Lemma 14.12. However, we provide the proofs of (14.111), (14.117), as some of the steps are useful later on. Under Assumption 14.2 (which allows us to choose $\hat{\alpha}$ continuously differentiable without loss of generality), Lemma 14.13, and the facts that $|\hat{w}(x,t)| \leq \int_0^1 |\hat{w}_x(x,t)|dx$ for all $x \in [0,1]$ (which follows from (14.31)) and that $f : C^2(\mathbb{R}^n \times \mathbb{R}; \mathbb{R}^n)$, $f(0,0)=0$, which allows us to conclude

$$|f(X,\omega)| \leq \alpha_5(|X|+|\omega|) \tag{14.118}$$

for some function $\alpha_5 \in \mathcal{K}_\infty \cap C^1$, we get from (14.40) that

$$|r(x,t)| \leq \frac{1}{\hat{D}}|\hat{w}_x(x,t)| + \alpha_r(\Lambda(t),R), \tag{14.119}$$

$$\Lambda(t) = |X(t)| + \int_0^1 \alpha^*(|\hat{w}(x,t)|)dx + \int_0^1 |\hat{w}_x(x,t)|dx \tag{14.120}$$

for some class $\mathcal{KC}_\infty$ function α_r continuously differentiable in its first argument. Analogously, differentiating (14.40) with respect to x and using (14.28) together with the fact that $\hat{D}r(x,t) = \hat{u}_x(x,t)$, it is shown that

$$\begin{aligned}|r_x(x,t)| \leq \frac{1}{\hat{D}}|\hat{w}_{xx}(x,t)| + \alpha_{1,r_x}(\Lambda(t),R) + \left(\mu^* + \alpha_{2,r_x}(\Lambda(t),R)\right)\\ \times |r(x,t)|\end{aligned} \tag{14.121}$$

for some positive constant μ^* and some functions $\alpha_{1,r_x}, \alpha_{2,r_x} \in \mathcal{KC}_\infty$ which are continuously differentiable with respect to their first argument. With the Cauchy–Schwarz inequality we get (14.111), (14.117). □

14.3.2 ▪ Proof of Lemma 14.5

It holds that $U(\theta) = U(t) - \int_\theta^t \dot{U}(s)ds$ for all $\phi(t) \leq \theta \leq t$, and hence, using $\int_\theta^t \dot{U}(s)ds = \int_{\frac{\sigma(\theta)-t}{\sigma(t)-t}}^1 u_x(x,t)dx$, (14.20), and (14.2) we get

$$\sup_{\phi(t)\leq\theta\leq t} |U(\theta)| \leq \hat{\alpha}\left(|\hat{p}(1,t)|\right) + \int_0^1 \left(|\hat{u}_x(x,t)| + |\tilde{u}_x(x,t)|\right)dx. \tag{14.122}$$

Lemma 14.11 and the Cauchy–Schwarz inequality give

$$\begin{aligned}\sup_{\phi(t)\leq\theta\leq t} |U(\theta)| \leq \alpha_4\Bigg(|X(t)| + \int_0^1 \alpha^*(|\hat{u}(x,t)|)dx\\ + \int_0^1 \hat{u}_x(x,t)^2dx + \int_0^1 \tilde{u}_x(x,t)^2dx\Bigg)\end{aligned} \tag{14.123}$$

for some class $\mathcal{K}_\infty$ function α_4. Using (14.22) and (14.118) from Lemma 14.15, conditions (14.17) for $0<c<1$ and (14.56) are satisfied if the following holds for all $\phi(t)\leq\theta\leq t$:

$$R_1 > c_1+\hat{\mu}(|P^*(\theta)|)+(c_1+\hat{\mu}(|P^*(\theta)|)) \times\alpha_5\left(|X(t)|+\sup_{\phi(t)\leq\theta\leq t}|U(\theta)|\right), \tag{14.124}$$

where $R_1=\min\{c,\hat{D}\}$. With Lemma 14.9, the right inequality in relation (14.64) of Lemma 14.8, and (14.123), the lemma is proved with $c^*=c_2^*$ and c_1 satisfying

$$R>\left(c_1+\hat{\mu}\left(\hat{\zeta}_1\left(\alpha_4\left(3\zeta_2\left(c_2^*,R\right)\right),R\right)\right)\right)\left(1+\alpha_5\left(\alpha_4\left(3\zeta_2\left(c_2^*,R\right)\right)\right)\right). \tag{14.125}$$

14.3.3 ▪ Proof of Lemma 14.6

Let c^* be the minimum of c_1^* and c_2^* defined in (14.106) from Lemma 14.13 and (14.125), respectively. Taking the derivative of V with $S^*=\sqrt{S}$ (Lemma 14.13) and using integration by parts together with (14.29)–(14.30), (14.35)–(14.36), (14.37)–(14.38), (14.41)–(14.44), and (14.104) from Lemma 14.13 we get

$$\begin{aligned}\dot{V}(t)\leq&-\frac{M_3}{2\sqrt{M_1}}|X(t)|+\frac{M_4L^*(R)}{2\sqrt{M_1}}|\hat{w}(0,t)|+\frac{M_4L^*(R)}{2\sqrt{M_1}}|\tilde{u}(0,t)|\\&-g_{11}\pi(0,t)|\tilde{u}(0,t)|-g_1g_{11}\int_0^1 e^{g_1x}\pi(x,t)|\tilde{u}(x,t)|dx\\&-g_{11}\pi_x(x,t)\int_0^1 e^{g_1x}|\tilde{u}(x,t)|dx+g_{11}\sup_{x\in[0,1]}\left|1-\hat{D}\pi(x,t)\right|\\&\times\int_0^1 e^{g_1x}|r(x,t)|dx+g_6e^{g_2}\left|\frac{1}{\pi(1,t)}-\hat{D}\right|^2r(1,t)^2\\&-g_6\pi(0,t)\tilde{u}_x(0,t)^2-g_2g_6\int_0^1 e^{g_2x}\pi(x,t)\tilde{u}_x(x,t)^2dx+g_6\pi_x(x,t)\\&\times\int_0^1 e^{g_2x}\tilde{u}_x(x,t)^2dx+2g_6\sup_{x\in[0,1]}\left|1-\hat{D}\pi(x,t)\right|\\&\times\int_0^1 e^{g_2x}|r_x(x,t)||\tilde{u}_x(x,t)|dx-g_7\hat{w}_x(0,t)^2-g_4g_7\\&\times\int_0^1 e^{g_4x}\hat{w}_x(x,t)^2dx-g_9g_{10}\int_0^1 e^{g_{10}x}|\hat{w}_x(x,t)|dx-g_9|w_x(0,t)|\\&+2g_7\int_0^1 e^{g_4x}|\hat{w}_x(x,t)||r_2(x,t)|dx\left|\tilde{f}(t)\right|+g_9\int_0^1 e^{g_{10}x}|r_2(x,t)|dx|\tilde{f}(t)|\\&+g_7e^{g_4}r_1(1,t)^2\left|\tilde{f}(t)\right|^2+g_9e^{g_{10}}|r_1(1,t)||\tilde{f}(t)|+g_8e^{g_5}\hat{w}_{xx}(1,t)^2\\&-g_8\hat{w}_{xx}(0,t)^2-g_8g_5\int_0^1 e^{g_5x}\hat{w}_{xx}(x,t)^2dx+2g_8\int_0^1 e^{g_5x}|r_3(x,t)|\\&\times|\hat{w}_{xx}(x,t)|dx\left|\tilde{f}(t)\right|-g_{12}\alpha^*\left(|\hat{w}(0,t)|\right)-g_{12}g_3\int_0^1\alpha^*\left(|\hat{w}(x,t)|\right)dx\end{aligned}$$

$$+ g_{12}\int_0^1 e^{g_3 x}|\alpha^{*\prime}(|\hat{w}(x,t)|)||r_1(x,t)|dx\left|\tilde{f}(t)\right| \tag{14.126}$$

for an increasing function $L^* \in C\left(\mathbb{R}_+;\mathbb{R}_+\right)$. Using (14.39) and Lemma 14.5 we get for all $x\in[0,1]$, $\frac{1}{(1+R)\left(\hat{D}+R\right)} \le \pi(x,t) \le \frac{1}{(1-R)\left(\hat{D}-R\right)}$, $|\pi_x(x,t)| \le \frac{\hat{c}}{(1-R)\left(\hat{D}-R\right)}$, where

$$\begin{aligned}\hat{c} = &\left(c_1 + \hat{\mu}\left(\hat{\zeta}_1\left(\alpha_4\left(3\zeta_2\left(\min\left\{c_1^*,c_2^*\right\},R\right)\right),R\right)\right)\right)\\ &\times\left(1+\alpha_5\left(\alpha_4\left(3\zeta_2\left(\min\left\{c_1^*,c_2^*\right\},R\right)\right)\right)\right).\end{aligned} \tag{14.127}$$

Moreover, since $\pi(x,t)$ is linear in x, it takes its maximum value either at $x=0$ or at $x=1$, and hence

$$\begin{aligned}\left|1-\hat{D}\pi(x,t)\right| &\le \max\left\{\left|1-\hat{D}\pi(0,t)\right|,\left|1-\hat{D}\pi(1,t)\right|\right\}\\ &\le M_2 Z(t),\end{aligned} \tag{14.128}$$

where

$$M_2 = \pi(0,t)+\pi(1,t) \le \frac{2}{(1-R)\left(\hat{D}-R\right)}, \tag{14.129}$$

$$\begin{aligned}Z(t) = \max\Bigg\{ &|\delta(\sigma(t),P^*(t))|,\left|\delta(\sigma(t),P^*(t))+\hat{D}\right|\\ &\times\left(|\delta_t(\sigma(t),P^*(t))|+|\nabla\delta(\sigma(t),P^*(t))f(P^*(t),U(t))|\right)\Bigg\}.\end{aligned} \tag{14.130}$$

Therefore, using (14.125) we have that

$$\sup_{x\in[0,1]}\left|1-\hat{D}\pi(x,t)\right| \le 2\hat{c}B^*(R), \tag{14.131}$$

$$B^*(R) = \frac{1}{\hat{D}-R}\times\frac{\hat{D}+2}{1-R}. \tag{14.132}$$

Since $f: C^2(\mathbb{R}^n\times\mathbb{R};\mathbb{R}^n)$, using relations (14.32), (14.45), and (14.46), with Lemma 3.1 from [75] and (14.57), we have that

$$\left|\tilde{f}(t)\right| + \left|\tilde{f}_{\hat{\rho}}(t)\right| + \left|\tilde{f}_{\hat{u}}(t)\right| \le \varkappa_1(R)\,|\tilde{u}(0,t)| \tag{14.133}$$

for an increasing function $\varkappa_1 \in C\left(\mathbb{R}_+;\mathbb{R}_+\right)$. Hence, from (14.42), (14.119) we conclude after using (14.133), (14.120) that

$$r(1,t)^2 \le \varkappa_2(R)|\tilde{u}(0,t)| + 2\alpha_r^2(\Lambda(t),R), \tag{14.134}$$

for an increasing function $\varkappa_2 \in C\left(\mathbb{R}_+;\mathbb{R}_+\right)$ (where we also used the fact that $|\tilde{f}|^2 \le c(R)|\tilde{f}|$, which follows from (14.32) and (14.57)). We are concerned next with $\hat{w}_{xx}(1,t)^2$. With Young's inequality and (14.111), from (14.44) we get that there exists an increasing function $\varkappa_3 \in C\left(\mathbb{R}_+;\mathbb{R}_+\right)$ such that (where we absorb the powers of $|\tilde{f}|$, $|\tilde{f}_{\hat{\rho}}|$, and $|\tilde{f}_{\hat{u}}|$ higher than one in $\varkappa_3$ based on (14.32), (14.45), (14.46), (14.57))

$$\begin{aligned}\hat{w}_{xx}(1,t)^2 \le\ &\varkappa_3(R)\left(\left|\tilde{f}(t)\right|+\left|\tilde{f}_{\hat{\rho}}(t)\right|+\left|\tilde{f}_{\hat{u}}(t)\right|\right)+\varkappa_3(R)\tilde{u}_x(0,t)^2+\varkappa_3(R)\hat{w}_x(0,t)\\ &+\left(\sup_{x\in[0,1]}\left|1-\hat{D}\pi(x,t)\right|\right)^2\varkappa_3(R)\alpha_r^2(\Lambda(t),R),\end{aligned} \tag{14.135}$$

where we used (14.120) and the fact that $r(0,t)^2 \leq \frac{2}{\hat{D}^2}\hat{w}_x(0,t)^2 + 2\alpha_r^2(\Lambda(t),R)$, which follows from (14.111). From (14.120), (14.121), and (14.57) we get

$$\int_0^1 r_x(x,t)^2 dx \leq \frac{3}{\hat{D}^2}\int_0^1 \hat{w}_{xx}(x,t)^2 dx + \varkappa_4(R)\left(\int_0^1 \hat{w}_x(x,t)^2 dx + \alpha_r^2(\Lambda(t),R)\right) + 3\alpha_{1,r_x}^2(\Lambda(t),R). \tag{14.136}$$

With relation (14.57) and Lemmas 14.7, 14.8, and 14.15, from (14.126) one can conclude that the terms that multiply $|\tilde{f}|$ are bounded by the quantity $(g_7 e^{g_4} + g_9 e^{g_{10}} + g_8 e^{g_5} + g_{12} e^{g_3})\varkappa_5(R)$. Choosing the parameters g_1 and g_2 as $g_1 = g_2 = (1+R)(\hat{D}+R)\left(1+\frac{R}{(1-R)(\hat{D}-R)}\right)$ and combining (14.134), (14.135), (14.136), (14.133) with Young's inequality, we get from (14.126), (14.119), (14.120), and (14.121)

$$\begin{aligned}
\dot{V}(t) \leq &-\frac{M_3}{2\sqrt{M_1}}|X(t)| + \frac{M_4 L^*(R)}{2\sqrt{M_1}}\left(|\hat{w}(0,t)| + |\tilde{u}(0,t)|\right) - g_{11}\int_0^1 |\tilde{u}(x,t)|dx \\
&- g_{12}\alpha^*\left(|\hat{w}(0,t)|\right) - g_{11}(1+R)^{-1}\left(\hat{D}+R\right)^{-1}|\tilde{u}(0,t)| - \tilde{u}_x(0,t)^2 \\
&\times\left(g_6(1+R)^{-1}\left(\hat{D}+R\right)^{-1} - g_8 e^{g_5}\varkappa_3(R)\right) - (g_7 - g_8 e^{g_5}\varkappa_3(R))\hat{w}_x(0,t)^2 \\
&- g_6\left(1 - 2\hat{c}B^*(R)\right)\int_0^1 e^{g_2 x}\tilde{u}_x(x,t)^2 dx - g_{12}g_3\int_0^1 \alpha^*\left(|\hat{w}(x,t)|\right)dx \\
&- g_8\hat{w}_{xx}(0,t)^2 - \left(g_7 g_4 - 2\hat{c}e^{g_2}g_6\left(\hat{D}\varkappa_4(R)+1\right)\hat{D}^{-1}B^*(R)\right) \\
&\times\int_0^1 e^{g_4 x}\hat{w}_x(x,t)^2 dx - \left(g_8 g_5 - g_6 e^{g_2}\hat{D}^{-2}6\hat{c}B^*(R)\right)\int_0^1 \hat{w}_{xx}(x,t)^2 dx \\
&+ \left(4g_6 e^{g_2}\left(\hat{D}+1\right)^2\left|1 - \hat{D}\pi(1,t)\right|^2 \varkappa_2(R)\right)|\tilde{u}(0,t)| + (g_7 e^{g_4} + g_9 e^{g_{10}} \\
&+ g_8 e^{g_5} + g_{12}e^{g_3})\varkappa_5(R)|\tilde{u}(0,t)| - \left(g_9 g_{10} - g_{11}e^{g_1}2\hat{c}B^*(R)\hat{D}^{-1}\right) \\
&\times\int_0^1 e^{g_{10}x}|\hat{w}_x(x,t)|dx + 6g_6 e^{g_2}\hat{c}B^*(R)\alpha_{1,r_x}^2(\Lambda(t),R) + 2g_6 e^{g_2}\hat{c}B^*(R) \\
&\times \varkappa_4(R)\alpha_r^2(\Lambda(t),R) - g_9|w_x(0,t)| + g_{11}\sup_{x\in[0,1]}\left|1 - \hat{D}\pi(x,t)\right| e^{g_1} \\
&\times \alpha_r(\Lambda(t),R) + \left(\sup_{x\in[0,1]}\left|1 - \hat{D}\pi(x,t)\right|\right)^2\left(8g_6 e^{g_2}(\hat{D}+1)^2 + g_8 e^{g_5}\varkappa_3(R)\right) \\
&\times \alpha_r^2(\Lambda(t),R).
\end{aligned} \tag{14.137}$$

From the proof of Lemma 14.15 we have that α_r and α_{1,r_x} are continuously differentiable and hence locally Lipschitz with respect to their first argument. Using (14.57) we can write $\alpha_r^2(s,R) \leq \alpha_{26}(R)\alpha_r(s,R) \leq L_1(R)\alpha_{26}(R)s$ and $\alpha_{1,r_x}^2(s,R) \leq \alpha_{26}(R)\alpha_r(s,R) \leq L_2(R)\alpha_{26}(R)s$ for every bounded s, some increasing functions $L_i \in C(\mathbb{R}_+;\mathbb{R}_+)$, $i = 1,2$, and some class $\mathcal{K}_\infty$ function α_{26}. Choosing $g_6 > (1+R)(\hat{D}+R)g_8 e^{g_5}\varkappa_3(R)$, $g_7 > g_8 e^{g_5}\varkappa_3(R)$, $g_{11} > (1+R)(\hat{D}+R)\big(4g_6 e^{g_2}(\hat{D}+1)^2\, 4B^{*2}(R)\varkappa_2(R) + (g_7 e^{g_4} + g_9 e^{g_{10}} + g_8 e^{g_5} + g_{12}e^{g_3})\,\varkappa_5(R) + \frac{M_4 L^*(R)}{2\sqrt{M_1}}\big)$ and also $g_4 > \frac{1}{g_7}\left(e^{g_2}g_6\left(\varkappa_4(R) + \frac{1}{\hat{D}}\right) \times 2B^*(R) + 1\right)$,

$g_3 = g_5 = g_8 = g_9 = g_{10} = g = 1$, and since from the proof of Lemma 14.13 $\alpha^*(s) > s$, choosing $g_{12} > \frac{M_4 L^*(R)}{2\hat{D}\sqrt{M_1}}$ we get

$$\begin{aligned}\dot{V}(t) \le & -\left(\frac{M_3}{2\sqrt{M_1}} - \hat{c}B\right)|X(t)| - g_{11}\int_0^1 |\tilde{u}(x,t)|dx - g_6\left(1 - \hat{c}B_2\right) \\ & \times \int_0^1 e^{g_2 x}\tilde{u}_x(x,t)^2 dx - \int_0^1 \hat{w}_x(x,t)^2 dx - \left(1 - \hat{c}\left(B_3 + B\right)\right)\int_0^1 |\hat{w}_x(x,t)|dx \\ & - \left(g_{12} - \hat{c}B\right)\int_0^1 \alpha^*\left(|\hat{w}(x,t)|\right)dx - \left(1 - \hat{c}B_1\right)\int_0^1 \hat{w}_{xx}(x,t)^2 dx, \end{aligned} \tag{14.138}$$

where we used (14.120) and

$$\begin{aligned} B(R) = & 2g_6 e^{g_2} B^*(R)\alpha_{26}(R)(3L_2(R) + \varkappa_4(R)L_1(R)) + 32 g_6 e^{g_2}\left(\hat{D} + 1\right)^2 L_1(R) \\ & \times \alpha_{26}(R) 4 B^{*2}(R) + 2L_1(R)B^*(R)\left(g_{11}e^{g_1} + 2\alpha_{26}(R)eB^*(R)\right), \end{aligned} \tag{14.139}$$

$$B_1(R) = 6 g_6 e^{g_2}\hat{D}^{-2}B^*(R), \tag{14.140}$$

$$B_2(R) = 2B^*(R), \tag{14.141}$$

$$B_3(R) = 2e\, g_{11} e^{g_1}\hat{D}^{-1}B^*(R). \tag{14.142}$$

Restricting $c^* = \min\left\{c_1^*, c_2^*\right\}$ and c_1 such that $\hat{c}$ in (14.127) satisfies $\hat{c} < \min\{R, \hat{c}_1\}$, with $\hat{c}_1 \max\{B_2, B_1, B_3 + B\} \le \frac{1}{2}\min\left\{\frac{M_3}{2\sqrt{M_1}}, g_{12}, 1\right\}$, we arrive at $\dot{V}(t) \le -\lambda V(t)$, with $\lambda = \frac{1}{2}\min\left\{\frac{M_3}{2\sqrt{M_1}}, 2g_{11}, g_6, 1, g_{12}\right\}$. With the comparison principle (see also Lemma B.4 from Appendix B) we get (14.61).

14.3.4 ▪ Proof of Lemma 14.7

Using (14.100), Lemma 14.12, the fact that for all $x \in [0,1]$, $|\tilde{u}(x,t)| \le \int_0^1 |\tilde{u}_x(x,t)|dx$ (which follows from (14.36)), the Cauchy–Schwarz inequality, and some routine class $\mathcal{K}$ calculations, the proof is immediate.

14.4 ▪ Notes and References

Robustness of linear time-varying delay predictors to delay uncertainties that depend on the input is studied by Bresch-Pietri and coauthors [22], [23].

Looking into the details of the proofs in this chapter, we note that in the case of nonlinear systems with state-dependent perturbations, there is a trade-off between the achievable region of attraction and the size of the perturbation and its rate, at the origin. With the available Lyapunov functional, one could study next the inverse optimal redesign problem of predictor feedback for nonlinear systems.

Chapter 15

State-Dependent Delays That Depend on Delayed States

In the previous chapters we developed a systematic methodology for the compensation of input delays that depend on the *current* state of the plant. In this chapter we consider a different problem in which the delay depends on *past* values of the state, at a time that depends on the delay itself. Since the delay is needed for computing the predictor state, the first design challenge that we resolve is computing the delay through this implicit relation. Since the prediction horizon, over which we design the predictor, depends on the state of the system, the second design challenge that we resolve is computing the predictor state. The principal difficulty of the problem is in the stability analysis of the closed-loop system. Due to an inherent limitation that the delay rate is larger than -1 (which ensures that the delay function is uniquely defined, and hence that the dynamical system which describes the dynamics of the plant is uniquely defined), and since the delay depends on the state, only local results are possible. We also consider nonlinear systems with state-dependent delays on the state, specifically systems in the strict-feedback form with a delay on the virtual input that depends on past values of the state of the system at which the control input enters.

We present the predictor feedback law for general nonlinear systems with state-dependent input delays that depend on delayed states in Section 15.1. In Section 15.2 we prove local asymptotic stability of the closed-loop system with the aid of a Lyapunov functional that we construct by introducing a backstepping transformation of the actuator state. We present a numerical example of a second order strict-feedforward system with a state-dependent input delay (Section 15.3). We also deal with nonlinear systems with state-dependent delays on the state in Section 15.4.

15.1 ▪ Predictor Feedback under Input Delays

We consider the plant

$$\dot{X}(t) = f(X(t), U(\phi(t))), \tag{15.1}$$

where $t \geq \phi(0)$, $U \in \mathbb{R}$, $f : \mathbb{R}^n \times \mathbb{R} \to \mathbb{R}^n$ is locally Lipschitz with $f(0,0) = 0$, and ϕ satisfies

$$t = \phi(t) + D(X(\phi(t))). \tag{15.2}$$

We impose the following assumptions on the delay function D and the plant (15.1).

Assumption 15.1. *$D \in C^1(\mathbb{R}^n;\mathbb{R}_+)$ and ∇D is locally Lipschitz.*[1]

Assumption 15.2. *The plant $\dot{X} = f(X,\omega)$ is strongly forward complete, that is, there exist a smooth positive definite function R and class $\mathcal{K}_\infty$ functions α_1, α_2, and α_3 such that*

$$\alpha_1(|X|) \le R(X) \le \alpha_2(|X|), \tag{15.3}$$

$$\frac{\partial R(X)}{\partial X} f(X,\omega) \le R(X) + \alpha_3(|\omega|) \tag{15.4}$$

for all $X \in \mathbb{R}^n$ and for all $\omega \in \mathbb{R}$.

This property differs from the standard forward-completeness [2] in that we assume that $f(0,0) = 0$ and hence $R(\cdot)$ is positive definite. Assumption 15.2 guarantees that system (15.1) does not exhibit finite escape time, that is, for every initial condition and every locally bounded input signal the corresponding solution is defined for all times.

Assumption 15.3. *The plant $\dot{X} = f(X,\varkappa(X)+\omega)$ is input-to-state stable with respect to ω, and the function $\varkappa$ is locally Lipschitz with $\varkappa(0) = 0$.*

We refer to the quantity $t - \phi(t) = D(X(\phi(t)))$ as "delay." This is the time interval that indicates how long ago the control signal that currently affects the plant was actually applied. Consequently, the delay D depends on the value of the state at the time the control was applied. The goal of our predictor-based design is to completely compensate for this input delay, that is, after the control signal reaches the plant, i.e., when $\phi(t) \ge 0$ (which happens for the first time at $t^* = D(X(0))$), to make the closed-loop system behave as if there were no delay at all. To achieve this we first have to appropriately define the predictor of the state X, that is, the signal that satisfies

$$P(\phi(t)) = X(t) \quad \text{for all } t \ge 0. \tag{15.5}$$

Assume for the moment that $\phi'(t) > 0$ for all $t \ge 0$ (we show later on, under a sufficient condition, which incorporates the delay function D and the initial conditions and solutions of the system, that this is true), which in particular implies that ϕ is invertible. Denoting

$$\sigma(\theta) = \phi^{-1}(\theta) \quad \text{for all } \phi(t) \le \theta \le t, \tag{15.6}$$

the predictor state P is $P(\theta) = X(\sigma(\theta))$ for all $\phi(t) \le \theta \le t$. With the help of (15.2) we have

$$\sigma(\theta) = \theta + D(X(\theta)) \quad \text{for all } \phi(t) \le \theta \le t. \tag{15.7}$$

Therefore, the predictor of X, $P(t) = X(\sigma(t))$ is

$$P(t) = X(t + D(X(t))) \quad \text{for all } t \ge 0. \tag{15.8}$$

Having defined the predictor of X we now need to compute this signal. This is not an easy task since P cannot be directly computed from relation (15.8), because P depends on the future values of X which are not available. In addition to that, the quantity $\sigma(t) - t = D(X(t))$, which from now on we refer to as the prediction horizon (this is the time

[1] To ensure uniqueness of solutions.

interval which indicates after how long an input signal that is currently applied affects the plant), depends on the state $X(t)$. Note here that the delay time $D(X(\phi(t)))$ is not equal to the prediction horizon $D(X(t))$.

We are now ready to compute P. Since the predictor state $P(\theta)$, $\phi(t) \leq \theta \leq t$, satisfies $P(\theta) = X(\sigma(\theta))$ we perform a change of variables $t = \sigma(\theta)$ in (15.1); using definition $\sigma(\theta) = \theta + D(X(\theta))$ we get that

$$\frac{dP(\theta)}{d\theta} = (1 + \nabla D(X(\theta)) f(X(\theta), U(\phi(\theta)))) f(P(\theta), U(\theta)). \tag{15.9}$$

Integrating this relation from $\phi(t)$ to t and using the fact that $P(\phi(t)) = X(t)$ we get for all $\phi(t) \leq \theta \leq t$

$$\begin{aligned} P(\theta) = X(t) \\ + \int_{\phi(t)}^{\theta} (1 + \nabla D(X(\sigma)) f(X(\sigma), U(\phi(\sigma)))) f(P(\sigma), U(\sigma)) d\sigma. \end{aligned} \tag{15.10}$$

From relation (15.10) one can observe that for computing $P(t)$, besides having available $X(t)$ and the history of the signals $X(\sigma)$, $U(\sigma)$, $U(\phi(\sigma))$, $P(\sigma)$ on the interval $[\phi(t), t)$, one needs to know the function $\phi(t)$. We compute next $\phi(t)$, which is defined implicitly through relation (15.2). We proceed analogously with the derivation of relation (15.10). We define the change of variables $t = \theta$ for all $t \leq \theta \leq t + D(X(t))$ and differentiate (15.2) to get

$$1 = \phi'(\theta) + \nabla D(X(\phi(\theta))) f(X(\phi(\theta)), U(\phi(\phi(\theta))))\phi'(\theta). \tag{15.11}$$

Solving for ϕ' and integrating backward the resulting relation starting at the known value $\phi(t + D(X(t))) = t$, which follows from (15.6) and (15.7), we arrive for all $t \leq \theta \leq t + D(X(t))$ at

$$\phi(\theta) = t - \int_{\theta}^{t+D(X(t))} \frac{d\sigma}{1 + \nabla D(X(\phi(\sigma))) f(X(\phi(\sigma)), U(\phi(\phi(\sigma))))}. \tag{15.12}$$

The predictor-based controller is now derived based on the delay-free design (see Assumption 15.3) as

$$U(t) = \varkappa(P(t)). \tag{15.13}$$

In an actual implementation of the predictor-based control law (15.13), at each time step one has to compute the predictor P using relation (15.10) for $\theta = t$. The integral in (15.10) is computed from the history of $X(s)$ for all $s \in [\phi(t), t)$ and of $U(s)$ for all $s \in [\phi(\phi(t)), t)$ using a method of numerical integration, with a total number of $N_P(t) = \frac{D(X(\phi(t)))}{h}$ points, where h is the discretization step. However, for computing P one has to first compute ϕ. This computation is performed by calculating the integral in (15.12) from the history of $X(s)$ for all $s \in (\phi(t), t]$ and of $U(s)$ for all $s \in (\phi(\phi(t)), \phi(t)]$[2] using a method of numerical integration with $N_\phi(t) = \frac{D(X(t))}{h}$ points. Alternatively, one could compute ϕ by numerically solving relation (15.2) with respect to ϕ.

[2]For computing the first value of $\phi(\theta)$, i.e., $\phi(t + D(X(t)) - h)$, one needs the value of $U(s)$ at $s = \phi(t)$. Since $\phi(t)$ is yet to be computed, one could apply a one-discretization step delay h to U and employ the value $U(s)$ at $s = \phi(t - h)$ instead.

One can observe from (15.12) that the denominator has to be nonzero. What is more, $\phi(\theta)$ has to be invertible for all $t \leq \theta \leq t + D(X(t))$. A sufficient condition, on the initial conditions and the solutions of the system, for these two facts to hold simultaneously is

$$\mathcal{F}_c: \quad c + \nabla D(X(\phi(\sigma))) f(X(\phi(\sigma)), U(\phi(\phi(\sigma)))) > 0 \quad \text{for all } \sigma \geq 0 \tag{15.14}$$

for some $c \in (0,1]$. We refer to $\mathcal{F}_1$ as the *feasibility condition* of the controller (15.10), (15.12), (15.13). Although the actual feasibility region of the controller is given by the initial conditions and solutions of the system satisfying the feasibility condition $\mathcal{F}_1$ in (15.14), it turns out that the stability analysis is simplified if one imposes a more restrictive condition (which guarantees the satisfaction of condition (15.14)). In the subsequent development we impose the following condition on the initial conditions and solutions of the system:

$$\mathcal{F}_c^*: \quad |\nabla D(X(\phi(\sigma))) f(X(\phi(\sigma)), U(\phi(\phi(\sigma))))| < c \quad \text{for all } \sigma \geq 0 \tag{15.15}$$

for some $0 < c < 1$. It is evident that if $\mathcal{F}_c^*$ is satisfied, then $\mathcal{F}_c$ is also satisfied.

Note that the requirement $\dot{D} = \nabla D f > -1$ is an inherent limitation of the plant and not a restriction of the control design. This condition guarantees that $0 < \phi' < \infty$, that is, it guarantees that ϕ is a single-valued function, which in turn ensures that the dynamical system which describes the dynamics of the plant is uniquely defined.

15.2 ▪ Stability Analysis under Input Delays

Theorem 15.4. *Consider the closed-loop system consisting of the plant* (15.1) *and the control law* (15.10), (15.12), (15.13). *Under Assumptions* 15.1, 15.2, *and* 15.3 *there exist a class $\mathcal{K}$ function ψ_{RoA}, a class $\mathcal{KC}_\infty$ function ρ, and a class $\mathcal{KL}$ function β such that for all initial conditions for which $X(\cdot)$, $U(\cdot)$, $U(\phi(\cdot))$ are locally Lipschitz on the interval $[\phi(0), 0)$, and which satisfy*

$$\Omega(0) < \psi_{\mathrm{RoA}}(c) \tag{15.16}$$

for some $0 < c < 1$, where

$$\Omega(t) = |X(t)| + \sup_{\phi(t) \leq \theta \leq t} |X(\theta)| + \sup_{\phi(\phi(t)) \leq \sigma \leq t} |U(\sigma)|, \tag{15.17}$$

there exists a unique solution to the closed-loop system with X Lipschitz on $[0,\infty)$, U Lipschitz on $(0,\infty)$, and

$$\Omega(t) \leq \beta(\rho(\Omega(0), c), t) \tag{15.18}$$

for all $t \geq 0$. Furthermore, there exists a class $\mathcal{K}$ function δ^* such that, for all $t \geq 0$,

$$\sup_{\phi(t) \leq \theta \leq t} D(X(\theta)) \leq D(0) + \delta^*(c), \tag{15.19}$$

$$\sup_{\phi(t) \leq \theta \leq t} \left|\dot{D}(X(\theta))\right| \leq c. \tag{15.20}$$

The proof of our main result is based on Lemmas 15.5–15.12, which are presented next. Note that the definition of class $\mathcal{KC}$ and $\mathcal{KC}_\infty$ functions is the one from Appendix C.1. We introduce first the backstepping transformation.

Lemma 15.5. *The infinite-dimensional backstepping transformation of the actuator state defined for all $\phi(\phi(t)) \leq \theta \leq t$ by*

$$W(\theta) = U(\theta) - \kappa(P(\theta)), \tag{15.21}$$

together with the control law (15.13), *transforms the plant* (15.1) *to the target system given by*

$$\dot{X}(t) = f(X(t), \kappa(X(t)) + W(\phi(t))) \quad \textit{for all } t \geq \phi(0), \tag{15.22}$$
$$W(t) = 0 \quad \textit{for all } t \geq 0. \tag{15.23}$$

Proof. Using (15.13) we get (15.23). Noting that for all $t \geq 0$, $P(\phi(t)) = X(t)$ and defining $P(\phi(t)) = X(t)$ also for $\phi(0) \leq t \leq 0$, from (15.21) we get $W(\phi(t)) = U(\phi(t)) - \kappa(X(t))$ for all $t \geq \phi(0)$, and hence from (15.1) we get (15.22). □

Lemma 15.6. *The inverse transformation of* (15.21) *is given for all $\phi(\phi(t)) \leq \theta \leq t$ by*

$$U(\theta) = W(\theta) + \kappa(\Pi(\theta)), \tag{15.24}$$

where for all $\phi(t) \leq \theta \leq t$

$$\Pi(\theta) = X(t) + \int_{\phi(t)}^{\theta} (1 + \nabla D(X(\sigma)) f(X(\sigma), \kappa(X(\sigma)) + W(\phi(\sigma)))) \times f(\Pi(\sigma), \kappa(\Pi(\sigma)) + W(\sigma)) d\sigma. \tag{15.25}$$

Proof. By direct verification we get that $\Pi(t) = P(t)$[3] for all $t \geq \phi(0)$. Defining $\Pi(\phi(t)) = X(t)$ for all $\phi(0) \leq t \leq 0$ we conclude that $\Pi(\sigma) = P(\sigma)$ for all $\phi(\phi(t)) \leq \sigma \leq t$, and hence using (15.21) we get (15.24). □

Lemma 15.7. *There exist a class $\mathcal{KL}$ function β^* and a class $\mathcal{KC}_\infty$ function ρ^* such that for all solutions of the system satisfying* (15.15) *for $0 < c < 1$, the following holds:*

$$\Xi(t) \leq \beta^*(\rho^*(\Xi(0), c), t) \tag{15.26}$$

for all $t \geq 0$, where

$$\Xi(t) = |X(t)| + |X(\phi(t))| + \sup_{\phi(\phi(t)) \leq \theta \leq t} |W(\theta)|. \tag{15.27}$$

Proof. Based on Assumption 15.3 and Theorem C.16, there exist a smooth function $S : \mathbb{R}^n \to \mathbb{R}_+$ and class $\mathcal{K}_\infty$ functions α_4, α_5, α_6, and α_7 such that for all $X \in \mathbb{R}^n$ and for all $\omega \in \mathbb{R}$ the following hold:

$$\alpha_4(|X(\theta)|) \leq S(X(\theta)) \leq \alpha_5(|X(\theta)|), \tag{15.28}$$
$$\frac{\partial S(X(\theta))}{\partial X} f(X(\theta), \kappa(X(\theta)) + \omega(\theta)) \leq -\alpha_6(|X(\theta)|) + \alpha_7(|\omega(\theta)|). \tag{15.29}$$

[3] The quantities P and Π are identical. However, we use two distinct symbols for the same quantity, because in one case P is expressed in terms of X and U, for the direct backstepping transformation, while in the other case Π is expressed in terms of X and W, for the inverse backstepping transformation.

Consider now the following Lyapunov functional for the target system given in (15.22)–(15.23):

$$V(t)=S(X(t))+S(X(\phi(t)))+\frac{2+\frac{1}{1-c}}{g}\int_0^{L(t)}\frac{\alpha_7(r)}{r}dr, \tag{15.30}$$

where

$$\begin{aligned}L(t)&=\sup_{\phi(t)\le\theta\le\sigma(t)}\left|e^{g(\sigma(\theta)-t)}W(\phi(\theta))\right|\\&=\lim_{n\to\infty}\left(\int_{\phi(t)}^{\sigma(t)}e^{2ng(\sigma(\theta)-t)}W(\phi(\theta))^{2n}d\theta\right)^{\frac{1}{2n}},\end{aligned} \tag{15.31}$$

with $g>0$. We now upper- and lower-bound $L(t)$ in terms of $\sup_{\phi(\phi(t))\le\theta\le t}|W(\theta)|$. From (15.6), (15.12), and (15.15) for $0<c<1$ we get for all $\phi(t)\le\theta\le\sigma(t)$ that $\frac{d\sigma(\theta)}{d\theta}=\frac{1}{\phi'(\sigma(\theta))}\le 2$. Integrating this relation from $\phi(t)$ to θ, and since $\sigma(\phi(t))=t$ and $\theta\le\sigma(t)$, we have

$$\sigma(\theta)-t\le 2(\sigma(t)-\phi(t)),\quad \phi(t)\le\theta\le\sigma(t). \tag{15.32}$$

Since $D\in C^1(\mathbb{R}^n;\mathbb{R}_+)$ there exists a function $\delta_1\in\mathcal{K}_\infty\cap C^1$ such that

$$D(X)\le D(0)+\delta_1(|X|). \tag{15.33}$$

Therefore, using (15.2) and (15.7) we arrive at

$$L(t)\le e^{4gD(0)}e^{4g\delta_1(|X(t)|+|X(\phi(t))|)}\sup_{\phi(\phi(t))\le\theta\le t}|W(\theta)|. \tag{15.34}$$

Moreover, since σ is increasing with $\sigma(\phi(t))=t$, based on (15.6) and (15.15), we get for all $\phi(t)\le\theta\le\sigma(t)$

$$0\le\sigma(\theta)-t. \tag{15.35}$$

Therefore, with the help of (15.35) we have that

$$L(t)\ge\sup_{\phi(\phi(t))\le\theta\le t}|W(\theta)|. \tag{15.36}$$

Taking the time derivative of $L(t)$, with (15.23) we get

$$\begin{aligned}\dot L(t)=\lim_{n\to\infty}\frac{1}{2n}&\left(\int_{\phi(t)}^{\sigma(t)}e^{2ng(\sigma(\theta)-t)}W(\phi(\theta))^{2n}d\theta\right)^{\frac{1}{2n}-1}\Bigg(-\phi'(t)W(\phi(\phi(t)))^{2n}\\&-2ng\int_{\phi(t)}^{\sigma(t)}e^{2ng(\sigma(\theta)-t)}W(\phi(\theta))^{2n}d\theta\Bigg).\end{aligned} \tag{15.37}$$

Using (15.15) we have $\phi'(t)>0$ and hence $\dot L(t)\le -gL(t)$. With this inequality and (15.29), taking the derivative of (15.30) we get with the help of (15.15)

$$\begin{aligned}\dot V(t)\le&-\alpha_6(|X(t)|)-\frac{1}{2}\alpha_6(|X(\phi(t))|)+\frac{1}{1-c}\alpha_7(|W(\phi(t))|)+\alpha_7(|W(\phi(\phi(t)))|)\\&-\left(2+\frac{1}{1-c}\right)\alpha_7(L(t)).\end{aligned} \tag{15.38}$$

With the help of (15.36) we get $\dot{V}(t) \le -\alpha_6(|X(t)|) - \frac{1}{2}\alpha_6(|X(\phi(t))|) - \alpha_7(L(t))$. Using (15.28), the definition of $L(t)$ in (15.31), and (15.30) we conclude that there exists a class $\mathcal{K}$ function γ_1 such that $\dot{V}(t) \le -\gamma_1(V(t))$. Using the comparison principle (Lemma B.7) and Lemma C.6, there exists a class $\mathcal{KL}$ function β_1 such that $V(t) \le \beta_1(V(0), t)$. Using (15.28), definition (15.30), and the properties of class $\mathcal{K}$ functions we get $|X(t)| + |X(\phi(t))| + L(t) \le \beta^*(\rho_1(|X(0)| + |X(\phi(0))| + L(0), c), t)$ for some class $\mathcal{KL}$ function β^* and some class $\mathcal{KC}_\infty$ function ρ_1. Using relations (15.34) and (15.36), the lemma is proved. □

Lemma 15.8. *There exists a class $\mathcal{K}_\infty$ function α_8 such that for all solutions of the system satisfying* (15.15) *for* $0 < c < 1$, *the following holds for all* $\phi(t) \le \theta \le t$:

$$|X(\theta)| + |P(\theta)| \le \alpha_8\left(|X(t)| + |X(\phi(t))| + \sup_{\phi(\phi(t)) \le s \le t} |U(s)|\right). \tag{15.39}$$

Proof. Setting in (15.4) $\omega = U(\theta)$, we get for all $\phi(t) \le \theta \le t$ that

$$\frac{dR(P(\theta))}{dP} f(P(\theta), U(\theta)) \le R(P(\theta)) + \alpha_3(|U(\theta)|). \tag{15.40}$$

Using (15.7) and (15.9), by multiplying (15.40) with $\frac{d\sigma(\theta)}{d\theta}$ and using (15.15) we get

$$\frac{dR(P(\theta))}{d\theta} \le 2(R(P(\theta)) + \alpha_3(|U(\theta)|)), \quad \phi(t) \le \theta \le t. \tag{15.41}$$

Using the comparison principle (Lemma B.7 in Appendix B) and (15.33) one gets for all $\phi(t) \le \theta \le t$

$$R(P(\theta)) \le e^{2(D(0)+\delta_1(|X(\phi(t))|))}\left(R(X(t)) + \sup_{\phi(t) \le s \le t} \alpha_3(|U(s)|)\right). \tag{15.42}$$

Similarly, since $X(\theta)$ satisfies $\frac{dX(\theta)}{d\theta} = f(X(\theta), U(\phi(\theta)))$, $\phi(t) \le \theta \le t$, setting $\omega = U(\phi(\theta))$ in (15.4) we get

$$R(X(\theta)) \le e^{2(D(0)+\delta_1(|X(\phi(t))|))} \times\left(R(X(\phi(t))) + \sup_{\phi(\phi(t)) \le s \le \phi(t)} \alpha_3(|U(s)|)\right). \tag{15.43}$$

With standard properties of class $\mathcal{K}_\infty$ functions we get (15.39), where the class $\mathcal{K}_\infty$ function α_8 is given as

$$\alpha_8(s) = 2\alpha_1^{-1}\left((\alpha_2(s) + \alpha_3(s))e^{2(D(0)+\delta_1(s))}\right). \qquad \square \tag{15.44}$$

Lemma 15.9. *There exists a class $\mathcal{K}$ function γ^* such that for all solutions of the system satisfying* (15.15) *for* $0 < c < 1$, *the following holds for all* $\phi(t) \le \theta \le t$:

$$|X(\theta)| + |\Pi(\theta)| \le \gamma^*\left(|X(t)| + |X(\phi(t))| + \sup_{\phi(\phi(t)) \le s \le t} |W(s)|\right). \tag{15.45}$$

Proof. Under Assumption 15.3 (see Definition C.14), there exists class $\mathcal{KL}$ function β_2 and class $\mathcal{K}$ function γ_1 such that

$$|Y(\tau)| \leq \beta_2(|Y(t_0)|, \tau - t_0) + \gamma_1\left(\sup_{t_0 \leq s \leq \tau} |\omega(s)|\right), \quad \tau \geq t_0, \tag{15.46}$$

where $Y(\tau)$ is the solution of $\dot{Y}(\tau) = f(Y(\tau), \varkappa(Y(\tau)) + \omega(\tau))$. Using the change of variables $y = \sigma(\theta)$ and definitions (15.7), (15.25), we have that

$$\frac{d\Pi(\phi(y))}{dy} = f(\Pi(\phi(y)), \varkappa(y, \Pi(\phi(y))) + W(\phi(y))). \tag{15.47}$$

Using (15.46) we have

$$|\Pi(\theta)| \leq \gamma_2(|X(t)|) + \gamma_1\left(\sup_{\phi(t) \leq s \leq t} |W(s)|\right), \quad \phi(t) \leq \theta \leq t, \tag{15.48}$$

where the class $\mathcal{K}$ function γ_2 is defined as $\gamma_2(s) = \beta_2(s, 0)$. Analogously, since $X(\theta)$ satisfies $\frac{dX(\theta)}{d\theta} = f(X(\theta), \varkappa(X(\theta)) + W(\phi(\theta)))$ for all $\phi(t) \leq \theta \leq t$, we get

$$|X(\theta)| \leq \gamma_2(|X(\phi(t))|) + \gamma_1\left(\sup_{\phi(\phi(t)) \leq s \leq \phi(t)} |W(s)|\right), \quad \phi(t) \leq \theta \leq t. \tag{15.49}$$

With the properties of class $\mathcal{K}$ functions we get (15.45), where $\gamma^*(s) = 2\gamma_1(s) + 2\gamma_2(s)$ is of class $\mathcal{K}$. □

Lemma 15.10. *There exist class $\mathcal{K}_\infty$ functions α_9 and α_{10} such that for all solutions of the system satisfying* (15.15) *for $0 < c < 1$, the following hold:*

$$\Omega(t) \leq \alpha_9(\Xi(t)), \tag{15.50}$$
$$\Xi(t) \leq \alpha_{10}(\Omega(t)) \tag{15.51}$$

for all $t \geq 0$, where Ω is defined in (15.17) *and Ξ is defined in* (15.27).

Proof. Using (15.45) we have that

$$\sup_{\phi(t) \leq \theta \leq t} |X(\theta)| \leq \gamma^*\left(|X(t)| + |X(\phi(t))| + \sup_{\phi(\phi(t)) \leq s \leq t} |W(s)|\right). \tag{15.52}$$

Under Assumption 15.3 (Lipschitzness of $\varkappa$ and $\varkappa(0) = 0$) there exists a class $\mathcal{K}_\infty$ function $\hat{\alpha}$ such that

$$|\varkappa(X)| \leq \hat{\alpha}(|X|). \tag{15.53}$$

With the inverse backstepping transformation (15.24) and relation (15.53) we arrive at

$$\begin{aligned}\sup_{\phi(\phi(t)) \leq \theta \leq t} |U(\theta)| \leq & \sup_{\phi(\phi(t)) \leq \theta \leq t} |W(\theta)| \\ & + \hat{\alpha}\left(\sup_{\phi(\phi(t)) \leq \theta \leq \phi(t)} |\Pi(\theta)| + \sup_{\phi(t) \leq \theta \leq t} |\Pi(\theta)|\right).\end{aligned} \tag{15.54}$$

Using relation (15.45) and definition $\Pi(\phi(\theta))=X(\theta)$, $\phi(t)\leq\theta\leq t$, we get (15.50) with

$$\alpha_9(s)=s+\gamma^*(s)+\hat{\alpha}(2\gamma^*(s)). \tag{15.55}$$

Analogously, using the direct backstepping transformation (15.21), relation (15.39), and definition $P(\phi(\theta)) = X(\theta)$, $\phi(t) \leq \theta \leq t$, we get (15.51) with $\alpha_{10}(s) = s + \hat{\alpha}(2\alpha_8(s))$. □

Lemma 15.11. *There exists a function α^* of class $\mathcal{K}_\infty$ such that all the solutions that satisfy*

$$\Omega(t)<\alpha^{*-1}(c), \quad t\geq 0, \tag{15.56}$$

for $0<c<1$ also satisfy (15.15), *where Ω is defined in* (15.17).

Proof. Since $D\in C^1(\mathbb{R}^n;\mathbb{R}_+)$ and $f:\mathbb{R}^n\times\mathbb{R}\to\mathbb{R}^n$ is locally Lipschitz with $f(0,0)=0$, there exist class $\mathcal{K}_\infty$ functions δ_2 and α_{11} such that

$$|\nabla D(X)|\leq|\nabla D(0)|+\delta_2(|X|), \tag{15.57}$$

$$|f(X,\omega)|\leq\alpha_{11}(|X|+|\omega|). \tag{15.58}$$

If a solution satisfies for all $t\geq 0$

$$(|\nabla D(0)|+\delta_2(|X(\theta)|))\,\alpha_{11}(|X(\theta)|+|U(\phi(\theta))|)<c, \quad \phi(t)\leq\theta\leq t, \tag{15.59}$$

for $0<c<1$, then it also satisfies (15.15). With the trivial inequalities $|X(\theta)|\leq\sup_{\phi(t)\leq\tau\leq t}|X(\tau)|$ and $|U(\phi(\theta))|\leq\sup_{\phi(\phi(t))\leq s\leq\phi(t)}|U(s)|$ for all $\phi(t)\leq\theta\leq t$, relation (15.59) is satisfied for $0<c<1$ for all $t\geq 0$ as long as (15.56) holds, where the class $\mathcal{K}_\infty$ function α^* is defined as

$$\alpha^*(s)=(|\nabla D(0)|+\delta_2(s))\,\alpha_{11}(s). \qquad \square \tag{15.60}$$

Lemma 15.12. *There exists a class $\mathcal{K}$ function ψ_{RoA} such that for all initial conditions of the closed-loop system* (15.1), (15.10), (15.12), (15.13) *that satisfy relation* (15.16), *the solutions of the system satisfy* (15.56) *for $0<c<1$ and hence satisfy* (15.15).

Proof. Using Lemma 15.10, with the help of (15.26), we have that

$$\Omega(t)\leq\alpha_9(\beta^*(\rho^*(\alpha_{10}(\Omega(0)),c),t)). \tag{15.61}$$

By defining the class $\mathcal{K}_\infty$ function α_{12} as $\alpha_{12}(s)=\alpha_9(\beta^*(s,0))$, we get

$$\Omega(t)\leq\alpha_{12}(\rho^*(\alpha_{10}(\Omega(0)),c)). \tag{15.62}$$

Hence, for all initial conditions that satisfy the bound (15.16) with any class $\mathcal{K}$ choice $\psi_{\mathrm{RoA}}(c)\leq\bar{\psi}^*_{\mathrm{RoA}}(\alpha^{*-1}(c),c)$, where $\bar{\psi}^*_{\mathrm{RoA}}(s,c)$ is the inverse of the class $\mathcal{KC}_\infty$ function $\psi^*_{\mathrm{RoA}}(s,c)=\alpha_{12}(\rho^*(\alpha_{10}(s),c))$ with respect to ψ^*_{RoA}'s first argument, the solutions satisfy (15.56). Furthermore, for all those initial conditions, the solutions verify (15.15) for all $\sigma\geq 0$. □

***Proof of Theorem* 15.4.** Using (15.61) we get (15.18) with $\beta(s,t)=\alpha_9(\beta^*(s,t))$ and $\rho(s,c)=\rho^*(\alpha_{10}(s),c)$. System (15.22), (15.23) guarantees the existence and uniqueness of

$X \in C^1(\sigma^*, \infty)$, where $\sigma^* = D(X(0))$. Consider now the case $t \in [0, \sigma^*)$. From (15.1) and (15.11) we have for all $t \in [0, \sigma^*)$ that

$$\dot{X}(t) = f(X(t), g_U(\phi(t))), \tag{15.63}$$

$$\dot{\phi}(t) = \frac{1}{1 + \nabla D(g_X(\phi(t))) f\left(g_X(\phi(t)), g_{U_d}(\phi(t))\right)}, \tag{15.64}$$

where the initial conditions for X, U, and $U(\phi)$ are defined as $X(s) = g_X(s)$, $U(s) = g_U(s)$, and $U(\phi(s)) = g_{U_d}(s)$ for all $\phi(0) \leq s < 0$. Lipschitzness of g_X, g_U, g_{U_d} on $[\phi(0), 0)$, Lipschitzness of f, and Assumption 15.1 (Lipschitzness of ∇D) guarantee that the right-hand side of (15.63) and of (15.64) are Lipschitz with respect to (X, ϕ), which guarantees, together with bound (15.15), the existence and uniqueness of $X \in C^1[0, \sigma^*)$. The boundedness of W and (15.22) guarantee that X is continuous at $t = \sigma^*$. By integrating (15.22) between any two time instants it is shown that X is Lipschitz on $[0, \infty)$ with a Lipschitz constant given by a uniform bound on the right-hand side of (15.22). The fact that $\Pi(t) = X(t + D(X(t)))$ for all $t \geq 0$ and Assumption 15.1 ($D \in C^1(\mathbb{R}^n; \mathbb{R}_+)$) guarantee that Π is Lipschitz for all $t \geq 0$. Since $U(t) = \kappa(\Pi(t))$, Assumption 15.3 (Lipschitzness of κ in both arguments) guarantees that U is Lipschitz in t on $(0, \infty)$. Using (15.15) and (15.33), we get (15.19), (15.20) with any class $\mathcal{K}$ function $\delta^*(c) \geq \delta_1(\alpha^{*-1}(c))$. □

15.3 ▪ Example

In this example we consider the following system in the feedforward form taken from [84]:

$$\dot{X}_1(t) = X_2(t) - X_2(t)^2 U(t - D(X(\phi(t)))), \tag{15.65}$$

$$\dot{X}_2(t) = U(t - D(X(\phi(t)))), \tag{15.66}$$

where

$$\phi(t) = t - D(X(\phi(t))), \tag{15.67}$$

$$D(X(\phi(t))) = \frac{1}{2}\sin(5X_2(\phi(t)))^2, \tag{15.68}$$

and hence

$$\sigma(t) = t + \frac{1}{2}\sin(5X_2(t))^2. \tag{15.69}$$

A nominal design for the delay-free plant is given in [84] as

$$U(t) = -X_1(t) - 2X_2(t) - \frac{1}{3}X_2(t)^2. \tag{15.70}$$

The predictor-based design is

$$U(t) = -P_1(t) - 2P_2(t) - \frac{1}{3}P_2(t)^2, \tag{15.71}$$

where

$$P_1(t)=X_1(t)+\int_{\phi(t)}^{t}(1+5\cos(5X_2(s))\sin(5X_2(s))\,U(\phi(s)))$$
$$\times\left(P_2(\theta)-P_2(\theta)^2U(\theta)\right)d\theta, \tag{15.72}$$
$$P_2(t)=X_2(t)+\int_{\phi(t)}^{t}(1+5\cos(5X_2(s))\sin(5X_2(s))\,U(\phi(s)))\,U(\theta)d\theta, \tag{15.73}$$

where

$$\phi(t)=t-\int_{t}^{t+\frac{1}{2}\sin(5X_2(t))^2}\frac{ds}{1+G(s)}, \tag{15.74}$$
$$G(s)=5\cos(5X_2(\phi(s)))\sin(5X_2(\phi(s)))\,U(\phi(\phi(s))). \tag{15.75}$$

We consider the initial conditions for the plant as $X_1(0)=1.3$, $X_2(s)=0$ for all $\phi(0)\leq s\leq 0$, and the initial conditions for the actuator state as $U(s)=U(\phi(s))=0$ for all $\phi(0)\leq s\leq 0$. With such an initial condition we get $\phi(0)=0$. Therefore, the control signal "kicks in" immediately, i.e., at $t=0$. In Figure 15.1 we show the response of the closed-loop system. As one can observe, the delay is immediately compensated for and X_1 and X_2 respond as if there were no delay at all. Yet the oscillations are evident in the control signal. The control signal oscillates in order to compensate for the oscillatory delay. In Figure 15.2 we show the delayed time ϕ and the prediction time σ.

15.4 ▪ Stabilization under State Delays

In the present section we consider the plant

$$\dot{X}(t)=f(X(t),\zeta(\phi(t))), \tag{15.76}$$
$$\dot{\zeta}(t)=U(t), \tag{15.77}$$

where $t\geq 0$, $U,\zeta\in\mathbb{R}$, $f:\mathbb{R}^n\times\mathbb{R}\to\mathbb{R}^n$ is locally Lipschitz with $f(0,0)=0$, and ϕ satisfies

$$t=\phi(t)+D(\zeta(\phi(t))). \tag{15.78}$$

Remark 15.13. *Let us highlight the importance of considering the class of nonlinear systems that satisfy* (15.76), (15.77), (15.78)*. First, this is a class of systems with state delay that depends on the past state of the system. This is different from system* (15.1)*, which has an input rather a state delay. Second, and more important, when one stabilizes system* (15.76), (15.77), (15.78) *then one can stabilize nonlinear systems with input delays that depend on the past input rather than the past state. To see this consider a nonlinear system with input delay that depends on past values of the input, i.e., consider the system*

$$\dot{X}(t)=f(X(t),V(\phi(t))), \tag{15.79}$$

where $t=\phi(t)+D(V(\phi(t)))$*. Then, by adding an integrator, one gets exactly equations* (15.76), (15.77), (15.78) *with* $\zeta=V$ *and* $\dot{V}=U$*, where* U *is designed in order to stabilize*

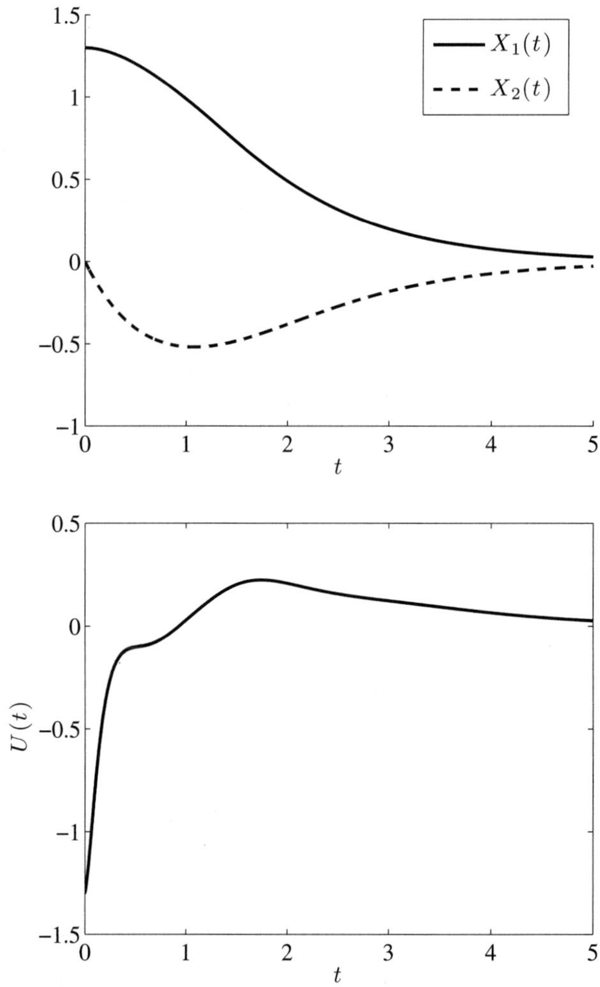

Figure 15.1. *The response of the closed-loop system* (15.65), (15.66) *with the predictor feedback law* (15.71)–(15.75) *(top) and the control effort* (15.71) *(bottom). The initial conditions are chosen as* $X_1(0) = 1.3$, $X_2(s) = 0$ *for all* $\phi(0) \le s \le 0$, *and* $U(s) = U(\phi(s)) = 0$ *for all* $\phi(0) \le s \le 0$. *For these initial conditions the control signal "kicks in" at* $t = 0$, *and hence the delay is immediately compensated for, resulting in identical, in the delay-free case, responses for* X_1, X_2. *The control signal oscillates to compensate for the oscillatory delay.*

the system $\dot{X}(t) = f(X(t), V(\phi(t)))$, $\dot{V}(t) = U(t)$. *Hence, stabilization of system* (15.76), (15.77), (15.78) *implies stabilization of system* (15.79).

We now make the following assumption regarding system (15.76).

Assumption 15.14. *There exists a function* $\mu \in C^1(\mathbb{R}^n;\mathbb{R})$, *with* $\mu(0) = 0$ *and* $\nabla \mu$ *locally Lipschitz,*[4] *such that the plant* $\dot{X}(t) = f(X(t), \mu(X(t)) + \omega(t))$ *is input-to-state stable with respect to* ω.

[4]To ensure uniqueness of solutions.

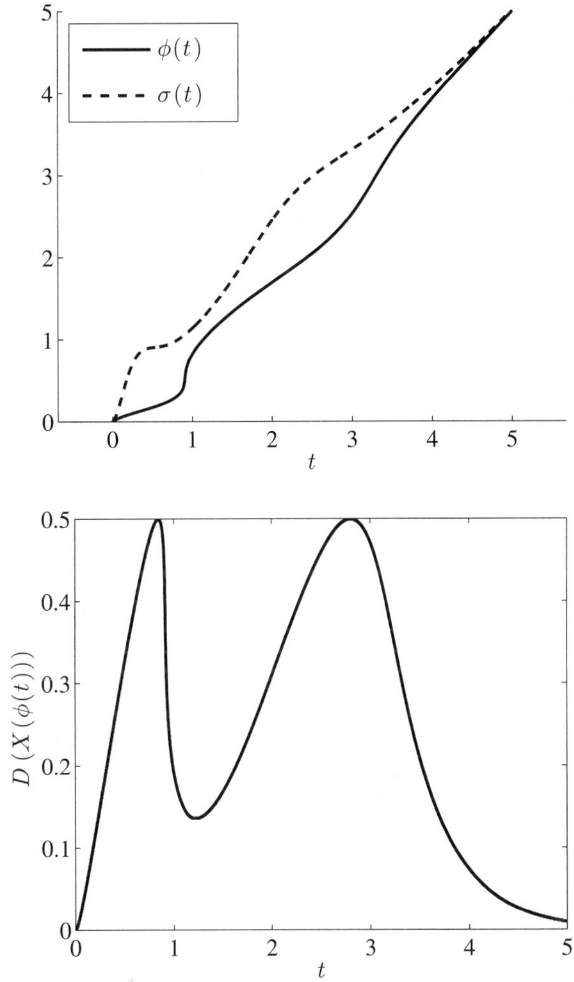

Figure 15.2. *The delayed time* (15.67) *and the prediction time* (15.69) *(top), and the delay* (15.68) *(bottom) of the closed-loop system* (15.65), (15.66) *with the predictor feedback law* (15.71)–(15.75). *The initial conditions are chosen as* $X_1(0) = 1.3$, $X_2(s) = 0$ *for all* $\phi(0) \leq s \leq 0$, *and* $U(s) = U(\phi(s)) = 0$ *for all* $\phi(0) \leq s \leq 0$. *For these initial conditions* $\phi(0) = \sigma(0) = 0$.

Note that we still assume that the delay function D (which is now defined in $\mathbb{R}$ rather than in $\mathbb{R}^n$) and the vector field f satisfy Assumptions 15.1 and 15.2, respectively. Assumption 15.14 is similar to Assumption 15.3 with the difference that in the present case the feedback law μ is assumed continuously differentiable rather than just locally Lipschitz (this regularity requirement for μ is a result of the backstepping procedure). Finally, note that the results of this section can be extended to the case in which the delay (15.78) depends also on $X(\phi(t))$. However, in order to keep the formulae of our design as simple as possible we do not consider this case.

From plant (15.76), (15.77) one can observe that the input signal reaches the state ζ at $t = 0$. However, it reaches X through a delayed integrator. Therefore, we need to define and derive an implementable form for the predictor of the state X, i.e., the signal that satisfies $P(\phi(t)) = X(t)$ for all $t \geq 0$. From relation (15.78) we get that

$$\begin{aligned}\phi^{-1}(\theta) &= \sigma(\theta) \\ &= \theta + D(\zeta(\theta)) \quad \text{for all } \phi(t) \leq \theta \leq t.\end{aligned} \tag{15.80}$$

Setting $t = \sigma(\theta)$ in (15.76), differentiating with respect to θ, and integrating the resulting expression from $\phi(t)$ to θ, with the help of the fact that $P(\phi(t)) = X(t)$, we get for all $\phi(t) \leq \theta \leq t$ that

$$P(\theta) = X(t) + \int_{\phi(t)}^{\theta} \left(1 + D'(\zeta(s))U(s)\right) f(P(s), \zeta(s))\, ds. \tag{15.81}$$

We compute next ϕ. Differentiating (15.78), and since $\phi(t + D(\zeta(t))) = \phi(\sigma(t)) = t$, we get that

$$\phi(\theta) = t - \int_{\theta}^{t+D(\zeta(t))} \frac{ds}{1 + D'(\zeta(\phi(s)))U(\phi(s))} \quad \text{for all } t \leq \theta \leq \sigma(t). \tag{15.82}$$

The predictor-based control law is based on a backstepping design on the delay plant and is given by

$$U(t) = \frac{\nabla\mu(P(t))f(P(t), \zeta(t)) - c_Z(\zeta(t) - \mu(P(t)))}{1 - \nabla\mu(P(t))f(P(t), \zeta(t))D'(\zeta(t))}, \tag{15.83}$$

where $c_Z > 0$ is arbitrary. Analogously to the case of input delay, in an actual implementation of the control law (15.83), (15.81), (15.82) one has to compute, at each time step, $\phi(t)$ by numerically computing the integral in (15.82) and using the history of ζ and U. Then one computes $P(t)$ using $\phi(t)$ and the history of ζ, P, and U. Finally, one calculates $U(t)$ from (15.83). However, in order to compute $\phi(t)$ one starts the integration at $\sigma(t) = t + D(\zeta(t))$. Yet the function inside the integral evaluated at $s = \sigma(t)$ depends on $U(t)$, i.e., on the current value of the input, which is yet to be computed. Therefore, since $\sigma(t)$ is strictly increasing, one can compute $\phi(t)$ by integrating (15.82) up to $s = \sigma(t - h)$, where h is the discretization step.

From (15.83) one can observe that besides a restriction that the denominator in (15.82) is positive, one has an additional condition that the denominator in (15.83) is also positive. Both conditions are satisfied when the following condition holds for all $\theta \geq \phi(0)$

$$\mathscr{G}_c: \quad \left|D'(\zeta(\theta))U(\theta)\right| + \left|\nabla\mu(P(\theta))f(P(\theta), \zeta(\theta))D'(\zeta(\theta))\right| < c, \tag{15.84}$$

for some $0 < c < 1$.

Theorem 15.15. *Consider the plant* (15.76)–(15.78) *together with the control law* (15.83), (15.81), (15.82). *Under Assumptions* 15.1, 15.2, *and* 15.14 *there exist a class $\mathscr{K}$ function ξ_{RoA}, a class $\mathscr{K}\mathscr{L}$ function $\hat{\beta}$, and a class $\mathscr{K}_\infty$ function σ_1 such that for all initial conditions for which ζ is locally Lipschitz on the interval $[\phi(0), 0]$, U is locally Lipschitz on the interval $[\phi(0), 0)$, and they satisfy* (15.77) *and*

$$\hat{\Omega}(0) < \xi_{\mathrm{RoA}}(c), \tag{15.85}$$

where

$$\hat{\Omega}(t) = |X(t)| + \sup_{\phi(t) \leq \theta \leq t} |\zeta(\theta)| + \sup_{\phi(t) \leq \theta \leq t} |U(\theta)| \tag{15.86}$$

for some $0 < c < 1$, *there exists a unique solution to the closed-loop system with* $X \in C^1[0,\infty)$, ζ *Lipschitz on* $[0,\infty)$, U *Lipschitz on* $(0,\infty)$, *and*

$$\hat{\Omega}(t) \leq \sigma_1 \left(1 + \frac{1}{1-c}\right) \hat{\beta}\left(\hat{\Omega}(0), t\right) \tag{15.87}$$

for all $t \geq 0$. *Furthermore, there exists a class* $\mathcal{K}$ *function* $\hat{\delta}^*$, *such that for all* $t \geq 0$ *the following hold:*

$$\sup_{\phi(t) \leq \theta \leq t} D(\zeta(\theta)) \leq D(0) + \hat{\delta}^*(c), \tag{15.88}$$

$$\sup_{\phi(t) \leq \theta \leq t} \left|\dot{D}(\zeta(\theta))\right| \leq c. \tag{15.89}$$

The proof of Theorem 15.15 is based on Lemmas 15.16–15.23, which are presented next.

Lemma 15.16. *The infinite-dimensional backstepping transformation of the state* ζ *defined by*

$$Z(\theta) = \zeta(\theta) - \mu(P(\theta)), \quad \phi(t) \leq \theta \leq t, \tag{15.90}$$

together with the predictor-based control law (15.83), (15.81), (15.82) *transforms the system* (15.76)–(15.77) *to the target system given by*

$$\dot{X}(t) = f\left(X(t), \mu(X(t)) + Z(\phi(t))\right), \tag{15.91}$$

$$\dot{Z}(t) = -c_Z Z(t). \tag{15.92}$$

Proof. Using (15.76) and the fact that $P(\phi(t)) = X(t)$ we get (15.91). Setting $\theta = t$ in (15.90) and taking the derivative with respect to t of the resulting equation we get (15.92) using (15.77), (15.81), and (15.83). □

Lemma 15.17. *The inverse of the infinite-dimensional backstepping transformation defined in* (15.90) *is*

$$\zeta(\theta) = Z(\theta) + \mu(\Pi(\theta)), \quad \phi(t) \leq \theta \leq t, \tag{15.93}$$

where

$$\begin{aligned}\Pi(\theta) = X(t) + \int_{\phi(t)}^{\theta} &\left(1 + D'\left(\mu(\Pi(s)) + Z(s)\right) U(s)\right) \\ &\times f\left(\Pi(s), \mu(\Pi(s)) + Z(s)\right) ds, \quad \phi(t) \leq \theta \leq t.\end{aligned} \tag{15.94}$$

Proof. The proof is by direct verification, noting also that $\Pi(\theta) = P(\theta)$ for all $\phi(t) \leq \theta \leq t$, where $\Pi(\theta)$ is driven by the transformed state $Z(\theta)$, whereas $P(\theta)$ is driven by the state $\zeta(\theta)$ for $\phi(t) \leq \theta \leq t$. □

Lemma 15.18. *There exists a class $\mathcal{K}_\infty$ function $\hat{\alpha}_8$ such that for all solutions of the system satisfying* (15.84) *for* $0 < c < 1$, *the following holds:*

$$|P(\theta)| \leq \hat{\alpha}_8\left(|X(t)| + \sup_{\phi(t)\leq\tau\leq t} |\zeta(\tau)|\right), \quad \phi(t) \leq \theta \leq t. \tag{15.95}$$

Proof. Under Assumption 15.2 we have that

$$\frac{dR(P(\theta))}{dP} f(P(\theta),\zeta(\theta)) \leq R(P(\theta)) + \alpha_3(|\zeta(\theta)|), \quad \phi(t) \leq \theta \leq t. \tag{15.96}$$

Multiplying both sides of (15.96) with $\dot{\sigma}(\theta) = 1 + D'(\zeta(\theta))U(\theta) > 0$, with (15.84) we get that

$$\frac{dR(P(\theta))}{d\theta} \leq 2(R(P(\theta)) + \alpha_3(|\zeta(\theta)|)), \quad \phi(t) \leq \theta \leq t. \tag{15.97}$$

Using relation (15.33) and the comparison principle, we have from (15.96) for all $\phi(t) \leq \theta \leq t$ that

$$R(P(\theta)) \leq e^{2(D(0)+\delta_1(|\zeta(\phi(t))|))}\left(R(X(t)) + \sup_{\phi(t)\leq\tau\leq t} \alpha_3(|\zeta(\tau)|)\right). \tag{15.98}$$

With standard properties of class $\mathcal{K}_\infty$ functions we get the statement of the lemma with $\hat{\alpha}_8 \in \mathcal{K}_\infty$ as

$$\hat{\alpha}_8(s) = \alpha_1^{-1}\left((\alpha_2(s) + \alpha_3(s))e^{2(D(0)+\delta_1(s))}\right). \qquad \square \tag{15.99}$$

Lemma 15.19. *There exists a class $\mathcal{K}$ function $\hat{\gamma}^*$ such that for all solutions of the system satisfying* (15.84) *for* $0 < c < 1$, *the following holds:*

$$|\Pi(\theta)| \leq \hat{\gamma}^*\left(|X(t)| + \sup_{\phi(t)\leq\tau\leq t} |Z(\tau)|\right), \quad \phi(t) \leq \theta \leq t. \tag{15.100}$$

Proof. Let $Y(s)$ be the solution of $\frac{dY(s)}{ds} = f(Y(s), \mu(Y(s)) + \omega(s))$ for $s \geq s_0 \geq 0$. Under Assumption 15.14 (see Definition C.14), there exist a class $\mathcal{KL}$ function $\hat{\beta}_2$ and a class $\mathcal{K}$ function $\hat{\gamma}_1$ such that

$$|Y(s)| \leq \hat{\beta}_2(|Y(s_0)|, s - s_0) + \hat{\gamma}_1\left(\sup_{s_0\leq r\leq s} |\omega(r)|\right) \quad \text{for all } s \geq s_0. \tag{15.101}$$

Using the change of variable $\theta = \phi(y)$ and definition (15.94), we have that

$$\frac{d\Pi(\phi(y))}{dy} = f(\Pi(\phi(y)), \mu(\Pi(\phi(y))) + Z(\phi(y))), \quad t \leq y \leq \sigma(t). \tag{15.102}$$

Since $\Pi(\phi(y))$ satisfies the same ODE in y as the ODE for $Y(s)$ in s, we have for all $t \leq y \leq \sigma(t)$ that

$$|\Pi(\phi(y))| \leq \hat{\beta}_2(|X(t)|, y - t) + \hat{\gamma}_1\left(\sup_{t\leq y\leq\sigma(t)} |Z(\phi(y))|\right). \tag{15.103}$$

With the fact that $\hat{\beta}(s,r) \leq \hat{\beta}(s,0)$ for all $r \geq 0$, we get from (15.103)

$$|\Pi(\theta)| \leq \hat{\beta}_2(|X(t)|,0) + \hat{\gamma}_1\left(\sup_{\phi(t)\leq\tau\leq t}|Z(\tau)|\right), \quad \phi(t) \leq \theta \leq t. \tag{15.104}$$

With the properties of class $\mathcal{K}$ functions we get (15.100), where $\gamma(s) = \hat{\beta}_2(s,0) + \hat{\gamma}_1(s)$. □

Lemma 15.20. *There exists a class $\mathcal{KL}$ function $\hat{\beta}^*$ such that for all solutions of the system satisfying* (15.84) *for $0 < c < 1$, the following holds for all $t \geq 0$:*

$$\hat{\Xi}(t) \leq \left(1 + \frac{1}{1-c}\right)\left(\hat{\beta}^*\left(\hat{\Xi}(0), t\right) + \hat{\beta}_4\left(\hat{\Xi}(0), \max\{0, t - \sigma(0)\}\right)\right), \tag{15.105}$$

where

$$\hat{\Xi}(t) = |X(t)| + \sup_{\phi(t)\leq\theta\leq t}|Z(\theta)| + \sup_{\phi(t)\leq\theta\leq t}|U(\theta)|. \tag{15.106}$$

Proof. Solving (15.92), we have that $Z(t) = Z(0)e^{-c_Z(t)}$ for all $t \geq 0$. Since $\phi(t)$ is increasing for all $t \geq 0$ we get

$$\sup_{\phi(t)\leq\theta\leq t}|Z(\theta)| \leq |Z(0)|e^{-c_Z\phi(t)} \quad \text{for all } t \geq \sigma(0). \tag{15.107}$$

Similarly, for all $0 \leq t \leq \sigma(0)$ we get

$$\sup_{\phi(t)\leq\theta\leq t}|Z(\theta)| \leq \sup_{\phi(0)\leq\theta\leq 0}|Z(\theta)| + \sup_{0\leq\theta\leq t\leq\sigma(0)}|Z(\theta)|, \tag{15.108}$$

and hence, combining (15.108) with (15.92), we get

$$\sup_{\phi(t)\leq\theta\leq t}|Z(\theta)| \leq 2\sup_{\phi(t_0)\leq\theta\leq t_0}|Z(\theta)| \quad \text{for all } 0 \leq t \leq \sigma(0). \tag{15.109}$$

Therefore, using (15.107), (15.109), and the fact that for all $t \leq \sigma(0)$, $\phi(t) \leq 0$ we get

$$\sup_{\phi(t)\leq\theta\leq t}|Z(\theta)| \leq 2\sup_{\phi(0)\leq\theta\leq 0}|Z(\theta)|e^{-c_Z\phi(t)} \quad \text{for all } t \geq 0. \tag{15.110}$$

Using (15.90) we get that $\phi(t) = t - D(\zeta(\phi(t))) = t - D(Z(\phi(t)) + \mu(X(t)))$, and hence the following holds for all $t \geq 0$:

$$\sup_{\phi(t)\leq\theta\leq t}|Z(\theta)| \leq 2\sup_{\phi(0)\leq\theta\leq 0}|Z(\theta)|e^{-c_Z t}e^{c_Z D(Z(\phi(t))+\mu(X(t)))}. \tag{15.111}$$

Using (15.33) we get that $D(Z(\phi(t)) + \mu(X(t))) \leq D(0) + \delta_1(2|Z(\phi(t))|) + \delta_1(2|X(t)|)$. Since for all $t \geq \sigma(0)$, $\phi(t) \geq 0$, from (15.92) we get that $|Z(\phi(t))| \leq |Z(0)|$ for all $t \geq \sigma(0)$. Moreover, for all $t \leq \sigma(0)$, $\phi(0) \leq \phi(t) \leq 0$. Hence, for all $t \geq 0$, $|Z(\phi(t))| \leq \sup_{\phi(0)\leq\theta\leq 0}|Z(\theta)|$. Therefore, (15.111) gives that for all $t \geq 0$

$$\begin{aligned}\sup_{\phi(t)\leq\theta\leq t}|Z(\theta)| \leq 2\sup_{\phi(0)\leq\theta\leq 0}|Z(\theta)|\,& e^{c_Z\left(D(0)+\delta_1\left(2\sup_{\phi(0)\leq\theta\leq 0}|Z(\theta)|\right)+\delta_1(2|X(t)|)\right)} \\ & \times e^{-c_Z t}.\end{aligned} \tag{15.112}$$

Under Assumption 15.14 (see Definition C.14) we get from (15.101) that

$$|X(t)| \le \hat{\beta}_2(|X(s)|, t-s) + \hat{\gamma}_1\left(\sup_{s\le\tau\le t} |Z(\phi(\tau))|\right), \quad t \ge s \ge 0. \tag{15.113}$$

Setting $s = 0$ we have that

$$|X(t)| \le \hat{\beta}_2(|X(0)|, t) + \hat{\gamma}_1\left(\sup_{\phi(0)\le\theta\le\phi(t)} |Z(\theta)|\right) \quad \text{for all } t \ge 0, \tag{15.114}$$

and hence, from (15.92),

$$|X(t)| \le \hat{\beta}_2(|X(0)|, 0) + \hat{\gamma}_1\left(2 \sup_{\phi(0)\le\theta\le 0} |Z(\theta)|\right) \quad \text{for all } t \ge 0. \tag{15.115}$$

Therefore, with (15.112) we arrive at

$$\sup_{\phi(t)\le\theta\le t} |Z(\theta)| \le \hat{\alpha}_{12}\left(|X(0)| + \sup_{\phi(0)\le\theta\le 0} |Z(\theta)|\right) e^{-c_Z t} \tag{15.116}$$

for all $t \ge 0$, where the class $\mathcal{K}_\infty$ function $\hat{\alpha}_{12}$ is defined as

$$\hat{\alpha}_{12}(s) = 2e^{c_Z D(0)} s e^{c_Z\left(\delta_1(s) + \delta_1(2\hat{\beta}_2(s,0) + 2\hat{\gamma}_1(s))\right)}. \tag{15.117}$$

Setting in (15.113) $s = \frac{t}{2}$ we get

$$|X(t)| \le \hat{\beta}_2\left(|X(0)|, \frac{t}{2}\right) + \hat{\gamma}_1\left(\sup_{\phi\left(\frac{t}{2}\right)\le\theta\le\phi(t)} |Z(\theta)|\right) \quad \text{for all } t \ge 0. \tag{15.118}$$

We estimate now $\sup_{\phi\left(\frac{t}{2}\right)\le\theta\le\phi(t)} |Z(\theta)|$. Solving (15.92) we get

$$\sup_{\phi\left(\frac{t}{2}\right)\le\theta\le\phi(t)} |Z(\theta)| \le 2 \sup_{\phi(0)\le\theta\le 0} |Z(\theta)| e^{-c_Z \phi\left(\frac{t}{2}\right)} \quad \text{for all } t \ge 2\sigma(0). \tag{15.119}$$

With the help of relations (15.92) and (15.119) we get

$$\begin{aligned}
\sup_{\phi\left(\frac{t}{2}\right)\le\theta\le\phi(t)} |Z(\theta)| &\le \sup_{\phi\left(\frac{t}{2}\right)\le\theta\le 0} |Z(\theta)| + \sup_{0\le\theta\le\phi(t)} |Z(\theta)| \\
&\le 2 \sup_{\phi(0)\le\theta\le 0} |Z(\theta)| \quad \text{for all } 0 \le t \le 2\sigma(0).
\end{aligned} \tag{15.120}$$

Hence, using the fact that $\phi\left(\frac{t}{2}\right) = \frac{t}{2} - D\left(\zeta\left(\phi\left(\frac{t}{2}\right)\right)\right)$ we get from (15.90)

$$\begin{aligned}
\sup_{\phi\left(\frac{t}{2}\right)\le\theta\le\phi(t)} |Z(\theta)| \le 2 \sup_{\phi(0)\le\theta\le 0} |Z(\theta)| e^{c_Z\left(D(0) + \delta_1\left(2\left|X\left(\frac{t}{2}\right)\right|\right) + \delta_1\left(2\left|Z\left(\phi\left(\frac{t}{2}\right)\right)\right|\right)\right)} \\
\times e^{-\frac{c_Z}{2} t}
\end{aligned} \tag{15.121}$$

for all $t \geq 0$. Setting $s = 0$ and replacing t by $\frac{t}{2}$ we get from (15.113) that

$$\left|X\left(\frac{t}{2}\right)\right| \leq \hat{\beta}_2\left(|X_1(0)|, \frac{t}{2}\right) + \hat{\gamma}_1\left(\sup_{\phi(0)\leq\theta\leq\phi\left(\frac{t}{2}\right)} |Z(\theta)|\right). \tag{15.122}$$

Since

$$\begin{aligned}\sup_{\phi(0)\leq\theta\leq\phi\left(\frac{t}{2}\right)} |Z(\theta)| &\leq \sup_{\phi(0)\leq\theta\leq 0} |Z(\theta)| + \sup_{0\leq\theta\leq\phi(t)} |Z(\theta)| \\ &\leq 2 \sup_{\phi(0)\leq\theta\leq 0} |Z(\theta)|,\end{aligned} \tag{15.123}$$

we arrive at

$$\left|X\left(\frac{t}{2}\right)\right| \leq \hat{\beta}_2(|X(0)|, 0) + \hat{\gamma}_1\left(2 \sup_{\phi(0)\leq\theta\leq 0} |Z(\theta)|\right) \quad \text{for all } t \geq 0. \tag{15.124}$$

Using also the fact that $|Z(\phi(\frac{t}{2}))| \leq \sup_{\phi(0)\leq\theta\leq 0} |Z(\theta)|$, combining (15.121), (15.124) we get from (15.118) that

$$|X(t)| \leq \hat{\beta}_3\left(|X(0)| + \sup_{\phi(0)\leq\theta\leq 0} |Z(\theta)|, t\right) \quad \text{for all } t \geq 0, \tag{15.125}$$

where the class $\mathcal{KL}$ function $\hat{\beta}_3$ is defined as

$$\begin{aligned}\hat{\beta}_3(s,t) &= \hat{\beta}_2\left(s, \frac{t}{2}\right) \\ &\quad + \hat{\gamma}_1\left(2s e^{c_Z D(0)} e^{c_Z\left(\delta_1(2s)+\delta_1\left(2\hat{\beta}_2(s,0)+2\hat{\gamma}_1(s)\right)\right)} e^{-c_Z \frac{t}{2}}\right).\end{aligned} \tag{15.126}$$

Using (15.83), (15.84) we get for all $\theta \geq 0$

$$|U(\theta)| \leq \frac{1}{1-c}|\nabla\mu(P(\theta)) f(P(\theta), \zeta(\theta))| + \frac{1}{1-c} c_Z |Z(\theta)|. \tag{15.127}$$

Since $f : \mathbb{R}^n \times \mathbb{R} \to \mathbb{R}^n$ is locally Lipschitz with $f(0,0) = 0$ and $\mu \in C^1(\mathbb{R}^n;\mathbb{R})$ with $\mu(0) = 0$, there exist class $\mathcal{K}_\infty$ functions $\hat{\alpha}_{11}$ and $\hat{\alpha}^*_{11}$ such that for all $(X, \omega) \in \mathbb{R}^{n+1}$

$$|\mu(X)| \leq \hat{\alpha}_{11}(|X|), \tag{15.128}$$
$$|\nabla\mu(X)| \leq |\nabla\mu(0)| + \hat{\alpha}_{11}(|X|), \tag{15.129}$$
$$|f(X, \omega)| \leq \hat{\alpha}^*_{11}(|X| + |\omega|). \tag{15.130}$$

Therefore, using (15.33), (15.90), and the fact that $P(\theta) = \Pi(\theta) = X(\sigma(\theta))$ we get for all $\theta \geq 0$ that

$$\begin{aligned}|U(\theta)| \leq{}& \frac{1}{1-c}(|\nabla\mu(0)| + \hat{\alpha}_{11}(|X(\sigma(\theta))|))\, \hat{\alpha}^*_{11}(|X(\sigma(\theta))| + |Z(\theta)| + \hat{\alpha}_{11}(|X(\sigma(\theta))|)) \\ &+ \frac{1}{1-c} c_Z |Z(\theta)|.\end{aligned} \tag{15.131}$$

Hence, with the help of (15.116), (15.125) we get for all $t \geq \sigma(0)$ that

$$\sup_{\phi(t)\leq\theta\leq t} |U(\theta)| \leq \frac{1}{1-c}\hat{\beta}_4\left(|X(0)| + \sup_{\phi(0)\leq\theta\leq 0} |Z(\theta)|, t\right), \tag{15.132}$$

where the class $\mathcal{KL}$ function $\hat{\beta}_4$ is defined as

$$\begin{aligned}\hat{\beta}_4(s,t) = &\left(|\nabla\mu(0)| + \hat{\alpha}_{11}\left(\hat{\beta}_3(s,t)\right)\right)\hat{\alpha}_{11}^*\left(\hat{\beta}_3(s,t) + \hat{\alpha}_{12}(s)e^{-c_Z t} + \hat{\alpha}_{11}\left(\hat{\beta}_3(s,t)\right)\right)\\ &+ c_Z\hat{\alpha}_{12}(s)e^{-c_Z t}.\end{aligned} \tag{15.133}$$

Moreover,

$$\sup_{\phi(t)\leq\theta\leq t} |U(\theta)| \leq \sup_{\phi(0)\leq\theta\leq 0} |U(\theta)| + \sup_{0\leq\theta\leq t} |U(\theta)| \quad \text{for all } t \leq \sigma(0), \tag{15.134}$$

and hence

$$\begin{aligned}\sup_{\phi(t)\leq\theta\leq t} |U(\theta)| \leq &\sup_{\phi(0)\leq\theta\leq 0} |U(\theta)|\\ &+\frac{1}{1-c}\hat{\beta}_4\left(|X(0)| + \sup_{\phi(0)\leq\theta\leq 0} |Z(\theta)|, 0\right)\end{aligned} \tag{15.135}$$

for all $t \leq \sigma(0)$. Combining (15.132), (15.135) and assuming without loss of generality that $\hat{\beta}_4(s,0) > s$ we arrive at

$$\sup_{\phi(t)\leq\theta\leq t} |U(\theta)| \leq \left(1 + \frac{1}{1-c}\right)\hat{\beta}_4\left(\hat{\Xi}(0), \max\{0, t-\sigma(0)\}\right) \tag{15.136}$$

for all $t \geq 0$. Combining (15.116), (15.125), (15.136) we get (15.105) with $\hat{\beta}^*(s,t) = \hat{\alpha}_{12}(s)e^{-c_Z t} + \hat{\beta}_3(s,t)$. □

Lemma 15.21. *There exist class $\mathcal{K}_\infty$ functions $\hat{\alpha}_9$, $\hat{\alpha}_{10}$ such that for all solutions of the system satisfying* (15.84) *for* $0 < c < 1$, *the following hold:*

$$\hat{\Omega}(t) \leq \hat{\alpha}_9\left(\hat{\Xi}(t)\right), \tag{15.137}$$

$$\hat{\Xi}(t) \leq \hat{\alpha}_{10}\left(\hat{\Omega}(t)\right) \tag{15.138}$$

for all $t \geq 0$, where $\hat{\Omega}$ is defined in (15.86) *and $\hat{\Xi}$ is defined in* (15.106).

Proof. Using the direct backstepping transformation (15.90) and bounds (15.95), (15.128) we get the bound (15.138) with $\hat{\alpha}_{10}(s) = s + \hat{\alpha}_{11}(\hat{\alpha}_8(s))$. Using the inverse backstepping transformation (15.93) and the bounds (15.100), (15.128) we get the bound (15.137) with $\hat{\alpha}_9(s) = s + \hat{\alpha}_{11}(\hat{\gamma}^*(s))$. □

Lemma 15.22. *There exists a function* $\hat{\delta}$ *of class* $\mathscr{K}_\infty$ *such that for all solutions of the system that satisfy*

$$\hat{\Omega}(t) < \hat{\delta}^{-1}(c) \quad \text{for all } t \geq 0 \tag{15.139}$$

for $0 < c < 1$, *they also satisfy* (15.84).

Proof. Using (15.57), (15.128) one can conclude that if a solution satisfies for all $t \geq 0$ and for all $\phi(t) \leq \theta \leq t$

$$\begin{aligned} c > & \left(\left|D'(0)\right| + \delta_2(|\zeta(\theta)|)\right) \\ & \times \left(|U(\theta)| + (|\nabla\mu(0)| + \hat{\alpha}_{11}(|P(\theta)|))\,\hat{\alpha}_{11}^*(|P(\theta)| + |\zeta(\theta)|)\right) \end{aligned} \tag{15.140}$$

for $0 < c < 1$, then it also satisfies (15.84). Using Lemma 15.18, (15.140) is satisfied for $0 < c < 1$ as long as the bound (15.139) holds, where the class $\mathscr{K}_\infty$ function $\hat{\delta}$ is given by

$$\hat{\delta}(s) = \left(\left|D'(0)\right| + \delta_2(s)\right)\left(s + (|\nabla\mu(0)| + \hat{\alpha}_{11}(\hat{\alpha}_8(s)))\,\hat{\alpha}_{11}^*(\hat{\alpha}_8(s) + s)\right). \qquad \square \tag{15.141}$$

Lemma 15.23. *There exists a class* $\mathscr{K}$ *function* ξ_{RoA} *such that for all initial conditions of the closed-loop system* (15.76)–(15.78), (15.83), (15.81), (15.82) *that satisfy* (15.85), *the solutions of the system satisfy* (15.139) *for* $0 < c < 1$ *and hence satisfy* (15.84).

Proof. Using Lemma 15.21, with the help of (15.105), we have that

$$\begin{aligned} \hat{\Omega}(t) \leq \hat{\alpha}_9 & \left(\left(1 + \frac{1}{1-c}\right)\hat{\beta}^*\left(\hat{\alpha}_{10}\left(\hat{\Omega}(0)\right), t\right)\right. \\ & \left. + \left(1 + \frac{1}{1-c}\right)\hat{\beta}_4\left(\hat{\alpha}_{10}\left(\hat{\Omega}(0)\right), \max\{0, t - \sigma(0)\}\right)\right). \end{aligned} \tag{15.142}$$

Hence, for all initial conditions that satisfy the bound (15.85) with any class $\mathscr{K}$ choice $\xi_{\text{RoA}}(c) \leq \bar{\xi}^*_{\text{RoA}}(\hat{\delta}^{-1}(c), c)$, where $\bar{\xi}^*_{\text{RoA}}(s, c)$ is the inverse of the class $\mathscr{K}\mathscr{C}_\infty$ function

$$\xi^*{}_{\text{RoA}}(s, c) = \hat{\alpha}_9\left(\left(1 + \frac{1}{1-c}\right)\left(\hat{\beta}^*(\hat{\alpha}_{10}(s), 0) + \hat{\beta}_4(\hat{\alpha}_{10}(s), 0)\right)\right) \tag{15.143}$$

with respect to ξ^*_{RoA}'s first argument, the solutions satisfy (15.139). Moreover, for all those initial conditions, the solutions verify (15.84) for all $\theta \geq \phi(0)$. $\square$

***Proof of Theorem* 15.15.** Using (15.33), (15.85), and $0 < c < 1$ we conclude that $\sigma(0) = D(\zeta(0)) \leq D(0) + \delta_1(\xi_{\text{RoA}}(1)) = \xi^*$. Hence, using Corollary 10 in [158] and relation (15.142) we get (15.87) with some class $\mathscr{K}_\infty$ function σ_1, where $\hat{\beta}(s, t) = \sigma_1(\hat{\beta}^*(\hat{\alpha}_{10}(s), t) + \hat{\beta}_4(\hat{\alpha}_{10}(s), \max\{0, t - \xi^*\}))$. Using relations (15.83), (15.94) and the fact that $P = \Pi$ we get for all $t \geq 0$ that

$$\frac{d\Pi(t)}{dt} = \frac{\left(1 - D'(\mu(\Pi(t)) + Z(t))c_Z Z(t)\right) f(\Pi(t), \mu(\Pi(t)) + Z(t))}{1 - \nabla\mu(\Pi(t)) f(\Pi(t), \mu(\Pi(t)) + Z(t)) D'(\mu(\Pi(t)) + Z(t))}. \tag{15.144}$$

Under Assumption 15.1 (Lipschitzness of D'), Assumption 15.14 (Lipschitzness of $\nabla\mu$), and relation (15.92) we conclude that the right-hand side of (15.92), (15.144) is Lipschitz with respect to (Z,Π), and hence, using also bound (15.100), there exists a unique solution $(Z(t),\Pi(t)) \in C^1(0,\infty)$. Using (15.93) we get the existence and uniqueness of $\zeta(t) \in C^1(0,\infty)$. The boundedness of U and (15.77) guarantee that ζ is continuous at $t=0$. By integrating (15.77) between any two time instants it is shown that ζ is Lipschitz on $[0,\infty)$ with a Lipschitz constant given by a uniform bound on U. With the fact that $\Pi = P$, relations (15.83), (15.84), and the Lipschitzness of D' and $\nabla\mu$ we get the existence and uniqueness of $U \in (0,\infty)$ and that U is locally Lipschitz in $(0,\infty)$. From (15.76) and (15.82) we have for all $t > \sigma(0)$ that

$$\dot{X}(t) = f(X(t),\zeta(\phi(t))), \tag{15.145}$$

$$\dot{\phi}(t) = \frac{1}{1+D'(\zeta(\phi(t)))U(\phi(t))}. \tag{15.146}$$

Since ζ is Lipschitz on $[0,\infty)$, U is Lipschitz on $(0,\infty)$, and D' is locally Lipschitz, one can conclude that the right-hand side of system (15.145)–(15.146) is Lipschitz with respect to (X,ϕ), and hence there exists a unique solution $(X(t),\phi(t)) \in C^1(\sigma(0),\infty)$. Similarly, the Lipschitzness of the initial conditions $\zeta(s)$ and $U(s)$ for $\phi(0) \le s < 0$ guarantees the existence and uniqueness of $(X(t),\phi(t)) \in C^1[0,\sigma(0))$. The boundness of the right-hand side of (15.145)–(15.146) guarantees that (X,ϕ) are continuous at $\sigma(0)$, and hence the Lipschitzness of ζ at 0 guarantees that the right-hand side of (15.145) is continuous at $\sigma(0)$. Therefore X is continuously differentiable also at $\sigma(0)$. With (15.84) we get bound (15.89), and with (15.33), (15.139) we get (15.88) with any class $\mathcal{K}$ function $\hat{\delta}^*(c) \ge \delta_1(\hat{\delta}^{-1}(c))$.

15.5 ▪ Notes and References

Delay that depends on past, rather than current, values of the state appear in electrodynamics and in networks. The time that is required for a relativistic particle to feel the electromagnetic force of another particle depends on the position of the particles at a past time instant [163]. The round-trip time of a signal in a network is a past value of the queuing delay at a time that depends on past values of the queuing delay itself [27], [30]. The results of this chapter can be directly extended to the case in which the delay is explicitly defined as a function of past values of the state, at a time instant that is a priori given. The results of this chapter seem also extendable to the case in which the delay might depend on past values of the state, but at a time instant that may be a function of the delay (rather than just identical to the delay).

Since in this chapter we deal with delays that depend on delayed states, whereas in Chapters 10–14 we deal with delays that depend on current states, it is reasonable to ask whether this methodology can be extended to the case in which the delay function depends both on delayed and current states. For designing the predictor feedback law for such a delay function, one has first to show the well-posedness of both the prediction and delay times (namely, σ and ϕ, respectively). To show this one has to study the existence and uniqueness of a two-point boundary value problem for the prediction and the delay times. For studying the existence and uniqueness of this problem one has to use fixed-point theory incorporating the properties of the dynamics of these two times and the properties of the solutions of the system. This study is far from trivial and is a topic for future research.

Appendix A
Basic Inequalities

Young's inequality (elementary version):

$$ab \leq \frac{\gamma}{2}a^2 + \frac{1}{2\gamma}b^2 \quad \textit{for all } \gamma > 0. \tag{A.1}$$

The Cauchy–Schwarz inequality:

$$\int_0^1 u(x)w(x)dx \leq \sqrt{\int_0^1 u(x)^2 dx}\sqrt{\int_0^1 w(x)^2 dx}. \tag{A.2}$$

Appendix B
Input-to-Output Stability

For a function $x : \mathbb{R}_+ \to \mathbb{R}^n$, we define the L_p norm, $p \in [1,\infty]$, as

$$||x||_p = \begin{cases} \left(\int_0^\infty |x(t)|^p dt \right)^{1/p}, & p \in [1,\infty), \\ \sup_{t \geq 0} |x(t)|, & p = \infty, \end{cases} \tag{B.1}$$

and the $L_{p,e}$ norm (truncated L_p norm) as

$$||x_t||_p = \begin{cases} \left(\int_0^t |x(\tau)|^p d\tau \right)^{1/p}, & p \in [1,\infty), \\ \sup_{\tau \in [0,t]} |x(\tau)|, & p = \infty. \end{cases} \tag{B.2}$$

Lemma B.1 (Hölder's inequality). *If* $p, q \in [1,\infty]$ *and* $\frac{1}{p} + \frac{1}{q} = 1$, *then*

$$||(f g)_t||_1 \leq ||f_t||_p ||g_t||_q \qquad \text{for all } t \geq 0. \tag{B.3}$$

We consider a linear time-invariant causal system described by the convolution

$$y(t) = h \star u = \int_0^t h(t-\tau) u(\tau) d\tau, \tag{B.4}$$

where $u : \mathbb{R}_+ \to \mathbb{R}$ is the input, $y : \mathbb{R}_+ \to \mathbb{R}$ is the output, and $h : \mathbb{R} \to \mathbb{R}$ is the system's impulse response, which is defined to be zero for negative values of its argument.

Theorem B.2 (Young's convolution theorem). *If* $h \in L_{1,e}$, *then*

$$||(h \star u)_t||_p \leq ||h_t||_1 ||u_t||_p, \qquad p \in [1,\infty]. \tag{B.5}$$

Proof. Let $y = h \star u$. Then, for $p \in [1,\infty)$, we have

$$
\begin{aligned}
|y(t)| &\leq \int_0^t |h(t-\tau)|\,|u(\tau)|d\tau \\
&= \int_0^t |h(t-\tau)|^{\frac{p-1}{p}}\,|h(t-\tau)|^{\frac{1}{p}}\,|u(\tau)|d\tau \\
&\leq \left(\int_0^t |h(t-\tau)|d\tau\right)^{\frac{p-1}{p}} \left(\int_0^t |h(t-\tau)|\,|u(\tau)|^p d\tau\right)^{\frac{1}{p}} \\
&= \|h_t\|_1^{\frac{p-1}{p}} \left(\int_0^t |h(t-\tau)|\,|u(\tau)|^p d\tau\right)^{\frac{1}{p}},
\end{aligned}
\tag{B.6}
$$

where the second inequality is obtained by applying Hölder's inequality. Raising (B.6) to power p and integrating from 0 to t, we get

$$
\begin{aligned}
\|y_t\|_p^p &\leq \int_0^t \|h_t\|_1^{p-1} \left(\int_0^\tau |h(\tau-s)|\,|u(s)|^p ds\right) d\tau \\
&= \|h_t\|_1^{p-1} \int_0^t \left(\int_s^t |h(\tau-s)|\,|u(s)|^p d\tau\right) ds \\
&= \|h_t\|_1^{p-1} \int_0^t \left(\int_0^t |h(\tau-s)|\,|u(s)|^p d\tau\right) ds \\
&= \|h_t\|_1^{p-1} \int_0^t |u(s)|^p \left(\int_0^t |h(\tau-s)|d\tau\right) ds \\
&\leq \|h_t\|_1^{p-1} \int_0^t |u(s)|^p \left(\int_0^t |h(\tau)|d\tau\right) ds \\
&\leq \|h_t\|_1^{p-1} \|h\|_1^1 \|u_t\|_1^p \\
&\leq \|h_t\|_1^p \|u_t\|_p^p,
\end{aligned}
\tag{B.7}
$$

where the second line is obtained by changing the sequence of integration, and the third line by using the causality of h. The proof for the case $p = \infty$ is immediate by taking a supremum of u over $[0,t]$ in the convolution. □

Lemma B.3. *Let v and ρ be real-valued functions defined on $\mathbb{R}_+$, and let b and c be positive constants. If they satisfy the differential inequality*

$$
\dot{v} \leq -cv + b\rho(t)^2, \qquad v(0) \geq 0,
\tag{B.8}
$$

(i) *then the following integral inequality holds:*

$$
v(t) \leq v(0)e^{-ct} + b\int_0^t e^{-c(t-\tau)}\rho(\tau)^2 d\tau.
\tag{B.9}
$$

(ii) *If, in addition, $\rho \in L_2$, then $v \in L_1$ and*

$$
\|v\|_1 \leq \frac{1}{c}\left(v(0) + b\|\rho\|_2^2\right).
\tag{B.10}
$$

Proof. (i) Upon multiplication of (B.8) by e^{ct}, it becomes

$$\frac{d}{dt}\left(v(t)e^{ct}\right) \le b\rho(t)^2 e^{ct}. \tag{B.11}$$

Integrating (B.11) over $[0,t]$, we arrive at (B.9).

(ii) By integrating (B.9) over $[0,t]$, we get

$$\begin{aligned}\int_0^t v(\tau)d\tau &\le \int_0^t v(0)e^{-c\tau}d\tau + b\int_0^t\left[\int_0^\tau e^{-c(\tau-s)}\rho(s)^2 ds\right]d\tau \\ &\le \frac{1}{c}v(0) + b\int_0^t\left[\int_0^\tau e^{-c(\tau-s)}\rho(s)^2 ds\right]d\tau. \end{aligned} \tag{B.12}$$

Noting that the second term is $b\|(h\star\rho^2)_t\|_1$, where

$$h(t)=e^{-ct}, \qquad t\ge 0, \tag{B.13}$$

we apply Theorem B.2. Since

$$\|h\|_1 = \frac{1}{c}, \tag{B.14}$$

we obtain (B.10). □

Lemma B.4. *Let v, l_1, and l_2 be real-valued functions defined on $\mathbb{R}_+$, and let c be a positive constant. If l_1 and l_2 are nonnegative and in L_1 and satisfy the differential inequality*

$$\dot v \le -cv + l_1(t)v + l_2(t), \quad v(0)\ge 0, \tag{B.15}$$

then $v\in L_\infty\cap L_1$ and

$$v(t) \le \left(v(0)e^{-ct} + \|l_2\|_1\right)e^{\|l_1\|_1}, \tag{B.16}$$

$$\|v\|_1 \le \frac{1}{c}\left(v(0)+\|l_2\|_1\right)e^{\|l_1\|_1}. \tag{B.17}$$

Proof. Using the facts that

$$v(t) \le w(t), \tag{B.18}$$

$$\dot w = -cw + l_1(t)w + l_2(t), \tag{B.19}$$

$$w(0) = v(0) \tag{B.20}$$

(the comparison principle), and applying the variation-of-constants formula, the differential inequality (B.15) is rewritten as

$$\begin{aligned} v(t) &\le v(0)e^{\int_0^t[-c+l_1(s)]ds} + \int_0^t e^{\int_\tau^t[-c+l_1(s)]ds} l_2(\tau)d\tau \\ &\le v(0)e^{-ct}e^{\int_0^\infty l_1(s)ds} + \int_0^t e^{-c(t-\tau)}l_2(\tau)d\tau e^{\int_0^\infty l_1(s)ds} \\ &\le \left[v(0)e^{-ct} + \int_0^t e^{-c(t-\tau)}l_2(\tau)d\tau\right]e^{\|l_1\|_1}. \end{aligned} \tag{B.21}$$

By taking a supremum of $e^{-c(t-\tau)}$ over $[0,\infty]$, we obtain (B.16). Integrating (B.21) over $[0,\infty]$, we get

$$\int_0^t v(\tau)d\tau \leq \left(\frac{1}{c}v(0)+\int_0^t\left[\int_0^\tau e^{-c(\tau-s)}l_2(s)ds\right]d\tau\right)e^{\|l_1\|_1}. \tag{B.22}$$

Applying Theorem B.2 to the double integral, we arrive at (B.17). □

Remark B.5. *An alternative proof that $v\in L_\infty\cap L_1$ in Lemma* B.4 *is to use the Gronwall lemma (Lemma* B.8*). However, with the Gronwall lemma, the estimates of the bounds* (B.16) *and* (B.17) *are more conservative:*

$$v(t)\leq\left(v(0)e^{-ct}+\|l_2\|_1\right)\left(1+\|l_1\|_1 e^{\|l_1\|_1}\right), \tag{B.23}$$

$$\|v\|_1\leq\frac{1}{c}\left(v(0)+\|l_2\|_1\right)\left(1+\|l_1\|_1 e^{\|l_1\|_1}\right), \tag{B.24}$$

because

$$e^x<(1+xe^x) \qquad \textit{for all } x>0. \tag{B.25}$$

Note that the ratio between the bounds (B.23) *and* (B.16) *and that between the bounds* (B.24) *and* (B.17) *are of the order $\|l_1\|_1$ when $\|l_1\|_1\to\infty$.*

For cases where l_1 and l_2 are functions of time that converge to zero but are not in L_p for any $p\in[1,\infty)$, we have the following lemma.

Lemma B.6. *Consider the differential inequality*

$$\dot v\leq-(c-\beta_1(r_0,t))v+\beta_2(r_0,t)+\rho, \quad v(0)=v_0\geq 0, \tag{B.26}$$

where $c>0$ and $r_0\geq 0$ are constants, and β_1 and β_2 are class $\mathcal{KL}$ functions. Then there exist a class $\mathcal{KL}$ function β_v and a class $\mathcal{K}$ function γ_v such that

$$v(t)\leq\beta_v(v_0+r_0,t)+\gamma_v(\rho), \quad t\geq 0. \tag{B.27}$$

Moreover, if

$$\beta_i(r,t)=\alpha_i(r)e^{-\sigma_i t}, \quad i=1,2, \tag{B.28}$$

where $\alpha_i\in\mathcal{K}$ and $\sigma_i>0$, then there exist $\alpha_v\in\mathcal{K}$ and σ_v such that

$$\beta_v(r,t)=\alpha_v(r)e^{-\sigma_v t}. \tag{B.29}$$

Proof. We start by introducing

$$\tilde v=v-\frac{\rho}{c} \tag{B.30}$$

and rewriting (B.26) as

$$\dot{\tilde v}\leq-[c-\beta_1(r_0,t)]\tilde v+\frac{\rho}{c}\beta_1(r_0,t)+\beta_2(r_0,t). \tag{B.31}$$

It then follows that

$$v(t) \le v_0 e^{\int_0^t [\beta_1(r_0,s)-c]ds} + \int_0^t \left[\frac{\rho}{c}\beta_1(r_0,\tau)+\beta_2(r_0,\tau)\right] e^{\int_\tau^t [\beta_1(r_0,s)-c]ds} d\tau + \frac{\rho}{c}. \quad \text{(B.32)}$$

We note that

$$e^{\int_\tau^t [\beta_1(r_0,s)-c]ds} \le k(r_0) e^{-\frac{c}{2}(t-\tau)} \qquad \text{for all } \tau \in [0,t], \quad \text{(B.33)}$$

where k is a positive, continuous, increasing function. To get an estimate of the overshoot coefficient $k(r_0)$, we provide a proof of (B.33). For each c, there exists a class $\mathcal{K}$ function $T_c : \mathbb{R}_+ \to \mathbb{R}_+$ such that

$$\beta_1(r_0,s) \le \frac{c}{2} \qquad \text{for all } s \ge T_c(r_0). \quad \text{(B.34)}$$

Therefore, for $0 \le \tau \le T_c(r_0) \le t$, we have

$$\begin{aligned}
\int_\tau^t [\beta_1(r_0,s)-c]ds &\le \int_\tau^{T_c(r_0)} [\beta_1(r_0,s)-c]ds + \int_{T_c(r_0)}^t \left(-\frac{c}{2}\right) ds \\
&\le (\beta_1(r_0,0)-c)(T_c(r_0)-\tau) - \frac{c}{2}(t-T_c(r_0)) \\
&\le T_c(r_0)\beta_1(r_0,0) - \frac{c}{2}(t-\tau),
\end{aligned} \quad \text{(B.35)}$$

so the overshoot coefficient in (B.33) is given by

$$k(r_0) \stackrel{\Delta}{=} e^{T_c(r_0)\beta_1(r_0,0)}. \quad \text{(B.36)}$$

For the other two cases, $t \le T_c(r_0)$ and $T_c(r_0) \le \tau$, getting (B.33) with $k(r_0)$ as in (B.36) is immediate. Now substituting (B.33) into (B.32), we get

$$v(t) \le v_0 k(r_0) e^{-\frac{c}{2}t} + k(r_0) \int_0^t \left[\frac{\rho}{c}\beta_1(r_0,\tau)+\beta_2(r_0,\tau)\right] e^{-\frac{c}{2}(t-\tau)} d\tau + \frac{\rho}{c}. \quad \text{(B.37)}$$

To complete the proof, we show that a class $\mathcal{KL}$ function β convolved with an exponentially decaying kernel is bounded by another class $\mathcal{KL}$ function:

$$\begin{aligned}
\int_0^t e^{-\frac{c}{2}(t-\tau)}\beta(r_0,\tau)d\tau &= \int_0^{t/2} e^{-\frac{c}{2}(t-\tau)}\beta(r_0,\tau)d\tau + \int_{t/2}^t e^{-\frac{c}{2}(t-\tau)}\beta(r_0,\tau)d\tau \\
&\le \beta(r_0,0)\int_0^{t/2} e^{-\frac{c}{2}(t-\tau)}d\tau + \beta(r_0,t/2)\int_{t/2}^t e^{-\frac{c}{2}(t-\tau)}d\tau \\
&\le \frac{2}{c}\left[\beta(r_0,0)e^{-\frac{c}{4}t} + \beta(r_0,t/2)\right].
\end{aligned} \quad \text{(B.38)}$$

Thus, (B.37) becomes

$$\begin{aligned}
v(t) \le k(r_0)\Bigg\{&\left[v_0 + \frac{2\rho}{c^2}\beta_1(r_0,0) + \frac{2}{c}\beta_2(r_0,0)\right] e^{-\frac{c}{4}t} \\
&+ \frac{2\rho}{c^2}\beta_1(r_0,t/2) + \frac{2}{c}\beta_2(r_0,t/2)\Bigg\} + \frac{\rho}{c}.
\end{aligned} \quad \text{(B.39)}$$

By applying the Young inequality to the terms

$$k(r_0)\frac{2\rho}{c^2}\beta_1(r_0,0)e^{-\frac{c}{4}t} \tag{B.40}$$

and

$$k(r_0)\frac{2\rho}{c^2}\beta_1(r_0,t/2), \tag{B.41}$$

we obtain (B.27) with

$$\begin{aligned}\beta_v(r,t) = k(r)\Bigg\{&\left[r+\frac{k(r)}{c^2}\beta_1(r,0)^2+\frac{2}{c}\beta_2(r,0)\right]e^{-\frac{c}{4}t}\\ &+\frac{k(r)}{c^2}\beta_1(r,t/2)^2+\frac{2}{c}\beta_2(r,t/2)\Bigg\},\end{aligned} \tag{B.42}$$

$$\gamma_v(r) = \frac{r}{c}+\frac{r^2}{c^2}. \tag{B.43}$$

The last statement of the lemma is immediate by substitution into (B.42). □

The proof of the following lemma can be found in [75] (Lemma 3.4).

Lemma B.7 (comparison principle). *Consider the scalar differential equation*

$$\dot{u} = f(t,u), \quad u(t_0) = u_0, \tag{B.44}$$

where $f(t,u)$ is continuous in t and locally Lipschitz in u for all $t \geq 0$ and all $u \in \mathbb{J} \subset \mathbb{R}$. Let $[t_0,T)$ (T could be infinity) be the maximal interval of existence of the solution $u(t)$, and suppose $u(t) \in \mathbb{J}$ for all $t \in [t_0,T)$. Let $v(t)$ be a continuous function whose upper right-hand derivative $D^+v(t)$ satisfies the differential inequality

$$D^+v(t) \leq f(t,v(t)), \quad v(t_0) \leq u(t_0), \tag{B.45}$$

with $v(t) \in \mathbb{J}$ for all $t \in [t_0,T)$. Then $v(t) \leq u(t)$ for all $t \in [t_0,T)$.

Now we give a version of Gronwall's lemma.

Lemma B.8 (Gronwall). *Consider the continuous functions $\lambda : \mathbb{R}_+ \to \mathbb{R}$, $\mu : \mathbb{R}_+ \to \mathbb{R}_+$, and $\nu : \mathbb{R}_+ \to \mathbb{R}_+$, where μ and ν are also nonnegative. If a continuous function $y : \mathbb{R}_+ \to \mathbb{R}$ satisfies the inequality*

$$y(t) \leq \lambda(t) + \mu(t)\int_{t_0}^{t} \nu(s)y(s)ds \qquad \text{for all } t \geq t_0 \geq 0, \tag{B.46}$$

then

$$y(t) \leq \lambda(t) + \mu(t)\int_{t_0}^{t} \lambda(s)\nu(s)e^{\int_s^t \mu(\tau)\nu(\tau)d\tau}ds \qquad \text{for all } t \geq t_0 \geq 0. \tag{B.47}$$

In particular, if $\lambda(t) \equiv \lambda$ is a constant and $\mu(t) \equiv 1$, then

$$y(t) \leq \lambda e^{\int_{t_0}^t \nu(\tau)d\tau} \qquad \text{for all } t \geq t_0 \geq 0. \tag{B.48}$$

Appendix C

Lyapunov Stability, Forward-Completeness, and Input-to-State Stability

C.1 ▪ Lyapunov Stability and $\mathcal{K}_\infty$ Functions

Consider the nonautonomous ODE system

$$\dot{x} = f(x,t), \tag{C.1}$$

where $f : \mathbb{R}^n \times \mathbb{R}_+ \to \mathbb{R}^n$ is locally Lipschitz in x and piecewise continuous in t.

Definition C.1. *The origin $x = 0$ is the equilibrium point for* (C.1) *if*

$$f(0,t) = 0 \qquad \textit{for all } t \geq 0. \tag{C.2}$$

Scalar *comparison functions* are important stability tools.

Definition C.2. *A continuous function $\gamma : [0,a) \to \mathbb{R}_+$ is said to belong to class $\mathcal{K}$ if it is strictly increasing and $\gamma(0) = 0$. It is said to belong to class $\mathcal{K}_\infty$ if $a = \infty$ and $\gamma(r) \to \infty$ as $r \to \infty$.*

Definition C.3. *A continuous function $\beta : [0,a) \times \mathbb{R}_+ \to \mathbb{R}_+$ is said to belong to class $\mathcal{KL}$ if, for each fixed s, the mapping $\beta(r,s)$ belongs to class $\mathcal{K}$ with respect to r and, for each fixed r, the mapping $\beta(r,s)$ is decreasing with respect to s and $\beta(r,s) \to 0$ as $s \to \infty$. It is said to belong to class $\mathcal{KL}_\infty$ if, in addition, for each fixed s, the mapping $\beta(r,s)$ belongs to class $\mathcal{K}_\infty$ with respect to r.*

Definition C.4. *We say that a continuous function $\rho : \mathbb{R}_+ \times (0,1) \mapsto \mathbb{R}_+$ belongs to class $\mathcal{KC}$ if, for each fixed c, the mapping $\rho(s,c)$ belongs to class $\mathcal{K}$ with respect to s and, for each fixed s, the mapping $\rho(s,c)$ is continuous with respect to c. It belongs to class $\mathcal{KC}_\infty$ if, in addition, for each fixed c, the mapping $\rho(s,c)$ belongs to class $\mathcal{K}_\infty$ with respect to s.*

The main list of stability definitions for ODE systems is given next.

Definition C.5 (Stability). *The equilibrium point $x = 0$ of* (C.1) *is*

- uniformly stable *if there exist a class-$\mathcal{K}$ function $\gamma(\cdot)$ and a positive constant c, independent of t_0, such that*

$$|x(t)| \leq \gamma(|x(t_0)|) \quad \textit{for all } t \geq t_0 \geq 0, \ \textit{for all } x(t_0) \textit{ s.t. } |x(t_0)| < c; \tag{C.3}$$

- uniformly asymptotically stable *if there exist a class-$\mathcal{KL}$ function $\beta(\cdot,\cdot)$ and a positive constant c, independent of t_0, such that*

$$|x(t)| \le \beta(|x(t_0)|, t - t_0) \quad \textit{for all } t \ge t_0 \ge 0, \ \textit{for all } x(t_0) \text{ s.t. } |x(t_0)| < c; \tag{C.4}$$

- exponentially stable *if* (C.4) *is satisfied with* $\beta(r,s) = kre^{-\alpha s}$, $k > 0$, $\alpha > 0$;
- globally uniformly stable *if* (C.3) *is satisfied with* $\gamma \in \mathcal{K}_\infty$ *for any initial state* $x(t_0)$;
- globally uniformly asymptotically stable *if* (C.4) *is satisfied with* $\beta \in \mathcal{KL}_\infty$ *for any initial state* $x(t_0)$*; and*
- globally exponentially stable *if* (C.4) *is satisfied for any initial state* $x(t_0)$ *and with* $\beta(r,s) = kre^{-\alpha s}$, $k > 0$, $\alpha > 0$.

The following lemma is proved in [75] (Lemma 4.4).

Lemma C.6. *Consider the scalar autonomous differential equation*

$$\dot{y} = -\alpha(y), \quad y(t_0) = y_0, \tag{C.5}$$

where α *is a locally Lipschitz class* $\mathcal{K}$ *function defined on* $[0,a]$. *For all* $0 \le y_0 < a$, *this equation has a unique solution* $y(t)$ *defined for all* $t \ge t_0$. *Moreover,*

$$y(t) = \sigma(y_0, t - t_0), \tag{C.6}$$

where σ *is a class* $\mathcal{KL}$ *function defined on* $[0,a) \times [0,\infty)$.

The main Lyapunov stability theorem is then formulated as follows.

Theorem C.7 (Lyapunov theorem). *Let* $x = 0$ *be an equilibrium point of* (C.1) *and let* $D = \{x \in \mathbb{R}^n \mid |x| < r\}$. *Let* $V : D \times \mathbb{R}^n \to \mathbb{R}_+$ *be a continuously differentiable function such that for all* $t \ge 0$ *and for all* $x \in D$,

$$\gamma_1(|x|) \le V(x,t) \le \gamma_2(|x|), \tag{C.7}$$

$$\frac{\partial V}{\partial t} + \frac{\partial V}{\partial x} f(x,t) \le -\gamma_3(|x|). \tag{C.8}$$

Then the equilibrium $x = 0$ *is*

- uniformly stable *if* γ_1 *and* γ_2 *are class* $\mathcal{K}$ *functions on* $[0,r)$ *and* $\gamma_3(\cdot) \ge 0$ *on* $[0,r)$;
- uniformly asymptotically stable *if* γ_1, γ_2, and γ_3 *are class* $\mathcal{K}$ *functions on* $[0,r)$;
- exponentially stable *if* $\gamma_i(\rho) = k_i\rho^\alpha$ *on* $[0,r)$, $k_i > 0$, $\alpha > 0$, $i = 1,2,3$;
- globally uniformly stable *if* $D = \mathbb{R}^n$, γ_1 *and* γ_2 *are class* $\mathcal{K}_\infty$ *functions, and* $\gamma_3(\cdot) \ge 0$ *on* $\mathbb{R}_+$;
- globally uniformly asymptotically stable *if* $D = \mathbb{R}^n$, γ_1 *and* γ_2 *are class* $\mathcal{K}_\infty$ *functions, and* γ_3 *is a class* $\mathcal{K}$ *function on* $\mathbb{R}_+$*; and*
- globally exponentially stable *if* $D = \mathbb{R}^n$ *and* $\gamma_i(\rho) = k_i\rho^\alpha$ *on* $\mathbb{R}_+$, $k_i > 0$, $\alpha > 0$, $i = 1,2,3$.

The following theorem is proved in [75] (Theorem 4.14).

Theorem C.8 (converse Lyapunov theorem for exponential stability). *Let $x = 0$ be an equilibrium point for the nonlinear system*

$$\dot{x} = f(t,x), \tag{C.9}$$

where $f : [0,\infty) \times D \to \mathbb{R}^n$ is continuously differentiable, $D = \{x \in \mathbb{R}^n | ||x|| < r\}$, and the Jacobian matrix $\left[\frac{\partial f}{\partial x}\right]$ is bounded on D, uniformly in t. Let κ, λ, and r_0 be positive constants with $r_0 < \frac{r}{k}$. Let $D_0 = \{x \in \mathbb{R}^n | ||x|| < r_0\}$. Assume that the trajectories of the system satisfy

$$||x(t)|| \leq k||x(t_0)||e^{-\lambda(t-t_0)} \quad \text{for all } x(t_0) \in D_0, \quad \text{for all } t \geq t_0 \geq 0. \tag{C.10}$$

Then there is a function $V : [0,\infty) \times D_0 \to \mathbb{R}^n$ that satisfies the inequalities

$$c_1||x||^2 \leq V(t,x) \leq c_2||x||^2, \tag{C.11}$$

$$\frac{\partial V}{\partial t} + \frac{\partial V}{\partial x} f(t,x) \leq -c_3||x||^2, \tag{C.12}$$

$$\left\| \frac{\partial V}{\partial x} \right\| \leq c_4||x|| \tag{C.13}$$

for some positive constant c_1, c_2, c_3, and c_4. Moreover, if $r = \infty$ and the origin is globally exponentially stable, then $V(t,x)$ is defined and satisfies the aforementioned inequalities on $\mathbb{R}^n$. Furthermore, if the system is autonomous, V can be chosen independent of t.

In adaptive control our goal is to achieve convergence to a set. For time-invariant systems, the main convergence tool is LaSalle's invariance theorem. For time-varying systems, a more refined tool is the LaSalle–Yoshizawa theorem. For pedagogical reasons, we introduce it via a technical lemma due to Barbalat. These key results and their proofs are of importance in guaranteeing that an adaptive system will fulfill its tracking task.

Lemma C.9 (Barbalat). *Consider the function $\phi : \mathbb{R}_+ \to \mathbb{R}$. If ϕ is uniformly continuous and $\lim_{t\to\infty} \int_0^\infty \phi(\tau)d\tau$ exists and is finite, then*

$$\lim_{t\to\infty} \phi(t) = 0. \tag{C.14}$$

Proof. Suppose that (C.14) does not hold, that is, either the limit does not exist or it is not equal to zero. Then there exists $\varepsilon > 0$ such that for every $T > 0$, one can find $t_1 \geq T$ with $|\phi(t_1)| > \varepsilon$. Since ϕ is uniformly continuous, there is a positive constant $\delta(\varepsilon)$ such that $|\phi(t) - \phi(t_1)| < \varepsilon/2$ for all $t_1 \geq 0$ and all t such that $|t - t_1| \leq \delta(\varepsilon)$. Hence, for all $t \in [t_1, t_1 + \delta(\varepsilon)]$, we have

$$\begin{aligned} |\phi(t)| &= |\phi(t) - \phi(t_1) + \phi(t_1)| \\ &\geq |\phi(t_1)| - |\phi(t) - \phi(t_1)| \\ &> \varepsilon - \frac{\varepsilon}{2} = \frac{\varepsilon}{2}, \end{aligned} \tag{C.15}$$

which implies that

$$\left| \int_{t_1}^{t_1+\delta(\varepsilon)} \phi(\tau)d\tau \right| = \int_{t_1}^{t_1+\delta(\varepsilon)} |\phi(\tau)|d\tau > \frac{\varepsilon\delta(\varepsilon)}{2}, \tag{C.16}$$

where the first equality holds since $\phi(t)$ does not change sign on $[t_1, t_1+\delta(\varepsilon)]$. Noting that $\int_0^{t_1+\delta(\varepsilon)} \phi(\tau)d\tau = \int_0^{t_1} \phi(\tau)d\tau + \int_{t_1}^{t_1+\delta(\varepsilon)} \phi(\tau)d\tau$, we conclude that $\int_0^t \phi(\tau)d\tau$ cannot converge to a finite limit as $t \to \infty$, which contradicts the assumption of the lemma. Thus, $\lim_{t\to\infty} \phi(t) = 0$. □

Corollary C.10. *Consider the function $\phi : \mathbb{R}_+ \to \mathbb{R}$. If $\phi, \dot{\phi} \in \mathscr{L}_\infty$, and $\phi \in \mathscr{L}_p$ for some $p \in [1,\infty)$, then*

$$\lim_{t\to\infty} \phi(t) = 0. \tag{C.17}$$

Theorem C.11 (LaSalle–Yoshizawa). *Let $x = 0$ be an equilibrium point of* (C.1) *and suppose f is locally Lipschitz in x uniformly in t. Let $V : \mathbb{R}^n \times \mathbb{R}_+ \to \mathbb{R}_+$ be a continuously differentiable function such that*

$$\gamma_1(|x|) \le V(x,t) \le \gamma_2(|x|), \tag{C.18}$$

$$\dot{V} = \frac{\partial V}{\partial t} + \frac{\partial V}{\partial x} f(x,t) \le -W(x) \le 0 \tag{C.19}$$

for all $t \ge 0$ and for all $x \in \mathbb{R}^n$, where γ_1 and γ_2 are class $\mathscr{K}_\infty$ functions and W is a continuous function. Then all solutions of (C.1) *are globally uniformly bounded and satisfy*

$$\lim_{t\to\infty} W(x(t)) = 0. \tag{C.20}$$

In addition, if $W(x)$ is positive definite, then the equilibrium $x = 0$ is globally uniformly asymptotically stable.

Proof. Since $\dot{V} \le 0$, V is nonincreasing. Thus, in view of the first inequality in (C.18), we conclude that x is globally uniformly bounded, that is, $|x(t)| \le B$ for all $t \ge 0$. Since $V(x(t),t)$ is nonincreasing and bounded from below by zero, we conclude that it has a limit V_∞ as $t \to \infty$. Integrating (C.19), we have

$$\begin{aligned}\lim_{t\to\infty} \int_{t_0}^t W(x(\tau))d\tau &\le -\lim_{t\to\infty} \int_{t_0}^t \dot{V}(x(\tau),\tau)d\tau \\ &= \lim_{t\to\infty} \{V(x(t_0),t_0) - V(x(t),t)\} \\ &= V(x(t_0),t_0) - V_\infty,\end{aligned} \tag{C.21}$$

which means that $\int_{t_0}^\infty W(x(\tau))d\tau$ exists and is finite. Now we show that $W(x(t))$ is also uniformly continuous. Since $|x(t)| \le B$ and f is locally Lipschitz in x uniformly in t, we see that for any $t \ge t_0 \ge 0$,

$$\begin{aligned}|x(t) - x(t_0)| = \left|\int_{t_0}^t f(x(\tau),\tau)d\tau\right| &\le L\int_{t_0}^t |x(\tau)|d\tau \\ &\le LB|t-t_0|,\end{aligned} \tag{C.22}$$

where L is the Lipschitz constant of f on $\{|x| \le B\}$. Choosing $\delta(\varepsilon) = \varepsilon/LB$, we have

$$|x(t) - x(t_0)| < \varepsilon \qquad \text{for all } |t-t_0| \le \delta(\varepsilon), \tag{C.23}$$

which means that $x(t)$ is uniformly continuous. Since W is continuous, it is uniformly continuous on the compact set $\{|x| \leq B\}$. From the uniform continuity of $W(x)$ and $x(t)$, we conclude that $W(x(t))$ is uniformly continuous. Hence, it satisfies the conditions of Lemma C.9, which then guarantees that $W(x(t)) \to 0$ as $t \to \infty$.

If, in addition, $W(x)$ is positive definite, there exists a class $\mathcal{K}$ function $\gamma_3(\cdot)$ such that $W(x) \geq \gamma_3(|x|)$. Using Theorem C.7, we conclude that $x = 0$ is globally uniformly asymptotically stable. □

In applications, we usually have $W(x) = x^{\mathrm{T}} Q x$, where Q is a symmetric positive-semidefinite matrix. For this case, the proof of Theorem C.11 simplifies using Corollary C.10 with $p = 1$.

C.2 ▪ Forward-Completeness

Definition C.12 (Forward-completeness). *A system*

$$\dot{x} = f(x, u), \tag{C.24}$$

with a locally Lipschitz vector field $f : \mathbb{R}^n \times \mathbb{R} \to \mathbb{R}^n$, is said to be forward complete if, for every initial condition $x(0) = \xi$ and every measurable locally essentially bounded input $u : \mathbb{R}_+ \to \mathbb{R}$, the corresponding solution is defined for all $t \geq 0$; i.e., the maximal interval of existence of solutions is $T_{\max} = +\infty$.

The following Lyapunov characterization of forward-completeness was proved in [2].

Theorem C.13. *System* (C.24) *is forward complete if and only if there exist a nonnegative-valued, radially unbounded, smooth function $V : \mathbb{R}^n \to \mathbb{R}_+$ and a class $\mathcal{K}_\infty$ function σ such that*

$$\frac{\partial V(x)}{\partial x} f(x, u) \leq V(x) + \sigma(|u|), \tag{C.25}$$

for all $x \in \mathbb{R}^n$ and all $u \in \mathbb{R}$.

C.3 ▪ Input-to-State Stability

The following definition of input-to-state stability was provided by Sontag [157].

Definition C.14 (Input-to-state stability). *The system*

$$\dot{x} = f(t, x, u), \tag{C.26}$$

where f is piecewise continuous in t and locally Lipschitz in x and u, is said to be input-to-state stable *(ISS) if there exist a class $\mathcal{KL}$ function β and a class $\mathcal{K}$ function γ such that for any $x(t_0)$ and for any input $u(\cdot)$ continuous and bounded on $[0, \infty)$, the solution exists for all $t \geq 0$ and satisfies*

$$|x(t)| \leq \beta(|x(t_0)|, t - t_0) + \gamma\left(\sup_{t_0 \leq \tau \leq t} |u(\tau)|\right) \tag{C.27}$$

for all t_0 and t such that $0 \leq t_0 \leq t$.

The following theorem establishes the connection between the existence of a Lyapunov-like function and the input-to-state stability.

Theorem C.15. *Suppose that for the system* (C.26), *there exists a* C^1 *function* $V:\mathbb{R}_+\times\mathbb{R}^n\to\mathbb{R}_+$ *such that for all* $x\in\mathbb{R}^n$ *and* $u\in\mathbb{R}^m$,

$$\gamma_1(|x|)\le V(t,x)\le\gamma_2(|x|), \tag{C.28}$$

$$|x|\ge\rho(|u|)\quad\Rightarrow\quad\frac{\partial V}{\partial t}+\frac{\partial V}{\partial x}f(t,x,u)\le-\gamma_3(|x|), \tag{C.29}$$

where γ_1, γ_2, *and* ρ *are class* $\mathcal{K}_\infty$ *functions and* γ_3 *is a class* $\mathcal{K}$ *function. Then system* (C.26) *is ISS with* $\gamma=\gamma_1^{-1}\circ\gamma_2\circ\rho$.

Proof. (Outline) If $x(t_0)$ is in the set

$$R_{t_0}=\left\{x\in\mathbb{R}^n\ \middle|\ |x|\le\rho\left(\sup_{\tau\ge t_0}|u(\tau)|\right)\right\}, \tag{C.30}$$

then $x(t)$ remains within the set

$$S_{t_0}=\left\{x\in\mathbb{R}^n\ \middle|\ |x|\le\gamma_1^{-1}\circ\gamma_2\circ\rho\left(\sup_{\tau\ge t_0}|u(\tau)|\right)\right\} \tag{C.31}$$

for all $t\ge t_0$. Define $B=[t_0,T)$ as the time interval before $x(t)$ enters R_{t_0} for the first time. In view of the definition of R_{t_0}, we have

$$\dot V\le-\gamma_3\circ\gamma_2^{-1}(V)\qquad\text{for all } t\in B. \tag{C.32}$$

Then there exists a class-$\mathcal{KL}$ function β_V such that

$$V(t)\le\beta_V(V(t_0),t-t_0)\qquad\text{for all } t\in B, \tag{C.33}$$

which implies

$$|x(t)|\le\gamma_1^{-1}(\beta_V(\gamma_2(|x(t_0)|),t-t_0))\overset{\Delta}{=}\beta(|x(t_0)|,t-t_0)\qquad\text{for all } t\in B. \tag{C.34}$$

On the other hand, by (C.31), we conclude that

$$|x(t)|\le\gamma_1^{-1}\circ\gamma_2\circ\rho\left(\sup_{\tau\ge t_0}|u(\tau)|\right)\overset{\Delta}{=}\gamma\left(\sup_{\tau\ge t_0}|u(\tau)|\right)\quad\text{for all } t\in[t_0,\infty]\setminus B. \tag{C.35}$$

Then, by (C.34) and (C.35),

$$|x(t)|\le\beta(|x(t_0)|,t-t_0)+\gamma\left(\sup_{\tau\ge t_0}|u(\tau)|\right)\qquad\text{for all } t\ge t_0\ge 0. \tag{C.36}$$

By causality, it follows that

$$|x(t)|\le\beta(|x(t_0)|,t-t_0)+\gamma\left(\sup_{t_0\le\tau\le t}|u(\tau)|\right)\qquad\text{for all } t\ge t_0\ge 0.\qquad\square \tag{C.37}$$

A function V satisfying the conditions of Theorem C.15 is called an *ISS-Lyapunov function*. The inverse of Theorem C.15 is introduced next, and an equivalent dissipativity-type characterization of input-to-state stability is also introduced.

Theorem C.16 (Lyapunov characterization of input-to-state stability). *For the system*

$$\dot{x} = f(t, x, u), \tag{C.38}$$

where f is periodic in t, the following properties are equivalent:

1. *the system is ISS;*
2. *there exists an ISS-Lyapunov function;*
3. *there exist a C^1 function $V : \mathbb{R}_+ \times \mathbb{R}^n \to \mathbb{R}_+$ and class $\mathcal{K}_\infty$ functions α_1, α_2, α_3, and α_4 such that the following hold for all $t \geq 0$, $x \in \mathbb{R}^n$, $u \in \mathbb{R}^m$:*

$$\alpha_1(|x|) \leq V(t,x) \leq \alpha_2(|x|), \tag{C.39}$$

$$\frac{\partial V(t,x)}{\partial t} + \frac{\partial V(t,x)}{\partial x} f(t,x,u) \leq -\alpha_3(|x|) + \alpha_4(|u|). \tag{C.40}$$

Moreover, if f is autonomous, then V can be chosen independent of t and C^∞.

The proof for the time-varying case can be found in [101], whereas the proof for the autonomous case can be found in [159].

The following lemma establishes a useful property that a cascade of two ISS systems is itself ISS.

Lemma C.17. *Suppose that in the system*

$$\dot{x}_1 = f_1(t, x_1, x_2, u), \tag{C.41}$$

$$\dot{x}_2 = f_2(t, x_2, u), \tag{C.42}$$

the x_1-subsystem is ISS with respect to x_2 and u, and the x_2-subsystem is ISS with respect to u; that is,

$$|x_1(t)| \leq \beta_1(|x_1(s)|, t-s) + \gamma_1\left(\sup_{s\leq\tau\leq t} \{|x_2(\tau)| + |u(\tau)|\}\right), \tag{C.43}$$

$$|x_2(t)| \leq \beta_2(|x_2(s)|, t-s) + \gamma_2\left(\sup_{s\leq\tau\leq t} |u(\tau)|\right), \tag{C.44}$$

where β_1 and β_2 are class $\mathcal{KL}$ functions and γ_1 and γ_2 are class $\mathcal{K}$ functions. Then the complete $x = (x_1, x_2)$-system is ISS with

$$|x(t)| \leq \beta(|x(s)|, t-s) + \gamma\left(\sup_{s\leq\tau\leq t} |u(\tau)|\right), \tag{C.45}$$

where

$$\begin{aligned}\beta(r,t) = {} & \beta_1(2\beta_1(r,t/2) + 2\gamma_1(2\beta_2(r,0)), t/2) \\ & + \gamma_1(2\beta_2(r,t/2)) + \beta_2(r,t),\end{aligned} \tag{C.46}$$

$$\gamma(r) = \beta_1(2\gamma_1(2\gamma_2(r) + 2r), 0) + \gamma_1(2\gamma_2(r) + 2r) + \gamma_2(r). \tag{C.47}$$

Proof. With $(s,t)=(t/2,t)$, (C.43) is rewritten as

$$|x_1(t)| \le \beta_1(|x_1(t/2)|, t/2) + \gamma_1\left(\sup_{t/2\le\tau\le t}\{|x_2(\tau)|+|u(\tau)|\}\right). \tag{C.48}$$

From (C.44), we have

$$\begin{aligned}\sup_{t/2\le\tau\le t}|x_2(\tau)| &\le \sup_{t/2\le\tau\le t}\left\{\beta_2(|x_2(0)|,\tau)+\gamma_2\left(\sup_{0\le\sigma\le\tau}|u(\sigma)|\right)\right\}\\ &\le \beta_2(|x_2(0)|,t/2)+\gamma_2\left(\sup_{0\le\tau\le t}|u(\tau)|\right),\end{aligned} \tag{C.49}$$

and from (C.43), we obtain

$$\begin{aligned}|x_1(t/2)| &\le \beta_1(|x_1(0)|,t/2)+\gamma_1\left(\sup_{0\le\tau\le t/2}\{|x_2(\tau)|+|u(\tau)|\}\right)\\ &\le \beta_1(|x_1(0)|,t/2)+\gamma_1\left(\sup_{0\le\tau\le t/2}\left\{\beta_2(|x_2(0)|,\tau)+\gamma_2\left(\sup_{0\le\sigma\le\tau}|u(\sigma)|\right)+|u(\tau)|\right\}\right)\\ &\le \beta_1(|x_1(0)|,t/2)+\gamma_1\left(\beta_2(|x_2(0)|,0)+\sup_{0\le\tau\le t/2}\{\gamma_2(|u(\tau)|)+|u(\tau)|\}\right)\\ &\le \beta_1(|x_1(0)|,t/2)+\gamma_1(2\beta_2(|x_2(0)|,0))\\ &\quad +\gamma_1\left(2\sup_{0\le\tau\le t/2}\{\gamma_2(|u(\tau)|)+|u(\tau)|\}\right),\end{aligned} \tag{C.50}$$

where in the last inequality we have used the fact that $\delta(a+b)\le\delta(2a)+\delta(2b)$ for any class $\mathcal{K}$ function δ and any nonnegative a and b. Then, substituting (C.49) and (C.50) into (C.48), we get

$$\begin{aligned}|x_1(t)| \le \beta_1\Bigg(&\beta_1(|x_1(0)|,t/2)+\gamma_1(2\beta_2(|x_2(0)|,0))\\ &+\gamma_1\left(2\sup_{0\le\tau\le t/2}\{\gamma_2(|u(\tau)|)+|u(\tau)|\}\right)\Bigg)\\ &+\gamma_1\left(\beta_2(|x_2(0)|,t/2)+\gamma_2\left(\sup_{0\le\tau\le t}|u(\tau)|\right)+\sup_{t/2\le\tau\le t}\{|u(\tau)|\}\right)\\ \le\ &\beta_1(2\beta_1(|x_1(0)|,t/2)+2\gamma_1(2\beta_2(|x_2(0)|,0)),t/2)\\ &+\gamma_1(2\beta_2(|x_2(0)|,t/2))\\ &+\beta_1\left(2\gamma_1\left(2\sup_{0\le\tau\le t}\{\gamma_2(|u(\tau)|)+|u(\tau)|\}\right),0\right)\\ &+\gamma_1\left(2\sup_{0\le\tau\le t}\{\gamma_2(|u(\tau)|)+|u(\tau)|\}\right).\end{aligned} \tag{C.51}$$

Combining (C.51) and (C.44), we arrive at (C.45) with (C.46)–(C.47). □

Since (C.41) and (C.42) are ISS, then there exist ISS-Lyapunov functions V_1 and V_2 and class $\mathcal{K}_\infty$ functions $\alpha_1, \rho_1, \alpha_2$, and ρ_2 such that

$$\frac{\partial V_1}{\partial x_1} f_1(t, x_1, x_2, u) \leq -\alpha_1(|x_1|) + \rho_1(|x_2|) + \rho_1(|u|), \tag{C.52}$$

$$\frac{\partial V_2}{\partial x_2} f_2(t, x_2, u) \leq -\alpha_2(|x_2|) + \rho_2(|u|). \tag{C.53}$$

The functions $V_1, V_2, \alpha_1, \rho_1, \alpha_2$, and ρ_2 can *always* be found such that

$$\rho_1 = \alpha_2/2. \tag{C.54}$$

Then the ISS-Lyapunov function for the complete system (C.41)–(C.42) can be defined as

$$V(x) = V_1(x_1) + V_2(x_2), \tag{C.55}$$

and its derivative

$$\dot{V} \leq -\alpha_1(|x_1|) - \frac{1}{2}\alpha_2(|x_2|) + \rho_1(|u|) + \rho_2(|u|) \tag{C.56}$$

establishes the ISS property of (C.41)–(C.42) by part 3 of Theorem C.16.

In some applications of input-to-state stability, the following lemma is useful, as it is much simpler than Theorem C.15.

Lemma C.18. *Let v and ρ be real-valued functions defined on $\mathbb{R}_+$, and let b and c be positive constants. If they satisfy the differential inequality*

$$\dot{v} \leq -cv + b\rho(t)^2, \qquad v(0) \geq 0, \tag{C.57}$$

then the following hold:

(i) *If $\rho \in \mathcal{L}_\infty$, then $v \in \mathcal{L}_\infty$ and*

$$v(t) \leq v(0)\mathrm{e}^{-ct} + \frac{b}{c}\|\rho\|_\infty^2. \tag{C.58}$$

(ii) *If $\rho \in \mathcal{L}_2$, then $v \in \mathcal{L}_\infty$ and*

$$v(t) \leq v(0)\mathrm{e}^{-ct} + b\|\rho\|_2^2. \tag{C.59}$$

Proof. (i) From Lemma B.3, we have

$$\begin{aligned} v(t) &\leq v(0)\mathrm{e}^{-ct} + b\int_0^t \mathrm{e}^{-c(t-\tau)}\rho(\tau)^2 d\tau \\ &\leq v(0)\mathrm{e}^{-ct} + b \sup_{\tau\in[0,t]} \{\rho(\tau)^2\} \int_0^t \mathrm{e}^{-c(t-\tau)} d\tau \\ &\leq v(0)\mathrm{e}^{-ct} + b\|\rho\|_\infty^2 \frac{1}{c}(1 - \mathrm{e}^{-ct}) \\ &\leq v(0)\mathrm{e}^{-ct} + \frac{b}{c}\|\rho\|_\infty^2. \end{aligned} \tag{C.60}$$

(ii) From (B.9), we have

$$\begin{aligned} v(t) &\leq v(0)\mathrm{e}^{-ct} + b \sup_{\tau\in[0,t]} \left\{\mathrm{e}^{-c(t-\tau)}\right\} \int_0^t \rho(\tau)^2 d\tau \\ &= v(0)\mathrm{e}^{-ct} + b\|\rho\|_2^2. \qquad \square \end{aligned} \tag{C.61}$$

Remark C.19. *From Lemma* C.18, *it follows that if*

$$\dot{v} \leq -cv + b_1\rho_1(t)^2 + b_2\rho_2(t)^2, \qquad v(0) \geq 0, \tag{C.62}$$

and $\rho_1 \in \mathscr{L}_\infty$ *and* $\rho_2 \in \mathscr{L}_2$, *then* $v \in \mathscr{L}_\infty$ *and*

$$v(t) \leq v(0)\mathrm{e}^{-ct} + \frac{b_1}{c}\|\rho_1\|_\infty^2 + b_2\|\rho_2\|_2^2. \tag{C.63}$$

This, in particular, implies the input-to-state stability with respect to two inputs: ρ_1 *and* $\|\rho_2\|_2$.

Appendix D

Parameter Projection

Our adaptive designs rely on the use of parameter projection in our identifiers. We provide a treatment of projection for a general convex parameter set. The treatment for some of our designs where projection is used for only a scalar estimate $\hat{D}$ is easily deduced from the general case.

Let us define the following convex set:

$$\Pi = \left\{ \hat{\theta} \in \mathbb{R}^p \,\middle|\, \mathscr{P}(\hat{\theta}) \leq 0 \right\}, \tag{D.1}$$

where by assuming that the convex function $\mathscr{P} : \mathbb{R}^p \to \mathbb{R}$ is smooth, we ensure that the boundary $\partial\Pi$ of Π is smooth. Let us denote the interior of Π by $\overset{\circ}{\Pi}$ and observe that $\nabla_{\hat{\theta}}\mathscr{P}$ represents an outward normal vector at $\hat{\theta} \in \partial\Pi$. The standard projection operator is

$$\operatorname{Proj}\{\tau\} = \begin{cases} \tau, & \hat{\theta} \in \overset{\circ}{\Pi} \text{ or } \nabla_{\hat{\theta}}\mathscr{P}^{\mathrm{T}}\tau \leq 0, \\ \left(I - \Gamma \dfrac{\nabla_{\hat{\theta}}\mathscr{P}\,\nabla_{\hat{\theta}}\mathscr{P}^{rmT}}{\nabla_{\hat{\theta}}\mathscr{P}^{\mathrm{T}}\Gamma\nabla_{\hat{\theta}}\mathscr{P}} \right)\tau, & \hat{\theta} \in \partial\Pi \text{ and } \nabla_{\hat{\theta}}\mathscr{P}^{\mathrm{T}}\tau > 0, \end{cases} \tag{D.2}$$

where Γ belongs to the set $\mathscr{G}$ of all positive definite symmetric $p \times p$ matrices. Although Proj is a function of three arguments, τ, $\hat{\theta}$, and Γ, for compactness of notation, we write only $\operatorname{Proj}\{\tau\}$.

The meaning of (D.2) is that when $\hat{\theta}$ is in the interior of Π or at the boundary with τ pointing inward, then $\operatorname{Proj}\{\tau\} = \tau$. When $\hat{\theta}$ is at the boundary with τ pointing outward, then Proj projects τ on the hyperplane tangent to $\partial\Pi$ at $\hat{\theta}$.

In general, the mapping (D.2) is discontinuous. This is undesirable for two reasons. First, the discontinuity represents a difficulty for implementation in continuous time. Second, since the Lipschitz continuity is violated, we cannot use standard theorems for the existence of solutions. Therefore, we sometimes want to smooth the projection operator. Let us consider the following convex set:

$$\Pi_\varepsilon = \left\{ \hat{\theta} \in \mathbb{R}^p \,\middle|\, \mathscr{P}(\hat{\theta}) \leq \varepsilon \right\}, \tag{D.3}$$

which is a union of the set Π and an $O(\varepsilon)$-boundary layer around it. We now modify (D.2) to achieve continuity of the transition from the vector field τ on the boundary of

Π to the vector field $\big(I-\Gamma\frac{\nabla_{\hat\theta}\mathscr{P}\,\nabla_{\hat\theta}\mathscr{P}^{\mathrm T}}{\nabla_{\hat\theta}\mathscr{P}^{\mathrm T}\Gamma\nabla_{\hat\theta}\mathscr{P}}\big)\tau$ on the boundary of Π_ε:

$$\operatorname{Proj}\{\tau\}=\begin{cases}\tau, & \hat\theta\in\mathring{\Pi}\text{ or }\nabla_{\hat\theta}\mathscr{P}^{\mathrm T}\tau\le 0,\\ \left(I-c(\hat\theta)\Gamma\dfrac{\nabla_{\hat\theta}\mathscr{P}\,\nabla_{\hat\theta}\mathscr{P}^{\mathrm T}}{\nabla_{\hat\theta}\mathscr{P}^{\mathrm T}\Gamma\nabla_{\hat\theta}\mathscr{P}}\right)\tau, & \hat\theta\in\Pi_\varepsilon\setminus\mathring{\Pi}\text{ and }\nabla_{\hat\theta}\mathscr{P}^{\mathrm T}\tau>0,\end{cases}\tag{D.4}$$

$$c(\hat\theta)=\min\left\{1,\frac{\mathscr{P}(\hat\theta)}{\varepsilon}\right\}.\tag{D.5}$$

It is helpful to note that $c(\partial\Pi)=0$ and $c(\partial\Pi_\varepsilon)=1$.

In our proofs of stability of adaptive systems, we use the following technical properties of the projection operator (D.4).

Lemma D.1 (projection operator). *The following are the properties of the projection operator* (D.4)*:*

(i) *The mapping* $\operatorname{Proj}:\mathbb{R}^p\times\Pi_\varepsilon\times\mathscr{G}\to\mathbb{R}^p$ *is locally Lipschitz in its arguments* $\tau,\hat\theta,$ *and* Γ.

(ii) $\operatorname{Proj}\{\tau\}^{\mathrm T}\Gamma^{-1}\operatorname{Proj}\{\tau\}\le\tau^{\mathrm T}\Gamma^{-1}\tau$ *for all* $\hat\theta\in\Pi_\varepsilon$.

(iii) *Let* $\Gamma(t),\tau(t)$ *be continuously differentiable and*

$$\dot{\hat\theta}=\operatorname{Proj}\{\tau\},\qquad\hat\theta(0)\in\Pi_\varepsilon.$$

Then, on its domain of definition, the solution $\hat\theta(t)$ *remains in* Π_ε.

(iv) $-\tilde\theta^{\mathrm T}\Gamma^{-1}\operatorname{Proj}\{\tau\}\le-\tilde\theta^{\mathrm T}\Gamma^{-1}\tau$ *for all* $\hat\theta\in\Pi_\varepsilon,\theta\in\Pi$.

Proof. (i) The proof of this point is lengthy but straightforward and is omitted here.

(ii) For $\hat\theta\in\mathring{\Pi}$ or $\nabla_{\hat\theta}\mathscr{P}^{\mathrm T}\tau\le 0$, we have $\operatorname{Proj}\{\tau\}=\tau$, and (ii) trivially holds with equality. Otherwise, a direct computation gives

$$\begin{aligned}\operatorname{Proj}\{\tau\}^{\mathrm T}\Gamma^{-1}\operatorname{Proj}\{\tau\}&=\tau^{\mathrm T}\Gamma^{-1}\tau-2c(\hat\theta)\frac{\left(\nabla_{\hat\theta}\mathscr{P}^{\mathrm T}\tau\right)^2}{\nabla_{\hat\theta}\mathscr{P}^{\mathrm T}\Gamma\nabla_{\hat\theta}\mathscr{P}}+c(\hat\theta)^2\frac{\left|\nabla_{\hat\theta}\mathscr{P}\nabla_{\hat\theta}\mathscr{P}^{\mathrm T}\tau\right|_\Gamma^2}{\left(\nabla_{\hat\theta}\mathscr{P}^{\mathrm T}\Gamma\nabla_{\hat\theta}\mathscr{P}\right)^2}\\&=\tau^{\mathrm T}\Gamma^{-1}\tau-c(\hat\theta)\left(2-c(\hat\theta)\right)\frac{\left(\nabla_{\hat\theta}\mathscr{P}^{\mathrm T}\tau\right)^2}{\nabla_{\hat\theta}\mathscr{P}^{\mathrm T}\Gamma\nabla_{\hat\theta}\mathscr{P}}\\&\le\tau^{\mathrm T}\Gamma^{-1}\tau,\end{aligned}\tag{D.6}$$

where the last inequality follows by noting that $c(\hat\theta)\in[0,1]$ for $\hat\theta\in\Pi_\varepsilon\setminus\mathring{\Pi}$.

(iii) Using the definition of the Proj operator, we get

$$\nabla_{\hat\theta}\mathscr{P}^{\mathrm T}\operatorname{Proj}\{\tau\}=\begin{cases}\nabla_{\hat\theta}\mathscr{P}^{\mathrm T}\tau, & \hat\theta\in\mathring{\Pi}\text{ or }\nabla_{\hat\theta}\mathscr{P}^{\mathrm T}\tau\le 0,\\ \left(1-c(\hat\theta)\right)\nabla_{\hat\theta}\mathscr{P}^{\mathrm T}\tau, & \hat\theta\in\Pi_\varepsilon\setminus\mathring{\Pi}\text{ and }\nabla_{\hat\theta}\mathscr{P}^{\mathrm T}\tau>0,\end{cases}\tag{D.7}$$

which, in view of the fact that $c(\hat{\theta}) \in [0,1]$ for $\hat{\theta} \in \Pi_\varepsilon \setminus \mathring{\Pi}$, implies that

$$\nabla_{\hat{\theta}} \mathscr{P}^{\mathrm{T}} \mathrm{Proj}\{\tau\} \leq 0 \text{ whenever } \hat{\theta} \in \partial \Pi_\varepsilon; \tag{D.8}$$

that is, the vector $\mathrm{Proj}\{\tau\}$ either points inside Π_ε or is tangential to the hyperplane of $\partial\Pi_\varepsilon$ at $\hat{\theta}$. Since $\hat{\theta}(0) \in \Pi_\varepsilon$, it follows that $\hat{\theta}(t) \in \Pi_\varepsilon$ as long as the solution exists.

(iv) For $\hat{\theta} \in \mathring{\Pi}$, (iv) trivially holds with equality. For $\hat{\theta} \in \Pi_\varepsilon \setminus \mathring{\Pi}$, since $\theta \in \Pi$ and $\mathscr{P}$ is a convex function, we have

$$(\theta - \hat{\theta})^{\mathrm{T}} \nabla_{\hat{\theta}} \mathscr{P} \leq 0 \text{ whenever } \hat{\theta} \in \Pi_\varepsilon \setminus \mathring{\Pi}. \tag{D.9}$$

With (D.9), we now calculate

$$\begin{aligned} -\tilde{\theta}^{\mathrm{T}} \Gamma^{-1} \mathrm{Proj}\{\tau\} &= -\tilde{\theta}^{\mathrm{T}} \Gamma^{-1} \tau \\ &\quad + \begin{cases} 0, & \hat{\theta} \in \mathring{\Pi} \text{ or } \nabla_{\hat{\theta}} \mathscr{P}^{\mathrm{T}} \tau \leq 0 \\ c(\hat{\theta}) \dfrac{(\tilde{\theta}^{\mathrm{T}} \nabla_{\hat{\theta}} \mathscr{P})(\nabla_{\hat{\theta}} \mathscr{P}^{\mathrm{T}} \tau)}{\nabla_{\hat{\theta}} \mathscr{P}^{\mathrm{T}} \Gamma \nabla_{\hat{\theta}} \mathscr{P}}, & \hat{\theta} \in \Pi_\varepsilon \setminus \mathring{\Pi} \text{ and } \nabla_{\hat{\theta}} \mathscr{P}^{\mathrm{T}} \tau > 0 \end{cases} \\ &\leq -\tilde{\theta}^{\mathrm{T}} \Gamma^{-1} \tau, \end{aligned} \tag{D.10}$$

which completes the proof. □

Bibliography

[1] Y. ALTINTAS, S. ENGIN, AND E. BUDAK, *Analytical stability prediction and design of variable pitch cutters*, ASME Journal of Manufacturing Science and Engineering, 121 (1999), pp. 173–178. (Cited on p. 177)

[2] D. ANGELI AND E. D. SONTAG, *Forward completeness, unboundedness observability, and their Lyapunov characterizations*, Systems & Control Letters, 38 (1999), pp. 209–217. (Cited on pp. 224, 244, 277)

[3] M. ARCAK, A. TEEL, AND P. V. KOKOTOVIC, *Robust nonlinear control of feedforward systems with unmodeled dynamics*, Automatica, 37 (2001), pp. 265–272. (Not cited)

[4] Z. ARTSTEIN, *Linear systems with delayed controls: A reduction*, IEEE Transactions on Automatic Control, 27 (1982), pp. 869–879. (Cited on pp. 2, 47, 49, 52, 61, 82, 128)

[5] T. S. AUCKENTHALER, *Modeling and Control of Three-Way Catalytic Converters*, Ph.D. dissertation, ETH, 2005. Available online from http://e-collection.library.ethz.ch/view/eth:27993. (Cited on pp. 89, 90, 98, 102)

[6] M. BALENOVIC, T. BACKX, AND T. DE BIE, *Development of a model-based controller for a three-way catalytic converter*, SAE, 2002, paper 2002- 01- 0475. (Cited on p. 102)

[7] N. BEKIARIS-LIBERIS, M. JANKOVIC, AND M. KRSTIC, *Adaptive stabilization of LTI systems with distributed input delay*, International Journal of Adaptive Control and Signal Processing, 27 (2013), pp. 46–65. (Not cited)

[8] ——, *Compensation of state-dependent state delay for nonlinear systems*, Systems & Control Letters, 61 (2012), pp. 849–856. (Not cited)

[9] ——, *PDE-based analysis and control of the oxygen storage level in three-way catalytic converters*, in Proceedings of the IEEE Conference on Decision and Control, 2012. (Not cited)

[10] N. BEKIARIS-LIBERIS AND M. KRSTIC, *Compensation of wave actuator dynamics for nonlinear systems*, IEEE Transactions on Automatic Control, under review, 2013. (Cited on p. 5)

[11] ——, *Nonlinear control under delays that depend on delayed states*, European Journal of Control, 11 (2013), pp. 389–398. (Not cited)

[12] ——, *Robustness of nonlinear predictor feedback laws to time- and state-dependent delay perturbations*, Automatica, 49 (2013), pp. 1576–1590. (Not cited)

[13] ——, *Compensation of state-dependent input delay for nonlinear systems*, IEEE Transactions on Automatic Control, 58 (2013), pp. 275–289. (Not cited)

[14] ——, *Compensation of time-varying input and state delays for nonlinear systems*, Journal of Dynamic Systems, Measurement, and Control, 134 (2012), paper 011009. (Not cited)

[15] ——, *Lyapunov stability of linear predictor feedback for distributed input delays*, IEEE Transactions on Automatic Control, 56 (2011), pp. 655–660. (Not cited)

[16] ——, *Compensating distributed effect of diffusion and counter-convection in multi-input and multi-output LTI systems*, IEEE Transactions on Automatic Control, 56 (2011), pp. 637-642. (Cited on p. 84)

[17] ——, *Compensating the distributed effect of a wave PDE in the actuation or sensing path of MIMO LTI Systems*, Systems & Control Letters, 59 (2010), pp. 713–719. (Cited on p. 84)

[18] ——, *Stabilization of linear strict-feedback systems with delayed integrators*, Automatica, 56 (2010), pp. 1902–1910. (Not cited)

[19] ——, *Delay-adaptive feedback for linear feedforward systems*, Systems & Control Letters, 59 (2010), pp. 277–283. (Not cited)

[20] K. P. BHAT AND H. N. KOIVO, *Model characterization of controllability and observability in time-delay systems*, IEEE Transactions on Automatic Control, 21 (1976), pp. 292–293. (Cited on p. 35)

[21] E. BRANDT, Y. WANG, AND J. GRIZZLE, *Dynamic modeling of a three-way catalyst for SI engine exhaust emission control*, IEEE Transactions on Control Systems Technology, 8 (2000), pp. 767–776. (Not cited)

[22] D. BRESCH-PIETRI, J. CHAUVIN, AND N. PETIT, *Invoking Halanay inequality to conclude on closed-loop stability of processes with input-varying delay*, in Proceedings of the 10th IFAC Workshop on Time Delay Systems, 2012. (Cited on p. 241)

[23] ——, *Prediction-based feedback control of a class of processes with input-varying delay*, in Proceedings of the American Control Conference, 2012. (Cited on p. 241)

[24] ——, *Adaptive control scheme for uncertain time-delay systems*, Automatica, 48 (2012), pp. 1536–1552. (Cited on p. 49)

[25] D. BRESCH-PIETRI and M. KRSTIC, *Delay-adaptive full-state predictor feedback for systems with unknown long actuator delay*, IEEE Transactions on Automatic Control, 55 (2010), pp. 2106–2112. (Cited on pp. 49, 72)

[26] ——, *Adaptive trajectory tracking despite unknown input delay and plant parameters*, Automatica, 45 (2009), pp. 2074–2081. (Cited on p. 49)

[27] C. BRIAT, H. HJALMARSSON, K. H. JOHANSSON, U. T. JONSSON, G. KARLSSON, AND H. SANDBERG, *Nonlinear state-dependent delay modeling and stability analysis of internet congestion control*, in Proceedings of the IEEE Conference on Decision and Control, 2010, pp. 1484–1491. (Cited on p. 264)

[28] F. CACACE, A. GERMANI, AND C. MANES, *An observer for a class of nonlinear systems with time varying observation delay*, Systems & Control Letters, 59 (2010), pp. 305–312. (Cited on p. 169)

[29] C. CALIFANO, L. A. MARQUEZ-MARTINEZ, AND C. H. MOOG, *Extended lie brackets for nonlinear time-delay systems*, IEEE Transactions on Automatic Control, 56 (2011), pp. 2213–2218. (Not cited)

[30] J. Y. CHOI, K. KOO, J. W. KIM, AND J. S. LEE, *Global asymptotic stability of fast TCP in the presence of link dynamics*, International Journal of Control, Automation and Systems, 7 (2009), pp. 809–816. (Cited on p. 264)

[31] N. CHOPRA, M. W. SPONG, S. HIRCHE, AND M. BUSS, *Bilateral teleoperation over the internet: The time varying delay problem*, in Proceedings of the American Control Conference, 2003, pp. 155–160. (Cited on p. 137)

[32] M. B. G. CLOOSTERMAN, N. VAN DE WOUW, W. P. M. H. HEEMELS, AND H. NIJMEIJER, *Stability of networked control systems with uncertain time-varying delays*, IEEE Transactions on Automatic Control, 54 (2009), pp. 1575–1580. (Cited on p. 152)

[33] K. L. COOKE AND W. HUANG, *On the problem of linearization for state-dependent delay differential equations*, Proceedings on the American Mathematical Society, 124 (1996), pp. 1417–1426. (Cited on p. 219)

[34] J.-M. CORON, *Control and Nonlinearity*, Mathematical Surveys and Monographs 136, AMS, Providence, RI, 2007. (Cited on pp. 59, 70)

[35] M. DIAGNE, F. COUENNE, AND B. MASCHKE, *Mass transport equation with moving interface and its control as input delay system*, in Proceedings of the IFAC Joint Conference, Grenoble, France, 2013. (Cited on p. 176)

[36] S. DIOP, I. KOLMANOVSKY, P. E. MORAAL, AND M. VAN NIEUWSTADT, *Preserving stability/performance when facing an unknown time-delay*, Control Engineering Practice, 9 (2001), pp. 1319–1325. (Cited on p. 49)

[37] L. DUGARD AND E. I. VERRIEST, *Stability and Control of Time-Delay Systems*, Springer, London, UK, 1997. (Cited on p. 1)

[38] H. A. EDWARDS, Y. LIN, AND Y. WANG, *On input-to-state stability for time varying nonlinear systems*, in Proceedings of the IEEE Conference on Decision and Control, Australia, 2000, pp. 3501–3506. (Cited on p. 112)

[39] K. ENGELBORGHS, M. DAMBRINE, AND D. ROOSE, *Limitations of a class of stabilization methods for delay systems*, IEEE Transactions on Automatic Control, 46 (2001), pp. 336–339. (Not cited)

[40] S. EVESQUE, A. M. ANNASWAMY, S.-I. NICULESCU, AND A. P. DOWLING, *Adaptive control of a class of time-delay systems*, ASME Transactions on Dynamics, Systems, Measurement, and Control, 125 (2003), pp. 186–193. (Cited on p. 49)

[41] Y. A. FIAGBEDZI AND A. E. PEARSON, *Feedback stabilization of linear autonomous time lag systems*, IEEE Transactions on Automatic Control, 31 (1986), pp. 847–855. (Cited on pp. 3, 47, 49)

[42] G. FIENGO, J. A. COOK, AND J. W. GRIZZLE, *Fore-aft oxygen storage control*, in Proceedings of the American Control Conference, 2002. (Cited on p. 102)

[43] A. GERMANI, C. MANES, AND P. PEPE, *A new approach to state observation of nonlinear systems with delayed output*, IEEE Transactions on Automatic Control, 47 (2002), pp. 96–101. (Not cited)

[44] ———, *Input-output linearization with delay cancellation for nonlinear delay systems: The problem of the internal stability*, International Journal of Robust and Nonlinear Control, 13 (2003), pp. 909–937. (Not cited)

[45] K. GU, V. L. KHARITONOV, AND J. CHEN, *Stability of Time Delay Systems*, Birkhäuser, Boston, MA, 2003. (Cited on p. 1)

[46] K. GU AND S.-I. NICULESCU, *Survey on recent results in the stability and control of time-delay systems*, Transactions of AMSE, 125 (2003), pp. 158–165. (Cited on p. 1)

[47] B.-Z. GUO, J.-M. WANG, AND G.-D. ZHANG, *Spectral analysis of a wave equation with Kelvin-Voigt damping*, ZAMM, 90 (2010), pp. 323–342. (Cited on p. 96)

[48] L. GUZZELLA AND C. ONDER, *Introduction to Modeling and Control of Internal Combustion Engine Systems*, Springer, Berlin, 2010. (Cited on pp. 88, 90, 102)

[49] J. K. HALE AND S. M. VERDUYN LUNEL, *Introduction to Functional Differential Equations*, Springer-Verlag, New York, 1993. (Cited on pp. 23, 160)

[50] M. HANSEN, J. STOUSTRUP, AND J.-D. BENDTSEN, *Modeling of nonlinear marine cooling systems with closed circuit flow*, in Proceedings of the 18th IFAC World Congress, Milan, 2011, pp. 5537–5542. (Cited on pp. 177, 216)

[51] G. H. HARDY, J. E. LITTLEWOOD, AND G. POLYA, *Inequalities*, 2nd ed., Cambridge University Press, Cambridge, 1959. (Cited on p. 76)

[52] D. HELBING, *Traffic and related self-driven many-particle systems*, Reviews of Modern Physics, 73 (2001), pp. 1067–1141. (Cited on p. 177)

[53] J. P. HESPANHA, P. NAGHSHTABRIZI, AND Y. XU, *A survey of recent results in networked control systems*, Proceedings of the IEEE, 95 (2007), pp. 138–162. (Cited on pp. 152, 176)

[54] P. F. HOKAYEM, D. M. STIPANOVIC, AND M. W. SPONG, *Suboptimal master-slave teleoperation control with delays*, in Proceedings of the American Control Conference, 2006, pp. 2028–2033. (Cited on p. 137)

[55] H. ITO, P. PEPE, AND Z.-P. JIANG, *A small-gain condition for iISS of interconnected retarded systems based on Lyapunov-Krasovskii functionals*, Automatica, 46 (2010), pp. 1646–1656. (Not cited)

[56] M. JANKOVIC, *Control Lyapunov-Razumikhin functions and robust stabilization of time delay systems*, IEEE Transactions on Automatic Control, 46 (2001), pp. 1048–1060. (Cited on p. 113)

[57] ——, *Control of nonlinear systems with time delay*, in Proceedings of the IEEE Conference on Decision and Control, 2003, pp. 4545–4550. (Cited on p. 113)

[58] ——, *Cross-term forwarding for systems with time delay*, IEEE Transactions on Automatic Control, 54 (2009), pp. 498–511. (Cited on p. 113)

[59] ——, *Forwarding, backstepping, and finite spectrum assignment for time delay systems*, Automatica, 45 (2009), pp. 2–9. (Cited on pp. 47, 48)

[60] ——, *Recursive predictor design for state and output feedback controllers for linear time delay systems*, Automatica, 46 (2010), pp. 510–517. (Cited on pp. 35, 47, 48, 49)

[61] E. JOBSON, O. HJORTSBERG, S. L. ANDERSSON, AND I. GOTTBERG, *Reactions over a double layer trimetal three-way catalyst*, SAE, 1996, paper 960801. (Cited on pp. 88, 102)

[62] I. KARAFYLLIS, *The non-uniform in time small-gain theorem for a wide class of control systems with outputs*, European Journal on Control, 10 (2004), pp. 307–323. (Cited on p. 166)

[63] —— *Finite-time global stabilization by means of time-varying distributed delay feedback*, SIAM Journal on Control and Optimization, 45 (2006), pp. 320–342. (Cited on pp. 47, 113, 152)

[64] ——, *Lyapunov theorems for systems described by retarded functional differential equations*, Nonlinear Analysis, 64 (2006) pp. 590–617. (Cited on p. 113)

[65] ——, *Stabilization by means of approximate predictors for systems with delayed input*, SIAM Journal on Control and Optimization, 49 (2011), pp. 1100–1123. (Cited on p. 113)

[66] I. KARAFYLLIS AND Z.-P. JIANG, *A small-gain theorem for a wide class of feedback systems with control applications*, SIAM Journal on Control and Optimization, 46 (2007), pp. 1483–1517. (Not cited)

[67] ——, *Necessary and sufficient Lyapunov-like conditions for robust nonlinear stabilization*, ESAIM Control, Optimization and Calculus of Variations, 16 (2010), pp. 887–928. (Cited on p. 28)

[68] ——, *Stability and Stabilization of Nonlinear Systems*, Springer, London, 2011. (Cited on pp. 1, 113)

[69] I. KARAFYLLIS AND M. KRSTIC, *Nonlinear stabilization under sampled and delayed measurements, and with inputs subject to delay and zero-order hold*, IEEE Transactions on Automatic Control, 57 (2012), pp. 1141–1154. (Cited on p. 113)

[70] ——, *Delay-robustness of linear predictor feedback without restriction on delay rate*, Automatica, 49 (2013), pp. 1761–1767. (Cited on p. 139)

[71] ——, *Predictor-based output feedback for nonlinear delay systems*, Automatica, submitted, 2011. Available online from http://arxiv.org/abs/1108.4499. (Not cited)

[72] I. KARAFYLLIS, P. PEPE, AND Z.-P. JIANG, *Input-to-output stability for systems described by retarded functional differential equations*, European Journal of Control, 6 (2008), pp. 539–555. (Cited on p. 113)

[73] ——, *Stability results for systems described by coupled retarded functional differential equations and functional difference equations*, Nonlinear Analysis, 71 (2009), pp. 3339–3362. (Cited on p. 113)

[74] I. KARAFYLLIS AND J. TSINIAS, *Non-uniform in time input-to-state stability and the small-gain theorem*, IEEE Transactions on Automatic Control, 49 (2004), pp. 196–216. (Not cited)

[75] H. KHALIL, *Nonlinear Systems*, 3rd ed., Prentice Hall, Upper Saddle River, NJ, 2002. (Cited on pp. 89, 93, 94, 130, 137, 201, 230, 235, 239, 272, 274, 275)

[76] V. KHARITONOV, *An extension of the prediction scheme*, Automatica, submitted, 2012. (Cited on p. 47)

[77] J. KLAMKA, *Controllability of linear systems with time-variable delays in control*, International Journal of Control, 24 (1976), pp. 869–878. (Not cited)

[78] ——, *Absolute controllability of linear systems with time variable delay in the control*, International Journal of Control, 26 (1977), pp. 57–63. (Not cited)

[79] ——, *Observer for linear feedback control of systems with distributed delays in controls and outputs*, Systems & Control Letters, 1 (1982), pp. 326–331. (Not cited)

[80] V. KOLMANOVSKII AND A. MYSHKIS, *Introduction to the Theory and Applications of Functional Differential Equations*, Kluwer Academic Publishers, Dordrecht, The Netherlands, 1999. (Cited on p. 1)

[81] G. KOLTSAKIS AND A. STAMATELOS, *Modeling dynamic phenomena in 3-way catalytic converters*, Chemical Engineering Science, 54 (1999), pp. 4567–4578. (Cited on p. 102)

[82] N. N. KRASOVSKII, *Stability of Motion*, Stanford University Press, Stanford, CA, 1963. (Cited on p. 1)

[83] C. KRAVARIS AND R. A. WRIGHT, *Deadtime compensation for nonlinear processes*, Journal of American Institute of Chemical Engineers, 35 (1989), pp. 1535–1542. (Cited on p. 177)

[84] M. KRSTIC, *Feedback linearizability and explicit integrator forwarding controllers for classes of feedforward systems*, IEEE Transactions on Automatic Control, 49 (2004), pp. 1668–1682. (Cited on p. 252)

[85] ——, *Lyapunov tools for predictor feedbacks for delay systems: Inverse optimality and robustness to delay mismatch*, Automatica, 44 (2008), pp. 2930–2935. (Cited on pp. 47, 83)

[86] ——, *On compensating long actuator delays in nonlinear control*, IEEE Transactions on Automatic Control, 53 (2008), pp. 1684–1688. (Cited on pp. 3, 23, 204)

[87] ——, *Delay Compensation for Nonlinear, Adaptive and PDE Systems*, Birkhäuser, Boston, MA, 2009. (Cited on pp. 1, 3)

[88] ——, *Lyapunov stability of linear predictor feedback for time-varying input delay*, IEEE Transactions on Automatic Control, 55 (2010), pp. 554–559. (Not cited)

[89] ——, *Input delay compensation for forward complete and feedforward nonlinear systems*, IEEE Transactions on Automatic Control, 55 (2010), pp. 287–303. (Cited on pp. 23, 113)

[90] M. KRSTIC AND A. BANASZUK, *Multivariable adaptive control of instabilities arising in jet engines*, Control Engineering Practice, 14 (2006), pp. 833–842. (Cited on p. 49)

[91] M. KRSTIC AND N. BEKIARIS-LIBERIS, *Compensation of infinite-dimensional input dynamics*, IFAC Annual Reviews, 34 (2010), pp. 233–244. (Cited on p. 84)

[92] M. KRSTIC, I. KANELLAKOPOULOS, AND P. V. KOKOTOVIC, *Nonlinear and Adaptive Control Design*, Wiley, New York, 1995. (Cited on pp. 49, 204)

[93] M. KRSTIC AND A. SMYSHLYAEV, *Backstepping boundary control for first-order hyperbolic PDEs and application to systems with actuator and sensor delays*, Systems & Control Letters, 57 (2008), pp. 750–758. (Cited on pp. 3, 47, 49)

[94] W. H. KWON AND A. E. PEARSON, *Feedback stabilization of linear systems with delayed control*, IEEE Transactions on Automatic Control, 25 (1980), pp. 266–269. (Cited on p. 47)

[95] X. LITRICO AND V. FROMION, *Analytical approximation of open-channel flow for controller design*, Applied Mathematical Modeling, 28 (2004), pp. 677–695. (Cited on pp. 152, 177)

[96] S. LIU, T. BASAR, AND R. SRIKANT, *Pitfalls in the fluid modeling of RTT variations in window-based congestion control*, in Proceedings of the IEEE INFOCOM, USA, 2005, pp. 1002–1012. (Cited on p. 176)

[97] H. LOGEMANN, R. REBARBER, AND G. WEISS, *Conditions for robustness and nonrobustness of the stability of feedback systems with respect to small delays in the feedback loop*, SIAM Journal on Control and Optimization, 34 (1996), pp. 572–600. (Not cited)

[98] J. J. LOISEAU, *Algebraic tools for the control and stabilization of time-delay systems*, Annual Reviews in Control, 24 (200), pp. 135–149. (Cited on p. 47)

[99] J. LOUISELL, *Delay differential systems with time-varying delay: New directions for stability theory*, Kybernetika, 37 (2001), pp. 239–252. (Not cited)

[100] J. M. MAHAFFY, J. BÉLAIR, AND M. MACKEY, *Hematopoietic model with moving boundary condition and state dependent delay: Applications in erythopoiesis*, Journal of Theoretical Biology, 190 (1998), pp. 135–146. (Cited on p. 177)

[101] M. MALISOFF AND F. MAZENC, *Further remarks on strict input-to-state stable Lyapunov functions for time-varying systems*, Automatica, 41 (2005), pp. 1973–1978. (Cited on p. 279)

[102] A. Z. MANITIUS AND A. W. OLBROT, *Finite spectrum assignment for systems with delays*, IEEE Transactions on Automatic Control, 24 (1979), pp. 541–553. (Cited on pp. 1, 2, 11, 47, 49, 52, 82)

[103] D. Q. MAYNE, *Control of linear systems with time delay*, Electronics Letters, 4 (1968), pp. 439–440. (Cited on pp. 1, 2, 11)

[104] F. MAZENC AND P.-A. BLIMAN, *Backstepping design for time-delay nonlinear systems*, IEEE Transactions on Automatic Control, 51 (2006), pp. 149–154. (Cited on p. 113)

[105] F. MAZENC AND H. ITO, *Lyapunov technique and backstepping for nonlinear neutral systems*, IEEE Transactions on Automatic Control, 58 (2013), pp. 512–517. (Not cited)

[106] F. MAZENC AND M. MALISOFF, *Stabilization of a chemostat model with Haldane growth functions and a delay in the measurement*, Automatica, 46 (2010), pp. 1428–1436. (Cited on p. 113)

[107] ———, *Asymptotic stabilization for feedforward systems with delayed feedbacks*, Automatica, 49 (2013), pp. 780–787. (Cited on p. 113)

[108] F. MAZENC, M. MALISOFF, AND Z. LIN, *Further results on input-to-state stability for nonlinear systems with delayed feedbacks*, Automatica, 44 (2008), pp. 2415–2421. (Cited on pp. 113, 139)

[109] F. MAZENC, S. MONDIE, AND R. FRANCISCO, *Global asymptotic stabilization of feedforward systems with delay in the input*, IEEE Transactions on Automatic Control, 49 (2004), pp. 844–850. (Cited on p. 113)

[110] F. MAZENC, S. MONDIE, AND S. I. NICULESCU, *Global asymptotic stabilization for chains of integrators with a delay in the input*, IEEE Transactions on Automatic Control, 48 (2003), pp. 57–63. (Cited on pp. 48, 113)

[111] F. MAZENC AND S.-I. NICULESCU, *Generating positive and stable solutions through delayed state feedback*, Automatica, 47 (2011), pp. 525–533. (Cited on p. 113)

[112] F. MAZENC, S.-I. NICULESCU, AND M. BEKAIK, *Backstepping for nonlinear systems with delay in the input revisited*, SIAM Journal on Control and Optimization, 49 (2011), pp. 2239–2262. (Cited on p. 113)

[113] ———, *Stabilization of time-varying nonlinear systems with distributed input delay by feedback of plant's state*, IEEE Transactions on Automatic Control, 58 (2013), pp. 264–269. (Not cited)

[114] F. MAZENC, S.-I. NICULESCU, AND M. KRSTIC, *Lyapunov-Krasovskii functionals and application to input delay compensation for linear time-invariant systems*, Automatica, 48 (2012), pp. 1317–1323. (Cited on pp. 47, 83)

[115] F. MAZENC AND L. PRALY, *Adding integrations, saturated controls, and stabilization of feedforward system*, IEEE Transactions on Automatic Control, 41 (2996), pp. 1559–1578. (Not cited)

[116] ———, *Asymptotic tracking of a reference state for systems with a feedforward structure*, Automatica, 36 (2000), pp. 179–187. (Not cited)

[117] W. MICHIELS AND S.-I. NICULESCU, *Stability and Stabilization of Time-Delay Systems: An Eigenvalue-Based Approach*, SIAM, Philadelphia, PA, 2007. (Cited on p. 1)

[118] W. MICHIELS, R. SEPULCHRE, AND D. ROOSE, *Stability of perturbed delay differential equations and stabilization of nonlinear cascade systems*, SIAM Journal on Control and Optimization, 40 (2001), pp. 661–680. (Not cited)

[119] L. MIRKIN, *On the approximation of distributed-delay control laws*, Systems & Control Letters, 51 (2004), pp. 331–342. (Cited on pp. 3, 47, 174)

[120] B. M. MIRKIN AND P. O. GUTMAN, *Adaptive control of linear time-delay systems*, in Proceedings of the American Control Conference, 2004, pp. 1241–1246. (Cited on p. 49)

[121] R. MOLLER, M. VOTSMEIER, C. ONDER, L. GUZZELLA, AND J. GIESHOFF, *Is oxygen storage in three-way catalysts an equilibrium controlled process?*, Applied Catalysis B: Environmental, 91 (2009), pp. 30–38. (Cited on pp. 88, 89, 90, 102)

[122] S. MONDIE AND W. MICHIELS, *Finite spectrum assignment of unstable time-delay systems with a safe implementation*, IEEE Transactions on Automatic Control, 48 (2003), pp. 2207–2212. (Cited on pp. 3, 47, 174)

[123] L. A. MONTESTRUQUE AND P. ANTSAKLIS, *Stability of model-based networked control systems with time-varying transmission lines*, IEEE Transactions on Automatic Control, 49 (2004), pp. 1562–1572. (Cited on p. 152)

[124] L. MOREAU, W. MICHIELS, D. AEYELS, AND D. ROOSE, *Robustness of nonlinear delay equations with respect to input perturbations: A trajectory based approach*, Mathematics of Control, Signals, and Systems, 15 (2002), pp. 316–335. (Not cited)

[125] H. MOUNIER AND J. RUDOLPH, *Flatness-based control of nonlinear delay systems: A chemical reactor example*, International Journal on Control, 71 (1998), pp. 871–890. (Cited on p. 177)

[126] K. R. MUSKE, J. C. P. JONES, AND E. M. FRANCESCHI, *Adaptive analytical model-based control for SI engines air-fuel ratio*, IEEE Transactions on Control Systems Technology, 16 (2008), pp. 763–768. (Cited on p. 92)

[127] S.-I. NICULESCU, *Delay Effects on Stability*, Springer-Verlag, London, 2001. (Cited on p. 1)

[128] S.-I. NICULESCU AND A. M. ANNASWAMY, *An adaptive Smith controller for time-delay systems with relative degree $n \leq 2$*, Systems & Control Letters, 49 (2003), pp. 347–358. (Cited on p. 49)

[129] G. NIEMEYER, AND J.-J. E. SLOTINE, *Toward bilateral internet teleoperation*, in Beyond Webcams: An Introduction to Online Robots, MIT Press, Cambridge, MA, 2001, pp. 193–213. (Cited on p. 177)

[130] M. NIHTILA, *Adaptive control of a continuous-time system with time-varying input delay*, Systems & Control Letters, 12 (1989), pp. 357–364. (Cited on p. 128)

[131] ———, *Finite pole assignment for systems with time-varying input delays*, in Proceedings of the IEEE Conference on Decision Control, 1992, pp. 927–928. (Cited on pp. 3, 128)

[132] A. W. OLBROT, *Stabilizability, detectability, and spectrum assignment for linear autonomous systems with general time delays*, IEEE Transactions on Automatic Control, 23 (1978), pp. 887–890. (Cited on p. 47)

[133] N. OLGAC AND R. SIPAHI, *An exact method for the stability analysis of time-delayed linear time-invariant (LTI) systems*, IEEE Transactions on Automatic Control, 47 (2002), pp. 793–797. (Cited on p. 33)

[134] G. OROSZ AND G. STEPAN, *Subcritical Hopf bifurcations in a car-following model with reaction-time delay*, Proceedings of the Royal Society of London, Series A, 462 (2006), pp. 2643–2670. (Cited on pp. 152, 177)

[135] R. ORTEGA AND R. LOZANO, *Globally stable adaptive controller for systems with delay*, International Journal of Control, 47 (1988), pp. 17–23. (Cited on p. 49)

[136] Z. J. PALMOR, *Time-delay compensation–Smith predictor and its modifications*, in The Control Handbook, W. S. Levine, ed., 1996, pp. 224–237. (Cited on p. 3)

[137] P. PEPE, *Input-to-state stabilization of stabilizable, time-delay, control-affine, nonlinear systems*, IEEE Transactions on Automatic Control, 54 (2009), pp. 1688–1693. (Not cited)

[138] P. PEPE AND H. ITO, *On saturation, discontinuities, and delays, in iISS and ISS feedback control redesign*, IEEE Transactions on Automatic Control, 57 (2012), pp. 1125–1140. (Not cited)

[139] P. PEPE AND Z.-P. JIANG, *A Lyapunov-Krasovskii methodology for ISS and iISS of time-delay systems*, Systems & Control Letters, 55 (2006), pp. 1006–1014. (Cited on p. 113)

[140] P. PEPE, I. KARAFYLLIS, AND Z.-P. JIANG, *On the Liapunov-Krasovskii methodology for the ISS of systems described by coupled delay differential and difference equations*, Automatica, 44 (2008), pp. 2266–2273. (Cited on p. 113)

[141] P. PEPE AND E. I. VERRIEST, *Liapunov criteria for stability in L_p norm of special neutral systems*, Automatica, 48 (2012), pp. 132–138. (Not cited)

[142] J.-B. POMET, *Explicit design of time-varying stabilizing control laws for a class of controllable systems without drift*, Systems & Control Letters, 18 (1992), pp. 147–158. (Cited on pp. 152, 190)

[143] L. PRALY, *Adaptive regulation: Lyapunov design with a growth condition*, International Journal of Adaptive Control and Signal Processing, 6 (1992), pp. 329–351. (Not cited)

[144] B. S. RAZUMIKHIN, *On the stability of systems with a delay*, Prikladnava Matematika i Mekhanika, 20 (1956), pp. 500–512. (Cited on p. 1)

[145] J.-P. RICHARD, *Time-delay systems: An overview of some recent advances and open problems*, Automatica, 39 (2003), pp. 1667–1694. (Cited on pp. 1, 12, 47)

[146] C. RODUNER, C. ONDER, AND H. GEERING, *Automated design of an air/fuel controller for an SI engine considering the three-way catalytic converter in the H^∞ approach*, in Proceedings of the Mediterranean Conference on Control and Systems, 1997. (Cited on p. 102)

[147] O. ROESCH, H. ROTH, AND S.-I. NICULESCU, *Remote control of mechatronic systems over communication networks*, in Proceedings of the International Conference in Mechatronics and Automation, 2005, pp. 1648–1653. (Cited on p. 82)

[148] R. SEPULCHRE, M. JANKOVIC, AND P. V. KOKOTOVIC, *Integrator forwarding: A new recursive nonlinear robust design*, Automatica, 33 (1997), pp. 979–984. (Cited on p. 82)

[149] ——, *Constructive Nonlinear Control*, Springer-Verlag, Berlin, 1997. (Not cited)

[150] E. SHAFAI, C. RODUNER, AND H. P. GEERING, *Indirect adaptive control of a three-way catalyst*, SAE, 1996, paper 961038. (Cited on p. 102)

[151] R. SIPAHI, *Delay modeling in supply network dynamics: Performance, synchronization, stability*, Paper 3, 2007. (Not cited)

[152] R. SIPAHI, F. M. ATAY, AND S.-I. NICULESCU, *Stability of traffic flow behavior with distributed delays modeling the memory effects of the drivers*, SIAM Journal on Applied Mathematics, 68 (2007), pp. 738–759. (Cited on pp. 152, 177)

[153] R. SIPAHI, S. LAMMER, S.-I. NICULESCU, AND D. HELBING, *On stability analysis and parametric design of supply networks under the presence of transportation delays*, in Proceedings of the ASME-IMECE Conference, 2006, pp. 135–144. (Cited on pp. 152, 177)

[154] G. R. SLEMON AND A. STRAUGHEN, *Electric Machines*, Addison-Wesley, Reading, MA, 1980. (Cited on pp. 230, 231)

[155] O. J. M. SMITH, *A controller to overcome deadtime*, ISA, 6 (1959), pp. 28–33. (Cited on pp. 1, 2)

[156] A. SMYSHLYAEV AND M. KRSTIC, *Adaptive Control of Parabolic PDEs*, Princeton University Press, Princeton, NJ, 2010. (Cited on pp. 49, 71)

[157] E. D. SONTAG, *Smooth stabilization implies coprime factorization*, IEEE Transactions on Automatic Control, 34 (1989), pp. 435–443. (Cited on p. 277)

[158] E. D. SONTAG, *Comments on integral variants of ISS*, Systems & Control Letters, 34 (1998), pp. 93–100. (Cited on p. 263)

[159] E. D. SONTAG AND Y. WANG, *On characterization of the input-to-state stability property*, Systems & Control Letters, 24 (1995), pp. 351–359 (Cited on p. 279)

[160] J. D. STERMAN, *Business Dynamics: Systems Thinking and Modeling for a Complex World*, McGraw-Hill, 2000. (Cited on pp. 152, 177)

[161] G. TADMOR, *The standard H^∞ problem in systems with a single input delay*, IEEE Transactions on Automatic Control, 45 (2000), pp. 382–397. (Not cited)

[162] A. TEEL, *Connections between Razumikhin-type theorems and the ISS nonlinear small-gain theorem*, IEEE Transactions on Automatic Control, 43 (1998), pp. 960–964. (Cited on p. 139)

[163] S. P. TRAVIS, *A one-dimensional two-body problem of classical electrodynamics*, SIAM Journal on Applied Mathematics, 28 (1975), pp. 611–632. (Cited on p. 264)

[164] Y. Z. TSYPKIN, *The systems with delayed feedback*, Avtomatika i Telemekhnika, 7 (1946), pp. 107–129. (Cited on p. 1)

[165] V. VAN ASSCHE, M. DAMBRINE, J. F. LAFAY, AND J. P. RICHARD, *Some problems arising in the implementation of distributed-delay control laws*, in Proceedings of the IEEE Conference on Decision and Control, 1999. (Not cited)

[166] R. VAZQUEZ AND M. KRSTIC, *Control of 1-D parabolic PDEs with Volterra nonlinearities – Part I: Design*, Automatica, 44 (2008), pp. 2778–2790. (Cited on p. 5)

[167] ——, *Control of 1-D parabolic PDEs with Volterra nonlinearities – Part II: Analysis*, Automatica, 44 (2008), pp. 2791–2803. (Cited on p. 5)

[168] E. I. VERRIEST, *Stability of systems with state-dependent and random delays*, IMA Journal of Mathematical Control and Information, 19 (2002), pp. 103–114. (Cited on p. 219)

[169] H.-O. WALTHER, *On a model for soft landing with state-dependent delay*, Journal of Dynamics and Differential Equations, 3 (2007), pp. 593–622. (Cited on pp. 217, 219)

[170] Q.-G. WANG, T.-H. LEE, AND K.-K. TAN, *Finite Spectrum Assignment for Time-Delay Systems*, Springer-Verlag, London, UK, 1998. (Cited on p. 3)

[171] K. WATANABE, *Finite Spectrum assignment and observer for multivariable systems with commensurate delays*, IEEE Transactions on Automatic Control, 31 (1986), pp. 543–550. (Cited on p. 3)

[172] K. WATANABE, E. NOBUYAMA, T. KITAMORI, AND M. ITO, *A new algorithm for finite spectrum assignment of single-input systems with time delay*, IEEE Transactions on Automatic Control, 37 (1992), pp. 1377–1383. (Cited on pp. 3, 47)

[173] E. WITRANT, C. C. DE-WIT, D. GEORGES, AND M. ALAMIR, *Remote stabilization via communication networks with a distributed control law*, IEEE Transactions on Automatic Control, 52 (2007), pp. 1480–1485. (Cited on pp. 152, 177)

[174] H. WU, *Adaptive stabilizing state feedback controllers of uncertain dynamical systems with multiple time delays*, IEEE Transactions on Automatic Control, 45 (2000), pp. 1697–1701. (Not cited)

[175] ——, *Robust adaptive control schemes for a class of uncertain dynamical systems with multiple time delays*, International Journal of Systems Science, 33 (2002), pp. 1241–1248. (Not cited)

[176] ——, *Adaptive robust tracking and model following of uncertain dynamical systems with multiple time delays*, IEEE Transactions on Automatic Control, 49 (2004), pp. 611–616. (Not cited)

[177] L. XIE, E. FRIDMAN, AND U. SHAKED, *Robust H_∞ control of distributed delay systems with application to combustion control*, IEEE Transactions on Automatic Control, 46 (2001), pp. 1930–1935. (Cited on p. 82)

[178] N. YEGANEFAR, P. PEPE, AND M. DAMBRINE, *Input-to-state stability of time-delay systems: A link with exponential stability*, IEEE Transactions on Automatic Control, 53 (2008), pp. 1526–1531. (Cited on p. 113)

[179] Y. YILDIZ, A. ANNASWAMY, I. V. KOLMANOVSKY, AND D. YANAKIEV, *Adaptive posicast controller for time-delay systems with relative degree $n^* \leq 2$*, Automatica, 46 (2010), pp. 279–289. (Cited on p. 49)

[180] T. ZHANG, H. WEN, AND Q. ZHU, *Adaptive control of time-delay nonlinear systems with unknown control direction*, in Proceedings of the Chinese Conference on Decision and Control, 2010, pp. 2057–2062. (Cited on p. 49)

[181] Q.-C. ZHONG, *On distributed delay in linear control laws–Part I: Discrete-delay implementation*, IEEE Transactions on Automatic Control, 49 (2006), pp. 2074–2080. (Cited on pp. 3, 47)

[182] ——, *Robust Control of Time-Delay Systems*, Springer, London, 2006. (Cited on pp. 1, 47)

[183] Q.-C. ZHONG AND L. MIRKIN, *Control of integral processes with dead time—Part 2: Quantitative analysis*, Proceedings of the IEEE, 149 (2002), pp. 291–296. (Not cited)

[184] B. ZHOU, H. GAO, Z. LIN, AND G.-R. DUAN, *Stabilization of linear systems with distributed input delay and input saturation*, Automatica 48 (2012), pp. 712–724. (Cited on p. 82)

[185] B. ZHOU, Z.-Y. LI, W. X. ZHENG, AND G.-R. DUAN, *Stabilization of some linear systems with both state and input delays*, Systems & Control Letters, 61 (2012), pp. 989–998. (Cited on pp. 47, 128)

[186] B. ZHOU, Z. LIN, AND G.-R. DUAN, *Truncated predictor feedback for linear systems with long time-varying input delays*, Automatica, 48 (2012), pp. 2387–2399. (Cited on p. 128)

[187] J. ZHOU, W. WANG, AND C. WEN, *Adaptive backstepping control of uncertain systems with unknown input time delay*, in Proceedings of the IFAC World Congress, 2008. (Not cited)

Index